AF361507

CITY POLITICS IN CANADA

CITY POLITICS IN CANADA

Forty Years of Continuity and Change

EDITED BY MARTIN HORAK, JACK LUCAS,
AND ZACK TAYLOR

UNIVERSITY OF TORONTO PRESS
Toronto Buffalo London

© University of Toronto Press 2025
Toronto Buffalo London
utppublishing.com
Printed in Canada

ISBN 978-1-4875-6933-4 (cloth)   ISBN 978-1-4875-6936-5 (EPUB)
ISBN 978-1-4875-6934-1 (paper)   ISBN 978-1-4875-6935-8 (PDF)

Library and Archives Canada Cataloguing in Publication

Title: City politics in Canada : forty years of continuity and change / edited by Martin Horak,
Jack Lucas, and Zack Taylor.
Other titles: City politics in Canada (2025)
Names: Horak, Martin, 1973– editor | Lucas, Jack, 1982– editor | Taylor, Zack, 1973– editor
Series: Political development: comparative perspectives.
Description: Series statement: Political development: comparative perspectives | Includes bibliographical references
    and index.
Identifiers: Canadiana (print) 20250276984 | Canadiana (ebook) 20250276992 | ISBN 9781487569334 (cloth) |
    ISBN 9781487569341 (paper) | ISBN 9781487569358 (PDF) | ISBN 9781487569365 (EPUB)
Subjects: LCSH: Municipal government – Canada – Case studies. | LCGFT: Case studies.
Classification: LCC JS1709 .C58 2025 | DDC 352.160971 – dc23

Cover design: John Beadle
Cover image: ENVATO

We wish to acknowledge the land on which the University of Toronto Press operates. This land is the traditional
territory of the Wendat, the Anishnaabeg, the Haudenosaunee, the Métis, and the Mississaugas of the Credit First
Nation.

The editors acknowledge the assistance of the J.B. Smallman Publication Fund, and the Faculty of Social Science,
The University of Western Ontario.

This book has been published with the help of a grant from the Federation for the Humanities and Social Sciences,
through the Awards to Scholarly Publications Program, using funds provided by the Social Sciences and Humanities
Research Council of Canada.

University of Toronto Press acknowledges the financial support of the Government of Canada, the Canada
Council for the Arts, and the Ontario Arts Council, an agency of the Government of Ontario, for its publishing
activities.

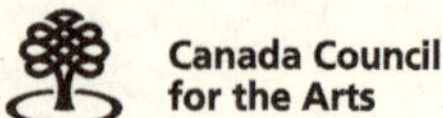

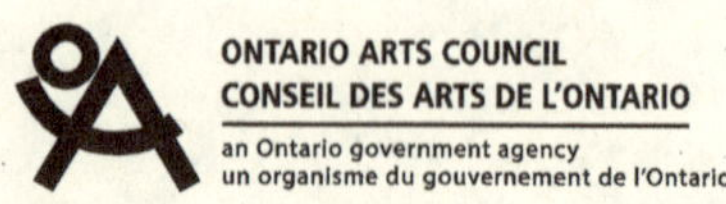

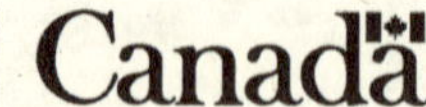

For Warren Magnusson, 1947–2025

Contents

Figures and Tables

Figures

Tables

Foreword

Andrew Sancton

Forty years after *City Politics in Canada* was published, it is being resurrected by three accomplished political scientists who specialize in Canadian urban politics. Many thanks to Martin Horak, Jack Lucas, and Zack Taylor for offering Warren Magnusson and me such a remarkable academic tribute.

Warren and I first met in the late 1960s when we were both pursuing graduate degrees at the University of Oxford. We ended up sharing the same supervisor, L. J. (Jim) Sharpe, for our doctoral work. Sharpe was a prominent scholar of English local government and politics, but he also had a great interest in Canada. He supervised doctoral research by a number of Canadians, two of the best-known being Keith Banting and the late Robert Young.

For one academic year, 1978–9, Warren was a member of the Political Science department at the University of Western Ontario, the place where I spent almost my entire academic career. It was during that year that we formulated our plan for *City Politics in Canada*. As we noted in the Preface, "The initial idea was to write a systematic comparison of city politics in Montreal and Toronto." I was an anglophone Montrealer whose research had been about that city, and Warren had done considerable work relating to urban politics in Toronto. As we thought more about the project, "We soon realized that, if we wanted our book to be widely read and used, we would have to go beyond the confines of central Canada. This we could not do ourselves."

Recruiting authors for the other cities was not easy, but in retrospect, it should have been a lot harder. Both of us were in the very early stages of our careers and most of our contributors were considerably more senior. They had no evidence that we could pull off this project, and we had virtually no experience with academic publishing. Somehow it all came together.

We knew that publishing was essential for academic careers, but more than anything, we wanted to edit a book that would help interested Canadians understand the politics of their major cities. I have no records of how many copies were sold or in how many courses it was adopted as a text. What I do know is that it seemed to attract a lot of attention, and not just from other scholars.

Today's young academics will no doubt be shocked to learn that my files contain twelve reviews of *City Politics* published in different academic and public policy journals. These were days when such publications paid much more attention to scholarly books.

Even more shocking by today's standards is that the book was reviewed or commented on in five different Canadian newspapers. Writing in the Books section of *The Globe and Mail* (20 August 1983), former Toronto mayor John Sewell (defeated in 1980 in his bid for re-election) used the occasion to provide his own reflections on Canadian urban politics. In discussing our book, he wrote that "It puts the past ten years in perspective, and gives a sense of continuity to the changes that were wrought and the changes that were passed over."

What Sewell and other journalistic reviewers wrote is less important than the fact that they were assessing an avowedly academic book in mainstream newspapers that was edited by little-known professors. Can we imagine such a thing happening today? What does this tell us about the current connections between university political science and our general political discourse? I would love to have my suspicions proven wrong by seeing a review of this new edition in one or two Canadian newspapers or magazines. But I am not holding my breath.

Some of the academic reviewers of the 1983 edition criticized us for being insufficiently theoretical. (Little did they know about what was to come from the work of Warren Magnusson!) Some praised the late Caroline Andrew's chapter on Ottawa-Hull for introducing the concept of collective consumption. But generally, we (and our authors) stuck to relatively straightforward narrative accounts, with critical analyses that did not require specialized jargon.

The current editors seem to have adopted much the same approach. Their concluding chapter, however, is much more theoretically sophisticated than was my conclusion in the 1983 edition. Building on their own work elsewhere and on the chapters in this book, they have much to say about theoretically important topics such as the role of ideology in Canadian city politics, the importance of group identities, urban political economy, and multilevel democracy.

Another big improvement in the current edition is the use of maps and graphics. Neither Warren nor I was as adept in such matters. In our defence, we did not have computers. Even if we had, I doubt that we would have had the courage to ask the University of Toronto Press to complicate their production process by going much beyond the printed word. We felt fortunate in being allowed to include an appendix of five pages, which was devoted to reproducing census data from our seven cities.

I wrote the chapter on Montreal in the 1983 edition. Despite having not lived in Montreal since 1977, I still have a great interest in the politics and government of the city and its wider metropolitan area. I maintain that students of urban politics outside Quebec should pay much more attention to Montreal. The authors of the Montreal chapter in this book describe well what happened with mergers and demergers during the early 2000s. But perhaps because they live with Montreal's peculiarities every day, they do not emphasize to outsiders the truly remarkable complexity of its local political institutions.

For example, the mayor of Montreal has major leadership roles in four different decision-making bodies: the borough council of Ville-Marie (i.e., the downtown central city), over which she presides; the Montreal city council; the agglomeration council; and the Montreal Metropolitan Council, over which she also presides.

The Montreal authors rightly point out that, after the amalgamation in 2002, there was only one mayor in the territory of the old Montreal Urban Community (i.e., the islands of Montreal, Ile-Bizard, and Dorval). Now there are 33 mayors, each with their own decision-making council. Some are mayors of de-amalgamated municipalities and some are mayors of boroughs within the city of Montreal. By reading the Montreal chapter, the rest of the country (and perhaps even observers of municipal amalgamations in other parts of the world) can discover how and why this remarkable development occurred.

It is true, as the authors of the Montreal chapter point out, that there is not much difference between the legal authority of a de-amalgamated municipality and a Montreal borough. However, voters in the former cannot vote in a Montreal city election. Because these voters live mainly in more anglophone and conservative areas, could it be that Valérie Plante and Projet Montréal might not have been nearly as politically successful had the de-amalgamations never happened? The authors of both the Montreal chapter and of the book's Conclusion claim that linguistic cleavages in Montreal city politics have significantly diminished. But perhaps the de-amalgamations are an important causal factor.

Montreal's boroughs and de-amalgamated municipalities provide an important opportunity to observe the possible effects of decentralization on housing policy, a subject of considerable concern for all the authors in this book. Everyone now acknowledges that municipal zoning is a key factor affecting the creation of new housing. Central parts of Montreal are well known for their residential density, typically provided by rows of attached duplexes and triplexes, many with the iconic outside staircases. Outside the centre, Montreal's boroughs and de-amalgamated municipalities control local zoning. Will they modify zoning rules about single-family housing as many municipalities in the country already have? Or will their relatively small populations make them more susceptible to local pressures to maintain the status quo? It is ironic indeed that the Montreal city council has scarcely any control over such matters.

Both editions of this book have been concerned with describing and analysing the developments within the municipal institutions of our major cities. In the first edition, we covered a longer historical period and were able to rely heavily on the work of urban historians. Authors in this edition are primarily concerned with events that have occurred over the last forty years. In the absence of academic historical accounts, they have had to rely primarily on local media and, in some cases, government documents. Both sources are now problematic. For example, there was no serious government study that investigated the need for the amalgamation of Toronto in 1998.

More seriously, we all know that local media, even in our big cities, are increasingly unable to provide the detailed and critical coverage of municipal decision making that is so necessary for a healthy local democracy. Where will the raw material come from to enable future scholars to provide narrative accounts of important developments within our major municipalities? I sometimes wonder if it will be necessary for them to attend municipal meetings themselves and act as their own reporters. Or maybe there will be more freelance municipal reporting on the internet. If so, will it be preserved in ways that future academics can easily access?

I noted at the beginning of this Foreword that the first edition of this book attracted the attention of journalists in major newspapers. Writing in the *Winnipeg Free Press* (18 June 1983), journalist Terence Moore reviewed our book and a book by Meyer Brownstone and T. J. Plunkett about the creation of Winnipeg's "Unicity" in 1972. Moore took issue with claims by Brownstone and Plunkett and by the authors of our Winnipeg chapter that the city would be better off if the mayor were not directly elected but was instead chosen by the city council. He criticizes these scholars for quoting each other about the need for coherent municipal political parties.

Terence Moore refers to the "Magnusson–Sancton reader on politics in seven Canadian cities" and states that "In the narrow world of urban studies in Canada, there is precious little for scholars to drink besides each other's bath water. The result is a swelling edifice of scholarly literature whose foundation of original research does not expand in proportion."

Hopefully no one can make such a claim today about the scholarly literature on Canadian urban politics, especially after the publication of the new edition of *City Politics in Canada*. More importantly though, I hope that informed observers outside "the narrow world of urban studies" will simply be paying attention. If no one except other scholars care about what we write, why are they writing?

Acknowledgments

This book would not be possible without the foundations laid by an earlier generation of scholars. We are especially grateful to Warren Magnusson and Andrew Sancton, whose 1983 book of the same title was our inspiration, and who agreed to write the book's foreword and afterword. Warren sadly passed away in April 2025. We dedicate this book to him. The contemporary study of urban politics and local government in Canada would not exist without their essential contributions. We are all Warren and Andrew's intellectual descendants. We also pay tribute to other Canadian scholars of their generation and earlier – some still active, others sadly no longer with us – who have shaped our field, influenced us, and sometimes mentored us, not least Caroline Andrew, Lionel Feldman, Frances Frisken, Katherine Graham, Pierre Hamel, Christopher Leo, James Lightbody, David Siegel, Richard Simeon, Richard Stren, Paul Tennant, Louise Quesnel, and Robert Young. We hope this tradition will be continued by a new generation of scholars.

We are very grateful to our chapter authors, each of whom gamely took up the challenge of interpreting the political development of their cities, participated in a lively manuscript workshop in March 2023, and revised their chapters to account for changing events. Thanks to them, we know much more about these cities than we did before. We also thank Daniel Quinlan for his enthusiasm for the project and efficiently shepherding it through the production process at University of Toronto Press. We are also grateful for the assistance of the J.B. Smallman Publication Fund and the Faculty of Social Science at the University of Western Ontario. We acknowledge the help of our research assistants at Western University: Amanda Gutzke, who helped to prepare the author information package, and Caleb Duffield, who assisted with the manuscript workshop. We are grateful to Frances Bula, Jill Grant, Anne Mévellec, Daniel Silver, and Eliot Tretter for taking the time to comment on chapter drafts.

Finally, we thank our families. Without the love and support of Amanda, Daphne, and Jenny, we could not do the work we do. We dedicate this book to our children, future urban citizens: Jacob, Oliver, Hannah, Zoe, and Sam.

Martin Horak, Jack Lucas, and Zack Taylor
September 2025

CITY POLITICS IN CANADA

Introduction: Bringing Local Politics Back in

Martin Horak, Jack Lucas, and Zack Taylor

Canada's big cities are at the forefront of social and economic change. They account for most of the population growth in the country, they are magnets for immigrants from all over the world, and they have led Canada's shift from an industrial to a post-industrial economy. Today – perhaps more than ever – Canada's cities are places where new policy problems, new political movements, and new demands for representation first emerge. While there is no shortage of academic work on specific aspects of urban politics in Canada, it has been forty years since we have had a broad survey of local politics in big Canadian cities. The first edition of *City Politics in Canada* (1983), coedited by Warren Magnusson and Andrew Sancton, remains perhaps our best single treatment of Canada's urban politics. The book analysed the roots and evolution of urban political conflict and the dynamics of local politics, focusing on the classic questions of whose interests prevail as well as when, why, and how. To do this, it presented holistic, developmental studies of politics in seven large Canadian cities, written by specialists on each city. The book was framed by an introductory chapter in which Magnusson presented what remains perhaps the definitive account of the history of Canadian urban politics and a conclusion in which Sancton interpreted the city case studies to draw broader conclusions about the character of city politics in Canada.

When the first edition of *City Politics in Canada* was published, many of Canada's big cities had recently experienced a period of tumultuous political change. Beginning in the 1960s, coalitions of neighbourhood groups and grassroots organizations – often led by urban middle-class professionals – mounted a spirited challenge to the post-war modernist urban development consensus. Pushing back against business leaders, developers, urban planners, and politicians, who had transformed the face of Canadian cities in the 1950s and 1960s with downtown redevelopment projects, highways, high-rise housing, and other major initiatives, these "reform" activists successfully captured power on many big-city councils with an agenda of neighbourhood preservation and resident participation in decision processes. Surveying these developments in *City Politics in Canada*, Andrew Sancton concluded that "The main line of cleavage in Canadian municipal politics involves attitudes towards development," and "Virtually all conflict in Canadian urban politics can be located on a pro-development–anti-development spectrum" (295). At the same time, Sancton noted that despite their self-conscious "progressiveness," reformers did not fundamentally change the focus of Canadian urban politics. He wrote, "Above all, municipal politics in Canada is about property" (296). Given the property-focused policy jurisdiction of local governments and their dependence on property tax revenues, even reform politicians pursued "boosterism" and local economic growth (292).

After forty years of social, economic, and institutional change, it is high time that we revisited Sancton's description of local politics in urban Canada. A focus on *local* politics may seem quaint when scholarship has shown that urban policy outcomes are shaped by social and economic shifts (immigration, globalization, deindustrialization) and government policies (provincial and federal) at broader spatial scales. Yet the multilevel character of urban governance does not make local politics in cities marginal or epiphenomenal. Indeed, local governments' combination of control over physical development and permeability to local interests makes them highly consequential sites of democratic representation and substantive decision making. The present book remains animated by our conviction that we need to bring the local level back into focus if we are to advance our understanding of urban politics in Canada as it operates today.

This book aims to address two big questions. First, how much has local politics in Canada's cities really changed over the last forty years? Specifically, is political conflict still structured by differing views on development and focused, above all, on property concerns, or have the transformative forces of immigration, deindustrialization, and globalization introduced new axes of conflict and new policy concerns? Second, has politics in different cities responded in different ways to the social, institutional, and economic changes of the last forty years, and if so, what explains those distinct trajectories? Inspired by the first edition of *City Politics in Canada*, this book aims to provide a set of holistic and comparative assessments of Canadian big-city politics that shed new light on these questions. Like its predecessor, it focuses on seven major Canadian cities: Toronto, Montreal, Vancouver, Halifax, Ottawa, Winnipeg, and Calgary[1] (see Figure 1.1).

The metropolitan and multilevel dimensions of Canadian city politics have been the subject of significant scholarly work in recent decades. We recognize the importance of the metropolitan and multilevel contexts for local politics, but our primary focus in this book is on the practice of local politics in the *central municipalities* of large urban areas in Canada – a subject that, as we will discuss later in this chapter, has not attracted much scholarly attention in recent years. While our authors do not ignore the metropolitan and multilevel contexts, they treat them as exactly that – *contexts* that set the stage for the main object of analysis, which is local politics. In this respect, too, this volume follows in the footsteps of the 1983 edition. Yet much has changed in the intervening decades, and as a result, the book you read today is also something entirely new – new authors, new themes, new data, and new arguments.

Figure 1.1: The seven cities

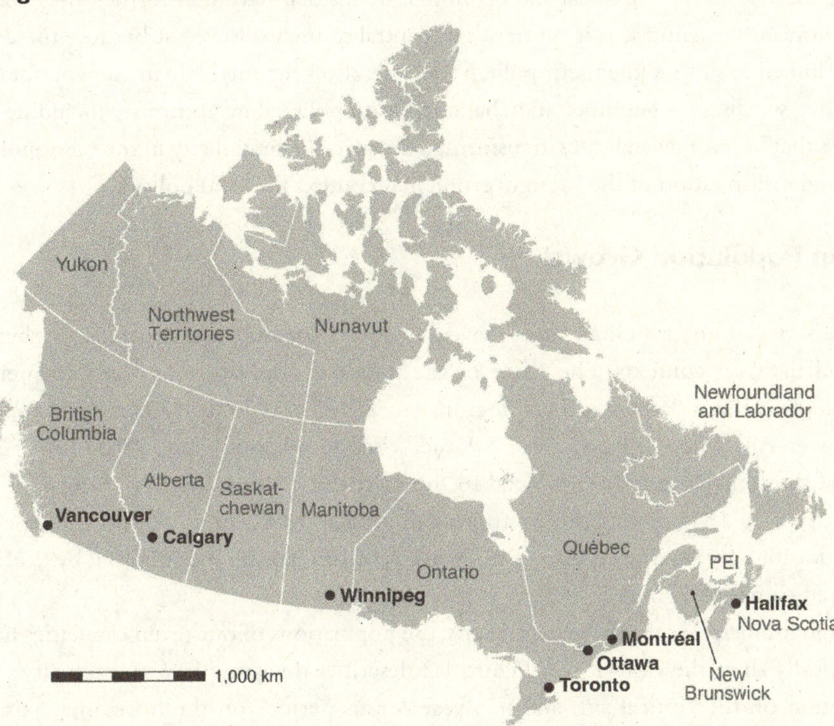

Before we turn to these new analyses, this chapter will set the stage. In the next section, we review the processes of economic and social transformation in urban Canada, with a focus on our seven cities, and describe changes to municipal, metropolitan, and multilevel political institutions since 1980. We then discuss developments in Canadian urban politics and governance scholarship over the last four decades, discussing advances in our understanding of multilevel politics, municipal elections, and urban policy. However, we also note the absence of – and the need for – the kind of local, comparative, and developmental analysis that we offer in this book. The chapter concludes with a brief overview of the rest of the volume.

FORTY YEARS OF SOCIAL AND ECONOMIC TRANSFORMATION IN URBAN CANADA

Contemporary Canadian urban politics takes place in a very different social and economic setting than it did forty years ago. As a result of a multiplicity of processes – immigration, post-industrialization, municipal amalgamation, suburbanization – Canada's major cities have become larger, more diverse, and more socio-economically polarized. To describe this new setting, we first survey the social and economic trends that have transformed the seven cities in this volume. Our main focus is on trends in central municipalities, but because these trends are closely linked to shifts at the metropolitan scale, we also keep this scale in view in our discussion. After that, we discuss continuity and change in local political institutions – including boundary changes that have, in several cases, transformed the place of central city in the metropolitan area – and the transformation of the intergovernmental context for local politics.

Urban Population Growth

Canada's largest metropolitan areas have become increasingly important in their provincial and national contexts. The share of the national population and employment growth in Canada's three largest urban regions – Toronto-Hamilton-Oshawa, Montreal, and Vancouver-Abbotsford – has increased decade over decade since the 1980s (Taylor 2024), and most of the remaining growth flowed to the next three largest regions: Calgary, Edmonton, and Ottawa. In 1981, the census metropolitan areas (CMAs) in which our seven central cities are located together accounted for 48 per cent of Canada's population; by 2021, this had increased to 57 per cent.

At the municipal scale of the central city, the populations of our seven case cities have grown dramatically since the early 1980s. Figure 1.2 describes this growth, with each city's municipal population on the vertical axis and five-year census periods on the horizontal axis. We mark amalgamations with vertical black lines and summarize each city's overall percentage growth since 1981 and 2006 in the bottom right corner of each panel. Each city's population trends upward since 1981, in some cases dramatically so.

These central cities' populations have grown steadily despite powerful forces of suburbanization, avoiding the hollowing out experienced in many American big cities. In some cases – Calgary, Halifax, Montreal, Ottawa, and Toronto – the apparent growth of municipal populations has occurred at least in part through the expansion of municipal boundaries, which we describe in more detail below. The amalgamation of municipalities to create new and larger central cities, or the annexation of neighbouring territory by central cities, has reinforced the relative weight of these central cities within their metropolitan, provincial, and national contexts. Vancouver and Winnipeg, by contrast, have essentially the same boundaries they did a generation ago.[2] While consistent long-term data are not available, the residential populations of downtowns and older core neighbourhoods of central cities have grown rapidly in recent years, even as suburban expansion has continued apace (Taylor et al. 2014; Gordon 2022; Pourali et al. 2022; Statistics Canada 2022). Given the fact of amalgamation or annexation in some cities but not in others, we cannot directly compare population growth rates over the full 1981–2021 period. Considering the past fifteen years, however, when boundaries have been stable, we see that some of our cities have grown faster than others. Calgary has grown the most (32 per cent) and Montreal the least (9 per cent). The other cities fall somewhere in between.

Figure 1.2: Municipal population change, 1981–2021, indicating amalgamation years and percentage change in population since 1981 and 2006

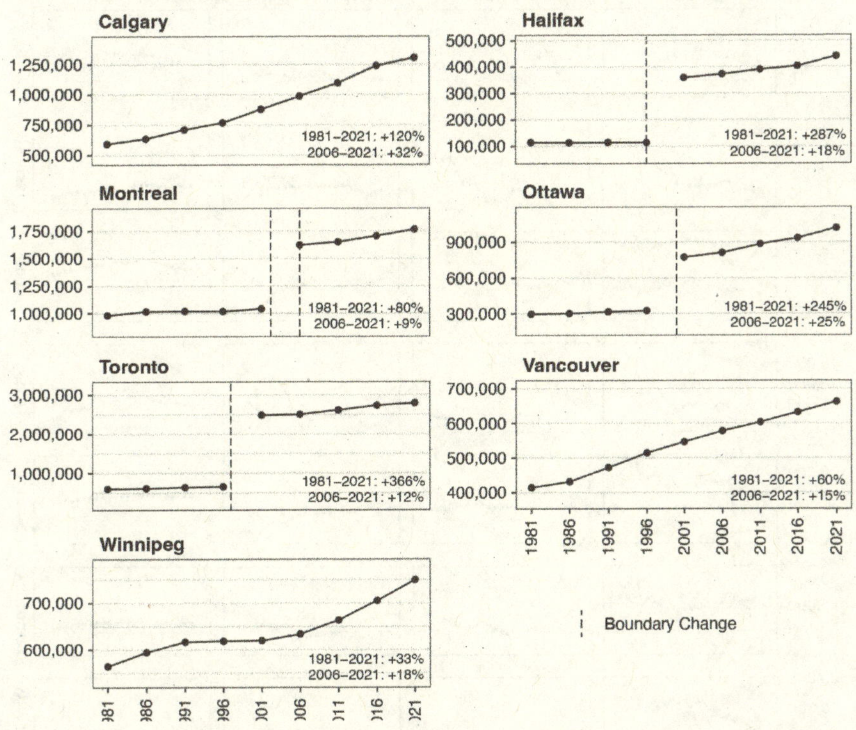

Notes: The vertical black lines indicate municipal amalgamations. Montreal experienced a two-step restructuring between the 2001 and 2006 censuses – a merger of all municipalities on the island in 2002, followed by a partial demerger in 2006. Calgary's incremental annexations of surrounding urbanizing countryside are not indicated, nor is the secession of tiny Headingley from Winnipeg in 1993.

Social Diversification through Migration

Urban population growth (or decline) can be caused by a variety of processes, including international migration, migration from other provinces, migration from other parts of the same province, and "natural" change due to births and deaths. Figure 1.3 summarizes these patterns at the metropolitan scale from 2002 to 2020. From left to right, the panels in each row show the annual amount of population growth or decline that occurred in each city as a result of these processes.

In the coming chapters, our authors have more to say about the significance of migration and natural population change for the practice of politics in their cities. For now, we merely highlight a few general patterns. Most importantly, the figure reinforces the importance of

Figure 1.3: Components of population growth by CMA, 2002–2020

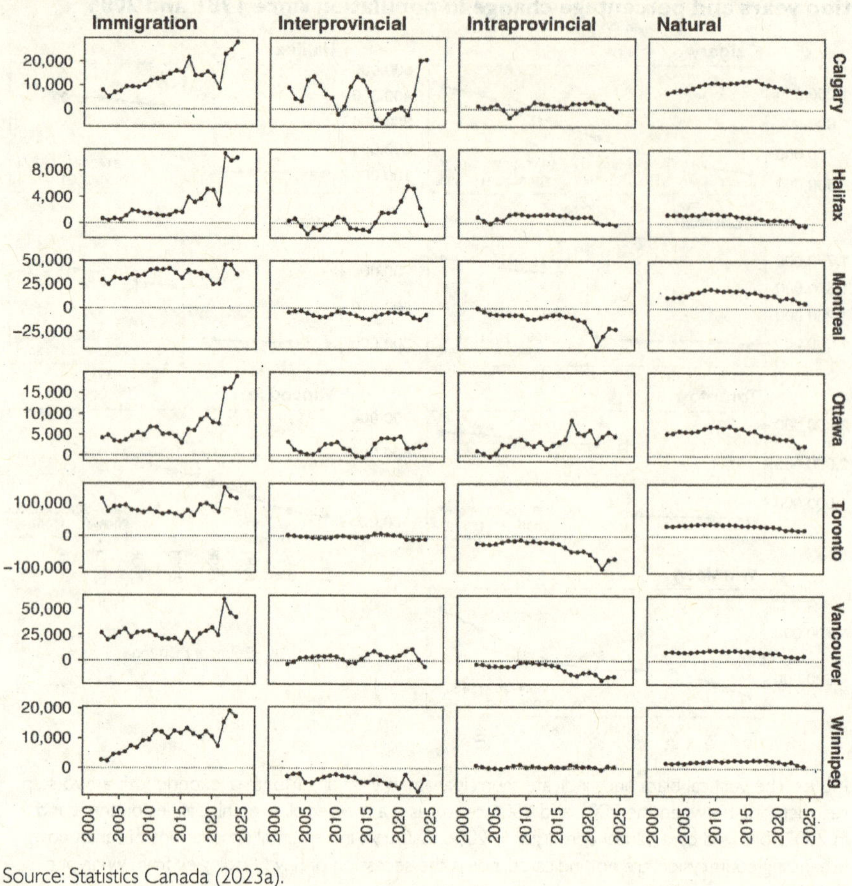

Source: Statistics Canada (2023a).

international migration for central city growth in Canada. Indeed, over the past twenty years, Toronto, Montreal, Vancouver, and Winnipeg have been net *exporters* of non-immigrant residents to other parts of Canada. Calgary, Halifax, and Ottawa have benefitted from domestic in-migration, although these flows have fluctuated with the business cycle and have generally been smaller than international immigration. Without immigration, Canada's major cities would shrink.

This influx of international migrants, largely from East and South Asia, but also from Latin America, the Caribbean, and Africa, has combined with net outflows of domestic-born Canadians to transform the demographic composition of Canada's central cities. Figure 1.4 summarizes the percentage of each city's population who are immigrants (grey lines) or who belong to visible minorities (black lines), which is the best available census category for measuring the presence of racialized residents (non-white, not including Indigenous). As a result of international

Figure 1.4: Growing diversity, 1981–2021

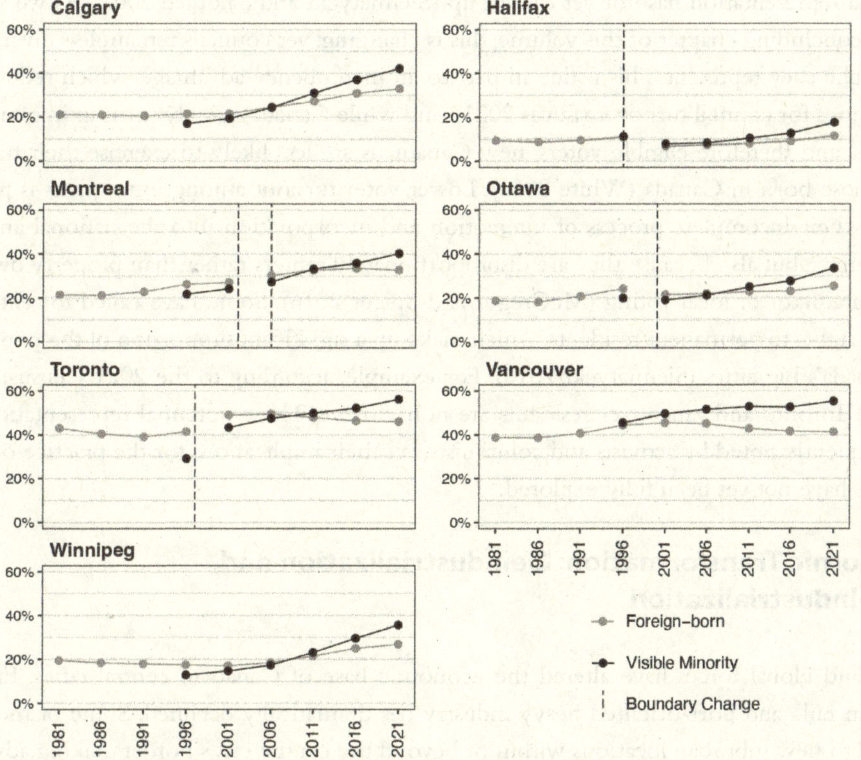

Notes: Visible minority data are available starting in 1996. The vertical black lines indicate municipal amalgamations. Montreal experienced a two-step restructuring between the 2001 and 2006 Censuses – a merger of all municipalities on the island in 2002, followed by a partial demerger in 2006. Calgary's incremental annexations of surrounding urbanizing countryside are not indicated, nor is the secession of tiny Headingley from Winnipeg in 1993.

immigration, the population of visible minorities has steadily increased in each of our central cities since 1996. In Toronto and Vancouver, visible minorities make up almost 60 per cent of the population, while over 40 per cent of the population is foreign-born. Calgary is not far behind, with visible minorities making up over 40 per cent of the population, while almost 35 per cent of the population is foreign-born. Winnipeg and Ottawa are similar: visible minorities making up 35 per cent, while 25 per cent is foreign-born. Halifax is the least diverse of the cities profiled in this volume: visible minorities make up 17 per cent of the population, while 12 per cent is foreign-born. There are of course important and politically salient variations within our cities – for instance, Halifax has a substantial historic Black population (4.7 per cent), and Winnipeg has the largest urban Indigenous population of any large Canadian central city (12.3 per cent).[3]

While rapid social diversification has quite literally changed the face of Canada's largest cities, political representation has not yet caught up (Siemiatycki and Saloojee 2002). As we will see in the concluding chapter of this volume, this is changing; yet councils remain less diverse than the public they represent. This is due in part to an incumbency advantage, which reduces opportunities for council turnover (Lucas 2021), and while Canada quickly turns immigrants into citizens, and therefore eligible voters, new Canadians are less likely to exercise their franchise than those born in Canada (White 2023). Lower voter turnout among immigrants is partially due to their incomplete process of integration and incorporation into the national and local community, but also because they are disproportionately renters rather than property owners – a key predictor of local voting (McGregor and Spicer 2016). Some have called for extending voting rights to permanent residents, which make up a significant proportion of the population of Canada's big cities (Siemiatycki 2010). For example, according to the 2021 Census, 17 per cent of Toronto and Vancouver residents are non-citizens. These potential representation gaps are frequently noted by activists and columnists, yet their implications for the practice of urban politics have not yet been fully explored.

Economic Transformation: Deindustrialization and Post-Industrialization

Local and global forces have altered the economic base of Canadian central cities. Employment in rail- and port-oriented heavy industry has dramatically declined. Some of these jobs moved to new suburban locations within or beyond the central city's borders, taking advantage of cheaper land, lower taxes, and proximity to highway infrastructure. Others moved outside Canada to locations elsewhere in the Americas or overseas, lured by cheap labour, tax breaks, and plummeting shipping costs. Many "old economy" jobs also disappeared with the advent of computers and automation in the 1970s. As Figure 1.5 shows, the trend in all of our municipalities has been towards long-term decline in manufacturing employment (marked in black), while jobs in high-value-added services in the finance, insurance, and real estate (FIRE) sector (marked in grey) have expanded.[4]

These employment shifts are symptomatic of broader socio-economic change, which has in turn reshaped the physical character of cities. In the late 1960s, a new middle class, characterized by higher educational attainment and professional employment in the emerging knowledge economy, began rehabilitating older housing in centrally located immigrant and working-class neighbourhoods (Bourne 1967; Ley 1994). To house new knowledge economy jobs, glass and steel office towers replaced low-rise stone and brick office, warehouse, and factory buildings in the downtowns. Renewed interest in urban liveability sparked public investments in downtown brownfield areas, often adjacent to waterfronts – for example, Historic Properties in Halifax, the CN Tower and Harbourfront in Toronto, the parliamentary precinct in Ottawa, Vieux Montréal, and False Creek in Vancouver. Often these were hitched to the creation of new tourism destina-

Figure 1.5: Post-industrialization: Change in employment by industry, 1981–2021

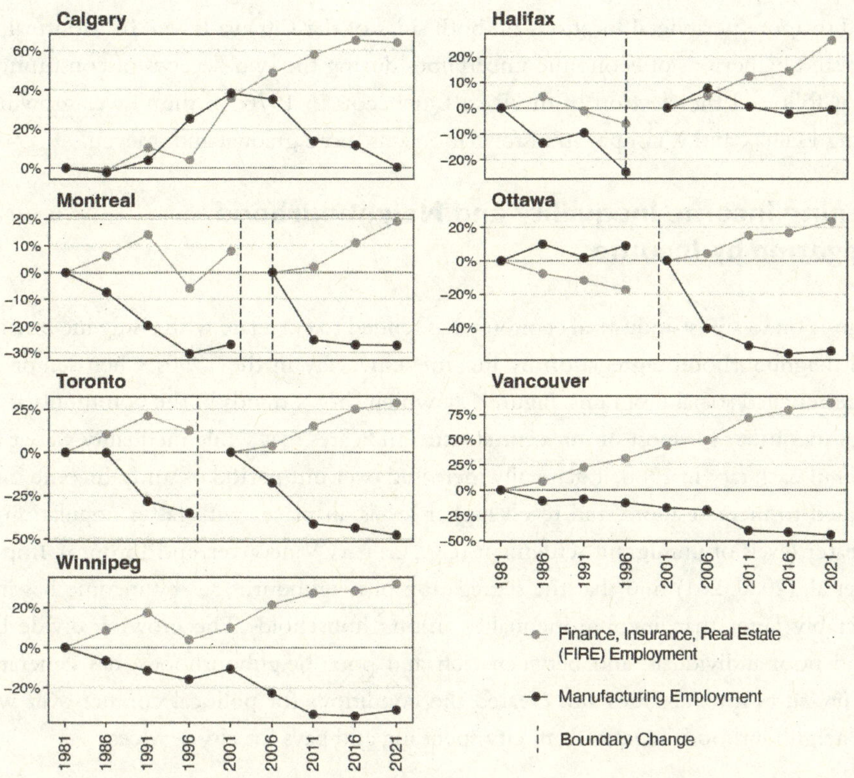

Notes: The values indicate the percentage change in employment in the finance, insurance, and real estate (FIRE) and manufacturing sectors relative to the start year. The start year is rebased after amalgamations. The vertical black lines indicate municipal amalgamations. Montreal experienced a two-step restructuring between the 2001 and 2006 Censuses – a merger of all municipalities on the island in 2002, followed by a partial demerger in 2006. Source: Employment by Industry, Census of Population, various years. Counts are by place of residence, not by place of work.

tions, new convention facilities and sports arenas, and bids for mega-events such as Olympics, world expositions, and the Canadian centennial in 1967. The urban liveability agenda, propelled into politics by the growing electoral power of the new middle class, also sought to protect residential areas adjacent to expanding central business districts and historic buildings within them. As we will see, the political consequences of this agenda remain important in many of our case cities.

In the larger, fast-growing cities – Calgary, Toronto, and Vancouver – these processes of economic change were transformational. Central business districts and core residential neighbourhoods in these cities looked very different in the 1980s and 1990s from four decades earlier. In Ottawa, redevelopment was constrained and redirected by the dominance of government functions, large-scale public land ownership, the centrality of the historical parliamentary precinct, and the planning of the National Capital Commission. Expanding federal office functions were directed to more peripheral locations on both sides of the Ottawa River. In Montreal, growth was arrested by periods of economic uncertainty during the two decades of constitutional debate that followed the election of the Parti Québécois in 1976. In the slower-growing cities, including Halifax and Winnipeg, transformation was more gradual and uneven.

Widening Income Inequality and Neighbourhood Segregation by Income

The transition to a post-industrial economy has tended to correlate with rising income inequality and neighbourhood segregation by income, especially in the 1990s when senior governments retrenched social programs. Figure 1.6, which shows trends in the commonly used Gini index of inequality for most of our central cities, indicates that while inequality was at a different level in each city in 1980, it generally increased over our period of study. Income inequality has tended to increase more, and reach higher levels, in places with faster population growth and greater levels of immigrant settlement (e.g., Calgary, Vancouver, and Toronto). Importantly, Grant et al. (2020, 261) find that the segregation of neighbourhoods by income has increased considerably faster than income inequality among households. The growing divide between rich and poor individuals, and between rich and poor neighbourhoods, has generated new policy pressures for our cities and created the conditions for political conflict over who, and whose neighbourhood, benefits from city spending and pays for city services.

CONTINUITY AND CHANGE IN LOCAL GOVERNMENT INSTITUTIONS

Demographic and economic changes have transformed Canada's central cities and their respective metropolitan areas. These changes are the backdrop for, and in some respects the drivers of, the many changes to local institutions and intergovernmental relations over the past several decades. The most visible changes have been related to geographic boundaries – the municipal amalgamations that took place in Ontario, Quebec, and Nova Scotia in the 1990s and early 2000s, as well as the emergence of new institutions of metropolitan coordination. Perhaps less appreciated, but nonetheless important, are changes to the legal foundations of local governance since the 1990s, which have altered the powers of municipalities and the discretion with which they may use them. A third important shift is the ebb and flow of federal government intervention in local affairs, principally through project funding, since the 1980s. Finally, we have seen incremental change in the composition of revenues and expenditures in our cities, generally in the context of fiscal constraint and inconsistent intergovernmental support.

Figure 1.6: Gini index of inequality for household income, 1980–2015

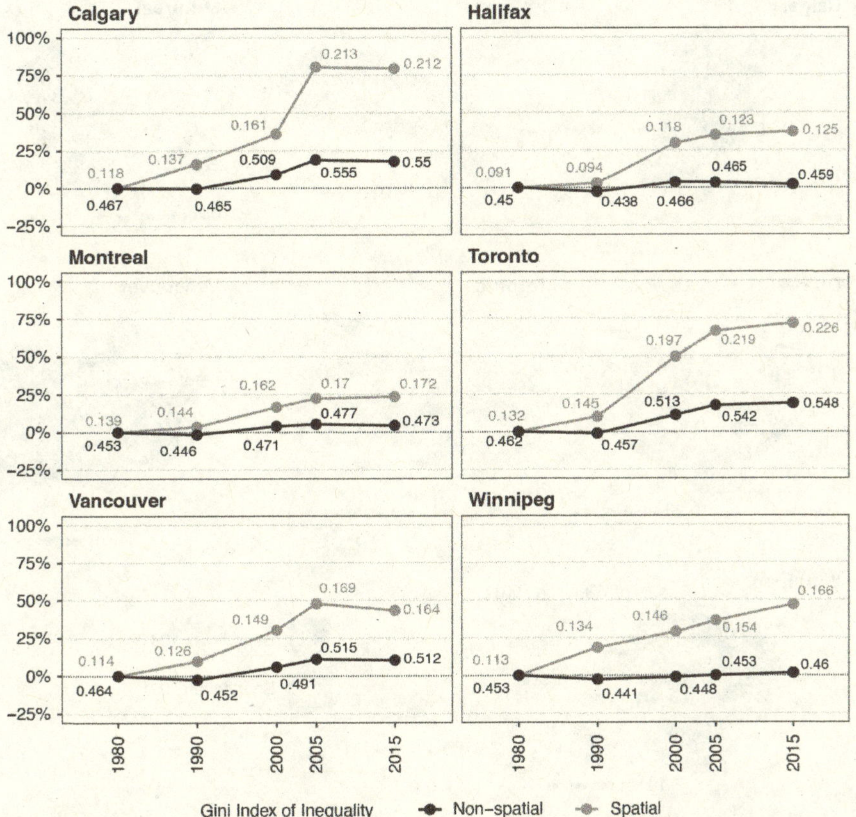

Notes: The non-spatial Gini index (shown in black) represents the level of income inequality among individuals. The spatial Gini index (shown in grey) represents the level of neighbourhood segregation by income. The higher the index value, the greater is the inequality or segregation level. To compare the rates of change of income inequality and income segregation, the vertical axis indicates percentage change relative to 1980. To enable comparison across cities in absolute levels of inequality and segregation, the points are labelled with the Gini index values. The Gini indices are calculated on total (before-tax) individual income for the Census Metropolitan Area. Source: Alan Walks, who prepared these data for Grant et al. (2020).

Municipal Restructuring and Metropolitan Governance

The restructuring of municipal institutions and territorial jurisdiction through amalgamation and annexation has profoundly influenced politics in almost all of our central cities. We map these changes in Figure 1.7, with 1981 municipal boundaries in dark grey and today's bounda-

Figure 1.7: Municipal boundary change, 1981–present

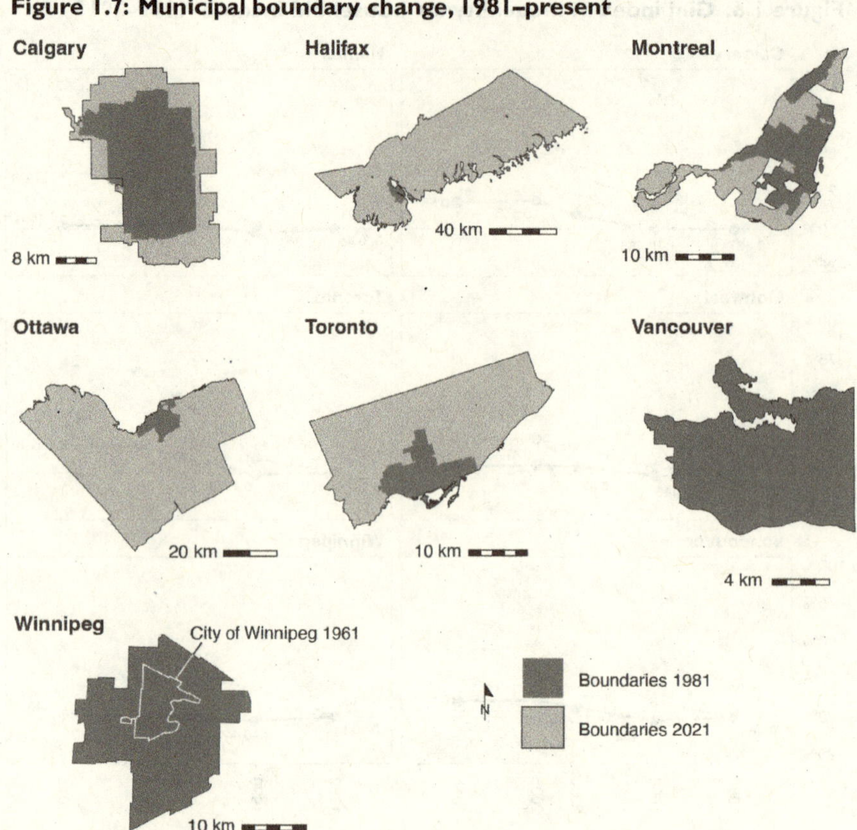

Notes: Dark grey = 1981; light grey = present day. White outline in Winnipeg indicates the boundaries of the former City of Winnipeg prior to the 1972 amalgamation.

ries in light grey. While the most recent round of restructuring, which occurred in Halifax in 1996, Toronto in 1997, Ottawa in 2000, and Montreal in 2002 (followed by a partial deamalgamation in 2006), reverberates even today, more distant institutional developments cast a long shadow in our other cities (Sancton 2000). Winnipeg experienced two major restructurings in earlier decades – the creation in 1960 of a two-tier metropolitan government encompassing 13 local municipalities and the consolidation of Metro Winnipeg into a large single-tier "Unicity" a decade later (Brownstone and Plunkett 1983). While Calgary never experienced a multi-municipal amalgamation, its physical and political development has been shaped by a policy of incremental annexation, whereby municipal boundaries were expanded outward prior to the servicing and development of peri-urban lands (Foran 2009; Taylor et al. 2014).

The motivations for amalgamations – in each case imposed by provincial governments, often over considerable opposition from local politicians and civil society – are complex. In some provinces, including Nova Scotia and Ontario, a discourse of efficiency prevailed: costs and tax burdens would be reduced, and fiscal and administrative capacity increased, by removing overlap and duplication. In Toronto and Montreal, the provincial government was also motivated by perceived competitive pressures – political consolidation would create a stronger international identity for the city and greater policy capacity to attract investment. Some have also argued that the primary motivation for municipal amalgamations was political, in the sense that provincial governments pursued amalgamation to deprive their political opponents of valuable institutional power bases (Keil 2000; but see Sancton 2006).

As chapters in this volume show, the integration of multiple communities with distinct identities and characteristics into a single, territorially expansive political unit has generated enduring axes of political disagreement. Amalgamations in Toronto and Winnipeg have pitted residents of older, denser urban cores against suburbs. The Halifax and Ottawa amalgamations combined not only dense pre-war urban cores with post-war automobile suburbs but also vast rural territories. Montreal's initial merger dissolved English-speaking communities into a majority francophone unit, which the *défusions* (deamalgamations) soon partially reversed. Calgary's politics is also shaped by incremental annexation, which steadily increased the suburbs' weight in civic affairs. Vancouver is the exception to the post-war pattern of incremental or disjunctive boundary change, having assumed its current boundaries in 1929, with the merger of the original City of Vancouver with neighbouring South Vancouver and Point Grey. Yet its politics features enduring geographic conflict between the neighbourhoods in the working-class and gentrifying east and the well-to-do west.

In some of our cities, amalgamation has created de facto metropolitan local governments (Taylor 2022). The Halifax and Ottawa amalgamations collapsed two-tier structures and associated special-purpose bodies into single-tier units of metropolitan scope. In Toronto and Montreal, amalgamation paralleled ongoing metropolitan governance debates, but it did not result in the creation of fully consolidated metropolitan municipalities. The proposed creation of a new upper-tier body for the Greater Toronto Area was rejected in favour of the province incrementally taking on a more muscular role in local land-use and infrastructure planning. The Quebec government took a different path, establishing "metropolitan communities" for the Montreal, Quebec City, and Gatineau regions – a third layer of local government with coordinating roles in transit, economic development, and environmental policy.

In western Canada, meanwhile, the trend has been towards provincially mandated bottom-up coordination. The Metro Vancouver Regional District, which was created in the 1960s from existing regional institutions established decades earlier (Taylor 2019), serves as the institutional template. Multipurpose metropolitan coordinating institutions have recently been created or substantially reformed in Calgary and Winnipeg. In sum, as shown in Table 1.1, amalgamations, where they occurred, increased the central city's share of the metropolitan population, sometimes creating local governments of near-metropolitan scope. Where restructuring did not occur, central cities have become smaller players on the metropolitan stage.

Table 1.1: Central city population as a percentage of CMA population, 1981 and 2021

	Central city population 1981	% of CMA population 1981	Central city population 2021	% of CMA population 2021
Montreal	967,470	34.2%	1,762,949	41.1%
Toronto	592,625	19.8%	2,794,356	45.1%
Halifax	113,430	40.8%	439,819	95.1%
Ottawa	291,850	53.3%	1,017,449	89.6%
Winnipeg	558,430	95.5%	749,607	89.9%
Vancouver	408,080	32.2%	662,248	25.2%
Calgary	587,020	99.0%	1,306,784	88.2%

Note: CMAs, or Census Metropolitan Areas, are defined by Statistics Canada based on commuting and other relationships. Sources: Taylor (2022); Statistics Canada (1981b, 1981a).

Ultimately, the geography of political conflict in Canada's big cities operates at two scales. *Intra*municipal conflict – which is the main focus in this volume – transpires within municipal institutions, while *inter*municipal conflict plays out between municipalities, often mediated by more or less elaborate and effective metropolitan institutions. The forces that give rise to distinct place-based political interests and agendas are the same; what differs are the local institutions through which these interests and agendas are channelled. As we will see, these institutional variations can result in the differential mobilization and demobilization of groups in society and the politicization or depoliticization of different issues and policy domains in city politics.

The Legal Status and Authority of Local Governments

To use a tired phrase, Canadian municipalities are "creatures of the province"; they have no independent existence outside provincial law, and they exercise powers delegated by the provinces. As provincial legislation sets out the range of voluntary and mandatory municipal functions and responsibilities, and defines institutional structures, these laws and regulations shape the practice of urban politics and local governance. It is therefore necessary to evaluate alterations to the legal frameworks, which have occurred over the past thirty years and which establish and empower the seven municipalities discussed in this book (for a review, see Taylor and Dobson 2020).

Since before Confederation until the 1990s, Canadian municipal legislation followed the legal doctrine of "express powers," whereby municipalities were given authority over an enumerated list of functions. Starting with Alberta in 1994 and spreading to most other provinces, modernized legislation now supplements express powers with permissive "spheres of jurisdiction," an expansive general welfare power, and natural person powers. At least in principle, these changes have given local officials more policy autonomy and discretion.

At the same time, provincial governments have established or updated special legislative arrangements for large cities. In 2006, the City of Toronto was detached from the general Municipal Act and now draws its authority from a separate City of Toronto Act. This act largely parallels the Municipal Act, but gives Toronto some additional taxing powers. Winnipeg's special legislation was updated in 2003, giving the power to establish tax increment finance districts and levy energy consumption taxes. Montreal is still governed by general legislation, but a separate act specifies idiosyncratic arrangements and powers for its borough governments. Alberta has taken a different approach for Calgary and Edmonton, giving those municipalities additional powers by regulation.

The effects of these legislative changes may be more symbolic than practical. It appears that these "charters" have done little to differentiate big cities from other municipalities, and neither big-city charters nor modernized general legislation alter the basic fact of municipal subordination to the provinces (Kitchen 2016; Sancton 2016). Indeed, in recent years, there have been major unilateral interventions by provincial governments across the country in local planning, infrastructure projects, and local institutional structures (Taylor 2024). For this reason, civil society groups continue to agitate for constitutional change that would enshrine a protected sphere of municipal jurisdiction (Charter City Toronto 2019; Good 2019; Good et al. 2020).

Multilevel Governance and Federal Involvement

Beyond these legal changes, the past four decades have also seen significant changes in intergovernmental policy delivery and the role of the federal government in particular. By the early 1980s, the federal government had retreated from its experiment with direct federal–municipal policy relations, the Ministry of State for Urban Affairs (Spicer 2011). Its enduring impacts were waterfront redevelopment projects involving federal lands. In the 1980s, the federal government initiated tri-level "urban development agreements" to address localized problems by flexibly mobilizing policy capacity and resources across governments (Doberstein 2011; Bradford 2020). A Winnipeg agreement focused on the economic malaise and social problems of the downtown area has been continually renewed since. Other agreements failed to take flight until the 1990s, when a new Liberal government established downtown-focused agreements in Edmonton, Victoria, and Vancouver. The negotiation of a Toronto agreement was abandoned following the election in 2006 of a new Conservative government led by Stephen Harper, which viewed federal involvement in urban affairs as an inappropriate incursion into provincial jurisdiction. This also spelled the end of the Liberal federal government's inchoate "New Deal for Cities and Communities," which promised local governments resources and influence over policies and projects (Bradford 2007; Horak 2012).

Two significant federal interventions persisted under Harper, however. The first is the so-called Gas Tax Fund, whereby the federal government, under agreements signed with provincial governments, municipalities, and municipal associations, transfers funds to local governments to support capital projects – mostly public transit – according to a formula (Bradford 2020, 19–22). Originally pegged to the value of the federal gas tax, it was renamed the Canada Community-Building Fund in 2021 in recognition of its expansion under the post-2015 Liberal government led by Justin Trudeau. The second is the maintenance and expansion, especially under the Trudeau government, of sectoral policy frameworks for federal–municipal collaboration, sometimes also involving provincial governments. These include multilevel funding and accountability agreements for place-based immigrant settlement services coordination, homelessness services, economic development, and services to off-reserve urban Indigenous populations (Taylor and Bradford 2020). While Canada does not have an explicit national urban policy, it has an implicit one through the accumulation of funding relationships and multilevel policy partnerships (Bradford 2018).

What impact have these federal programs had on politics and policy at the municipal level? Perhaps most directly, they have reduced municipal revenue and borrowing pressure by infusing federal money into local capital budgets. This has given local governments, especially in the largest cities, greater fiscal capacity to maintain and expand transit and other infrastructure systems. Less visibly, municipal participation in multilevel partnerships has given them a seat at the table when devising interventions in response to difficult social problems with far-reaching consequences, including the growing homelessness and drug addiction crises. This has led them to engage with policy areas beyond local governments' traditional domains of providing property-related services and promoting economic development.

The Fiscal Position of Large Urban Municipalities

Our story so far has been one of dramatic change: population growth, economic and demographic transformation, institutional and legal reform, and new patterns of intergovernmental relations. Yet when it comes to the fiscal position of our cities, the story is largely one of continuity. As Figure 1.8 illustrates with respect to the operating budget, the mix of municipal revenue sources by broad category has been fairly stable over the past twenty years in most of our cities.[5] Property tax remains the single largest revenue source, especially in Halifax, and user fees, mostly comprising transit fares and solid waste fees, represent an important secondary source.[6] Transfers from the provincial and federal governments play a larger role in Ontario, where municipalities have delegated responsibility for social assistance on a cost-shared basis. In sum, Canada's big cities exercise a high degree of fiscal autonomy in the sense that the lion's share of their operating budgets are funded by locally collected revenues. At the same time, dependence on the property tax limits their ability to capture the benefits of population and economic growth. This has generated political demands for new own-source revenues, such as road tolls and taxes on property transactions, as well as greater operating

Figure 1.8: Operating revenue sources, 2003–2019

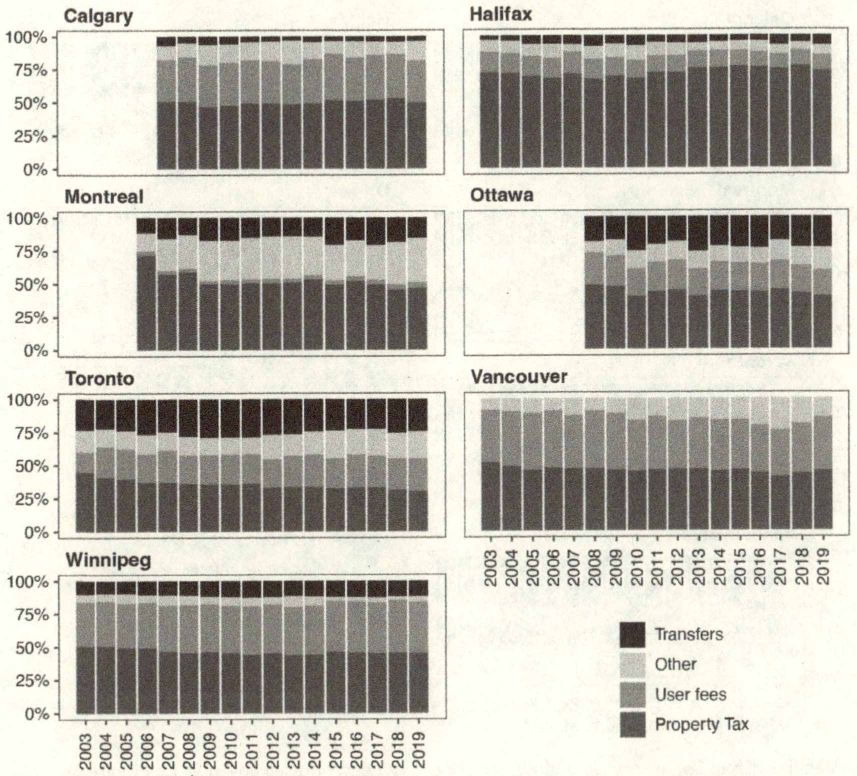

Notes: The property tax category includes payments in lieu of taxes from other governments. In Vancouver, the property tax category also includes revenue from penalties and interest, and intergovernmental transfers could not be differentiated from other sources and are included in the "Other" category. In Montreal, revenues collected by other municipalities associated with agglomeration cost sharing and from the regional transit authority for local transit are not shown. In Toronto, revenues from its unique Municipal Land Transfer Tax are not shown. Data are derived from municipal budget documents and annual financial returns over the past twenty years or post-amalgamation, where available. The series ends in 2019, before the COVID-19 pandemic upended public finance.

support from the federal and provincial governments in the form of program-specific transfers and general revenue sharing.

There has also been considerable stability on the expenditures side (see Figure 1.9). The major spending items (in descending order of importance) are protective services (police and fire, sometimes also including land ambulance), utilities and environmental services (water, sewer, and solid waste management), transportation (public transit and roads), and parks, recreational services, and libraries. There are important idiosyncrasies in some cities. Public transit is region-

Figure 1.9: Operating expenditures, 2003–2019

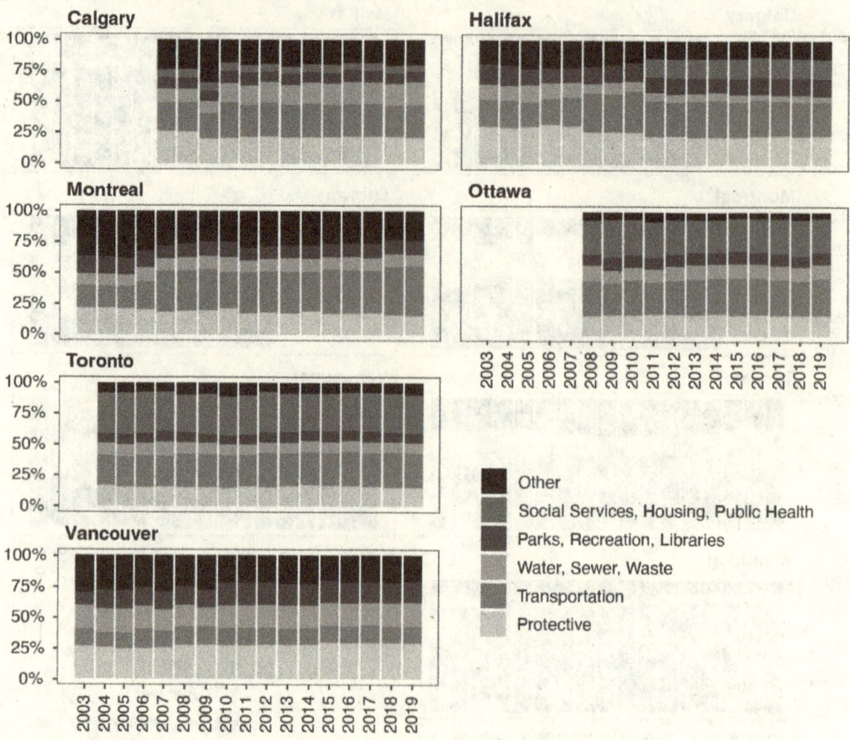

Notes: In Halifax, "Social Services and Housing" includes education starting in 2015. Montreal's expenditure mix shifted in 2006 following the de-amalgamations and the establishment of agglomeration cost sharing. Winnipeg is not included because its expenditure categories were not comparable. Data are derived from municipal budget documents and annual financial returns over the past twenty years or post-amalgamation, where available. The series ends in 2019, before the COVID-19 pandemic upended public finance.

alized in Vancouver. In Ontario, the municipalities deliver social, housing, and public health services, which are entirely provincial in most other provinces, and do so within highly prescriptive policy frameworks. Nova Scotia requires Halifax Regional Municipality to contribute funding for local public education from its property tax. The expenditure mix has remained fairly consistent over time in each of our cities. The big-ticket items are focused on property and mobility rather than people, with the partial exception of Ontario's delegated administration of provincial social and health services. Focusing on expenditures, however, does not reveal what are perhaps the most important policy levers possessed by municipalities: their regulatory powers over land use and business activities, as well as their control over municipally owned land and infrastructure.

POLITICAL SCIENCE AND POLITICS IN THE CHANGING CANADIAN CITY

Central cities across Canada have clearly experienced dramatic social, economic, and institutional changes since *City Politics in Canada* was first published. Our aim in this book is to illuminate important questions about the character and evolution of city politics in light of these changes. How has local politics in central cities responded to new economic and social realities? What are the issues that now shape political conflict in our cities? Is this conflict still centred primarily around differing approaches to development, as Sancton argued four decades ago? What are the key interests, and who are the most important and powerful local political actors?

How have these elements – issues, cleavages, interests, actors – changed over time? Can we still usefully generalize about the character of big-city politics in Canada, or have different cities taken fundamentally different paths since the 1980s?

Despite forty years of fruitful research on local and urban politics in Canada, we do not yet have clear answers to such questions. In part, this is because very few holistic, developmental studies of local political dynamics in individual cities have been produced – studies like Tom Urbaniak's *Her Worship* (2009), a political biography of Mississauga mayor Hazel McCallion that also describes thirty years of politics in an important and fast-growing Canadian city. Unlike *Her Worship*, most Canadian urban politics scholarship since the 1980s has fallen into one of three types: a *multilevel* focus that examines how non-local economic, institutional, and demographic forces shape urban politics; an *electoral* focus that explores voting behaviour and electoral politics in municipalities; and *policy-sectoral* research that focuses on politics and governance in specific urban policy fields. Each of these strands of research give us some insight into continuity and change in local politics in Canada's big cities in recent decades, but each also leaves many questions unanswered.

The Multilevel Focus

The fiscal restructuring, intergovernmental downloading, and forced amalgamations of the 1990s and early 2000s highlighted the weak legal position of even the largest urban governments in Canada (Good 2019). These provincial government actions, together with a new focus on place-based policy and infrastructure investment at the federal level, prompted many Canadian researchers to turn from a *municipal* to a *multilevel* perspective on urban politics. This perspective has afforded valuable insights into Canadian urban politics. Despite the increasing economic and demographic weight of urban Canada, local governments are only "*part* of a multi-level system of government for cities" (Sancton 2008, 32), and local institutions have in recent decades been deeply shaped by (sometimes unilateral) provincial action (Downey and Williams 1998; Vojnovic 2000; Sancton 2006; Good 2019). However, complex urban policy challenges – ranging from building large-scale infrastructure to supporting the social, cultural, and economic inclusion of new immigrants to building affordable housing – require action by more than one level of government, and multilevel interaction and coordination efforts are common across many urban policy fields (Horak and Young 2012; Lucas and Smith 2019). Local governments in Canada's big cities can be significant players in these multilevel policy processes, but their de facto influence varies widely depending on their administrative capacity and political inclination to engage, as well as by policy field (Horak and Young 2012).

Another strand of "supramunicipal" urban politics research in Canada has explored how global capitalism and neoliberal ideology have reshaped the terrain of urban politics, pushing all levels of government to prioritize economic competitiveness, often at the expense of equity and inclusion (Kipfer and Keil 2002). This work has connected the study of Canadian urban politics and policy to a vast interdisciplinary literature in critical urban studies; of all the branches of Canadian urban politics research, it has arguably had the broadest scholarly impact in terms of connecting the study of Canadian cities to broader debates about the changing character of urban politics and governance. However, while this research has demonstrated that global competitiveness rhetoric is used to frame and justify local boosterism, it may also overstate the novelty of the focus on growth and competitiveness, which – as the first edition of *City Politics in Canada* showed – is a long-standing and institutionally embedded feature of urban politics in Canada. Relatedly, this work's tendency to present local politics as epiphenomenal to changes in the global political economy of capitalism limits its potential as a theoretical frame for understanding important *variation* in local politics and policy processes in cities across Canada (Horak 2013).

The Electoral Focus

Alongside a new emphasis on multilevel approaches to Canadian urban politics and policy, several other aspects of local politics have received increased attention since the 1980s. The most recent area of growth has been in the study of municipal elections and electoral representation. Over the past decade, local election studies have become a major component of Canadian urban politics research, and of research on Canadian elections and voting more generally, with large-scale survey-based studies, modelled on national election studies, exploring turnout, voting behaviour, and policy attitudes in Canadian municipalities (McGregor et al. 2016; Lucas and McGregor 2021; McGregor et al. 2021; Bélanger et al. 2022). This work is often implicitly or explicitly comparative, integrating Canadian municipal elections and voting into studies of voting behaviour and public attitudes in other countries or levels of government – a significant development in a subfield that until recently had been dominated by single-city case studies.

One important insight from these election studies has been that despite the non-partisan character of municipal politics in most Canadian cities, along with municipal governments' tightly circumscribed policy responsibilities, local election dynamics are in many ways similar to those at other levels of government (Lucas and McGregor 2021). For instance, while both practitioners and researchers have argued that Canadian local government is (or ought to be) "non-ideological" in character, recent research has found strong evidence that left–right ideology does structure policy attitudes, vote choice, and political representation in Canadian municipalities (Lucas 2022, 2023; Lucas et al. 2023). Furthermore, contrary to long-standing claims that the smaller scale and absence of partisan "gatekeeping" would make local politics more open to under-represented groups, research on political candidacy and electoral success for women and racialized minorities in Canadian cities suggests that they face many of the same barriers at the local level as they do at other levels of government (Siemiatycki 2011; Tremblay and Mévellec 2013; Breux et al. 2019).

Yet, most researchers in this emerging literature acknowledge that elections in Canadian municipalities do differ in important ways from those at other levels. Most obviously, the absence of political parties from most municipal elections in Canada means that municipal voters have access to very different kinds of information about candidates (and often much less information overall), creating distinctive patterns of accountability and responsiveness at the local level (Breux and Couture 2018). One consequence of this low-information environment is that incumbent candidates dominate municipal elections to a degree that is unparalleled in Canadian provincial or federal politics, often winning more than 90 per cent of the races they contest in big-city elections (Lucas 2021; Lucas et al. 2021). Moreover, as we will see throughout this book, Canada's cities are not demographic or economic microcosms of the country as a whole, or even of the provinces in which they are located. This creates locally distinctive political cleavages and coalitions in municipal elections and politics (Doering et al. 2021).

Unsurprisingly, the salient issues in municipal elections also differ from the issues that dominate provincial or federal elections – and perhaps also from the issues that animated municipal elections in previous decades (McGregor et al. 2016; Silver et al. 2020; Lucas and McGregor 2021). Many chapters in this book suggest that the pro-development versus anti-development divide identified by Sancton four decades ago is not as evident in more recent big-city elections, even as many prominent issues in local elections – transportation and transit, property taxes and economic development, housing, and land use – remain closely tied to the politics of property. In addition, property owners are more likely to vote in local elections than those who do not own property (McGregor and Spicer 2016). While there is undoubtedly a wide variety of salient issues in big-city elections today (e.g., policing, social justice, climate change policy, immigrant settlement), many electorally important municipal policy debates continue to revolve around property-related concerns.

Despite its many contributions, this research on municipal elections has weaknesses that a more general and developmental perspective can help to remedy. Perhaps most importantly, recent municipal elections research does not provide a clear picture of how local electoral politics has been *reshaped* by the changes we described earlier – social change, economic transformation, institutional reform – because, with few exceptions (Doering et al. 2021;

Lucas 2021), electoral research has focused on point-in-time analyses. In addition, municipal elections are only one component of urban politics more generally, and we must be careful not to draw general conclusions about urban politics from analyses of election-specific behaviour or attitudes. Compared with the provincial and federal levels, local policy priorities and solutions in big cities are more amenable to public influence (Sancton 2015, 258–9), but they are also more likely to be shaped by bureaucrats than politicians, especially in technically complex fields (Siegel 2015). Furthermore, the fiscal dependence on property-based revenues in Canadian cities may lead to a gap between electoral promises and substantive responses to many emerging issues in Canadian cities. In other words, as Clarence Stone noted in his classic work on urban regimes in the United States, elections by themselves do not deliver the power to govern a city (Stone 2008).

The Policy-Sectoral Focus

A third important area of urban political science research in recent decades has focused on politics and policymaking in particular policy domains. While this research has been voluminous, it has largely consisted of "orphans and islands," as Taylor and Eidelman (2010) memorably put it – discrete bodies of theory and analysis that are weakly connected to each other, if at all. These orphans and islands engage many different theoretical traditions – urban political economy, critical political economy, policy studies, governance studies, and social movement research – and focus on a variety of specific policy fields. The few cross-field studies that do exist suggest that governance arrangements and policy dynamics differ widely by policy field (Horak and Young 2012; Lucas 2017a). In addition, a disproportionate amount of Canadian urban policy research is grounded in the Toronto context, with much less research in many other city contexts. We know little about how well these findings travel to other Canadian cities.

While these limitations make it difficult to draw broader conclusions about the character and evolution of big-city politics and policy from this literature, we can still advance some tentative insights. Since the 1980s, for example, the "traditional" core urban politics fields of planning, development, and physical infrastructure have seen the rise in a variety of new policy approaches and priorities. In the 1990s, as Canadian cities became post-industrial nodes in a globalized and liberalized economy, local economic development efforts increasingly shifted to emphasize competitiveness in post-industrial sectors, with an accompanying focus on attracting human capital and the "creative class" (Kipfer and Keil 2002; Darchen and Tremblay 2013). In recent years, urban planners have also emphasized the benefits of mixed-use development and well-designed public spaces. However, resident opposition to new development in existing neighbourhoods is common (Moore 2013), and while new research suggests that such opposition rarely succeeds in completely derailing new development (Moore and Caporale 2023), it has encouraged planners and politicians to concentrate new development in existing urban contexts in small nodes of high density in former industrial and commercial zones. Meanwhile, increasing pressure on existing transportation systems, together with climate change concerns and renewed federal infrastructure funding, have fuelled the development of new transit and active transportation infrastructure in many big cities (Assunçao-Denis and Tomalty 2019; Horak 2021).

Beyond the realms of urban development and infrastructure, increased social and ethno-cultural diversity and heightened social inequality have pushed a range of new issues onto local policy agendas. Since the 1980s, urban governments have, to varying degrees, developed initiatives to address the needs of recent immigrants (Good 2009; Siemiatycki 2011), 2SLGBTQI+ residents (Podmore 2015), Indigenous residents (Peters 2012), and other marginalized groups. Increasingly visible urban homelessness since the 1990s has also sparked local efforts to mount a policy response (Smith 2022), and a steep rise in housing costs since the 2010s has left local politicians and policymakers scrambling to find ways to strengthen social and supportive housing systems weakened by three decades of low intergovernmental support (Suttor 2016).

New policy demands related to social diversity and inequality have changed the landscape of local politics in Canadian cities and have, in some cases, resulted in notable policy innovations. Nonetheless, the research suggests that responses to identity-based and redistributive concerns are in many ways limited and constrained, even under ostensibly progressive local councils (Lynch et al. 2013). As we have seen, the property-based revenue systems and core policy responsibilities of local government have not changed much since the 1980s. In this context, local politicians often face little incentive to offer more than symbolic responses to identity-based and redistributive claims. Funding allocations for programs to support marginalized populations are typically small, and even the most promising policy initiatives often fail to secure sustained resources (Peters 2012; Horak and Moore 2015). Successful programs and initiatives often rely on complex governance arrangements that lean heavily on the resources of local non-profit groups and other non-governmental actors (Smith 2022). In short, it seems that the political demands that emerge from the social transformation of big cities run up against the structural logic of Canadian local government institutions.

The Local-Comparative-Developmental Perspective

Urban politics scholars have learned a great deal about city politics over the past four decades from multilevel, electoral, and policy-sectoral research. Even so, we see considerable value in returning once again to the analytical approach that characterized the first edition of *City Politics in Canada*. We characterize this approach as *local, comparative*, and *developmental*.

Following the model of the first edition of *City Politics in Canada*, our authors take an explicitly *local* approach in their chapters, focusing on the practice of politics within a single central city municipality. As the multilevel and policy-sectoral approaches have taught us, there are important drawbacks to the "methodological localism" that can characterize such single-city approaches (Brenner 2009) – the notion that what happens in a city must be caused by processes contained within that city. In our view, however, a *local* analysis need not imply a *localist* explanation. Instead, our authors explore the practice of politics in their case cities in rich detail without neglecting the metropolitan, provincial, federal, and even global processes that shape this practice. Adopting this local approach allows our authors to remain aware of multilevel influences while maintaining an analytical focus on concrete local political action.

Our approach is also *comparative*. Most obviously, our chapters are comparative in the explicit contrasts that our authors draw between their cities and other cities in the book. But our approach is also comparative in a more implicit or "emergent" sense: by placing our seven case cities alongside one another in a single volume, and by asking our authors to structure their analysis in similar ways in each chapter, our book is intended to enable comparison between cities by readers themselves. This implicit comparison not only helps to avoid the dangers of excessively localist explanation but also allows readers to compare among the cities that most interest them.

Finally, our approach is *developmental*, with chapters tracing the political development of the practice of politics in each case city through more than forty years of urban history. Since the publication of *City Politics in Canada*, the subfield of urban political development has blossomed, with numerous studies tracing what political development scholars call "durable shifts in governing authority" through time (see Lucas 2016; Taylor 2019). While we did not ask our authors to adopt the theoretical machinery of the political development approach (Dilworth 2009; Lucas 2017b; Dilworth 2020), the chapters in this book embody a key strength of political-developmental scholarship: carefully tracing how complex processes – social, economic, institutional – operating at and across multiple scales, have observable effects in the concrete practices of local politics in Canada's big cities.

CITY POLITICS IN CANADA: NEW PERSPECTIVES, FORTY YEARS ON

Our brief synthetic review of Canadian urban politics research since the 1980s provides some tantalizing partial answers to the broad questions that animate this volume. However, much remains unexplained, both in terms of how city politics in Canada has evolved over the last four decades and in terms of commonalities and differences among individual cities. The dynamics of urban political change over time are largely unmapped. While there is some work on the politics of urban development in Canadian cities (Leo and Brown 2000; Moore 2013; Young 2017), it is unclear whether – and if so, how – the broader battle lines between developers and residents that characterized Canadian urban politics in the 1970s have shifted. To what extent is the political cleavage between pro-development and anti-development forces that dominated Canadian city politics in the 1970s still salient? What other cleavages have emerged? To what extent have the terms of local politics – both during and between elections – been reshaped by the social and economic transformations of the past forty years?

While city politics across Canada appears to share some basic commonalities, there are also significant differences. For instance, individual cities differ significantly in their policy responses to issues ranging from immigrant settlement and homelessness to transportation infrastructure (Good 2009; Horak 2021; Smith 2022). Research on public participation mechanisms suggests that some cities provide more access and opportunity than others – compare, for instance, Montreal (Latendresse 2005) to Toronto (Joy and Vogel 2015) – but it is not clear why. How have major institutional shocks, such as amalgamations, shaped these distinct trajectories? How have mayors and other key local figures influenced the trajectory of city politics? What underlying pressures and tensions have shaped and constrained how local politics in different cities has responded to social and economic transformation?

The chapters in this volume provide valuable new answers to these questions. To enable readers to compare patterns across cities, we have asked our chapter authors to follow a similar structure in discussing their city's political development since 1980. Each chapter begins with an overview of *social*, *economic*, and *institutional* change in its city, adding important detail and context to the overview we provided in this introduction, and discussing the importance of these changes to local politics. Each chapter then turns to an in-depth discussion of the *practice of politics*, providing a chronological treatment of important elections, policy debates, political leaders, and other events in the city. In these sections – which we see as the "heart and soul" of the book – our authors provide a synthetic description of politics in their cities based on in-depth analyses of secondary materials, personal observations as participants in or close observers of their city's politics, and, in many cases, original qualitative research for this book. Each chapter then offers a concluding section, summarizing key arguments and findings.

While this volume can certainly be read cover to cover, readers can also feel free to move through the chapters in whichever order suits their interests. We encourage readers with interests in a specific city to read at least one additional chapter as well; reading the chapters alongside one another will, we believe, illuminate important areas of similarity and difference across Canadian urban politics more generally.

Our chapters are ordered chronologically by each city's incorporation date. We thus begin, in chapter two, with Montreal (incorporated 1832). In this chapter, Laurence Bherer, Sandra Breux, and Sophie L. Van Neste provide a rich treatment of Montreal's dramatic institutional and political evolution since 1980, with a particular focus on its distinctive party politics and institutional changes in the early 2000s. Bherer et al. offer a critical but ultimately optimistic picture of the consequences of metropolitan reform for local democracy in Montreal.

In chapter three, Martin Horak and Zack Taylor offer a rather less optimistic perspective on the longer-term political consequences of the 1998 amalgamation in Toronto (incorporated 1834). Horak and Taylor describe two enduring crises facing the City of Toronto since amalgamation – a crisis of institutional and fiscal capacity, and a crisis of "institutional intermediation." Horak and Taylor describe a city that, despite continued growth and a capable local bureaucracy, continues to struggle to address gaps in democratic governance caused by imposed institutional reforms.

In chapter four, Robert Finbow provides an analysis of Halifax (incorporated 1842), Atlantic Canada's largest city. Here, too, we find a city grappling with the ongoing consequences of a dramatic municipal amalgamation. Finbow describes these institutional changes and the challenges of governance in a city whose boundaries now include everything from diverse urban cores to remote fishing villages. Finbow also surveys contentious recent debates related to systemic racism, legacies of colonialism, and other pressing policy challenges.

Chapter five turns our attention to Ottawa (incorporated 1855). Luc Turgeon begins by noting that important continuities in Ottawa's local politics have persisted since the 1980s. Yet, Turgeon also notes – again – how a forced municipal amalgamation has reshaped the salient policy debates and political cleavages in contemporary Ottawa, introducing a territorial cleavage and contentious policy debates into what had once been Ottawa's staid middle-class municipal politics.

In chapter six, by Ursula Stelman and Aaron Moore, we move westward to Winnipeg (incorporated 1873). Stelman and Moore describe the practice of politics in a city with a legacy of strong local class divisions, focusing particular attention on the interaction between the city's political and administrative elites. Stelman and Moore valuably highlight the importance of civic administrative institutions for democracy and representation in Canada's cities.

In chapter seven, Ian Bushfield and Stewart Prest describe developments in Vancouver (incorporated 1886), perhaps Canada's most institutionally and politically distinctive big city. Bushfield and Prest discuss Vancouver's fascinating institutional structure – its historic local party system, at-large elections, and complex metropolitan governance – as well as the consequences of these institutions for the city's electoral politics and policy debates. Bushfield and Prest put special emphasis on the ways that the politics of *land* continue to animate Vancouver politics – and have begun to fracture long-standing political coalitions.

Jack Lucas discusses Calgary (incorporated 1894), Canada's third-largest municipality, in chapter eight. Lucas emphasizes how Calgary's politics and policy agenda are repeatedly upended by economic forces far outside the city's municipal boundaries, but also notes how Calgary's expansive municipal boundaries themselves, which have long incorporated both urban and suburban communities, provide Calgary with a strong dose of contentious and ideological local politics.

In our concluding chapter, we draw together insights from our individual city chapters, building on their rich accounts of political development and political conflict. In doing so, we return to the two "big questions" that we posed at the start of this chapter and offer some answers. We argue that the development of local politics in our cities over the past forty years has been shaped by the juxtaposition of far-reaching social transformation one hand and continuity in the basic institutions of local government on the other hand. The structural and functional foundations of local government are not very different than they were in the 1980s, so the issues that they deal with – property development, services to property, planning and zoning – also remain the same. However, urban societies and economies have changed dramatically, so the terms of political debate over these issues have changed as well.

The dominant mid-century political cleavage that pitted proponents of modernist urban development against defenders of neighbourhood values and public engagement, we suggest, has been transformed into a new divide that pits older middle-class homeowners – now often characterized as NIMBYs – against proponents of what we call "twenty-first century urbanism." Notwithstanding long-standing claims that Canadian city politics is "non-ideological," we also argue that the past decades have seen the emergence of a durable ideological divide between an urban right espousing low taxes and business principles in local government, and an urban left that embraces a variety of environmental and social goals. Finally, we suggest that long-standing ethnic, linguistic, and religious group identity cleavages in city politics have faded, and despite (or perhaps because of) the socio-cultural hyper-diversity of many large Canadian cities, ethno-racial cleavages are generally not dominant in Canadian city politics.

Such common changes have not, however, meant that intercity differences in local politics have disappeared. On the contrary, politics in different cities has evolved in distinct directions. For instance, post-amalgamation Toronto is characterized by centralized decision-making processes that are quite inaccessible to ordinary citizens, whereas Montreal has experimented with a variety of innovations in participatory democracy. Politics in some cities features a strong divide between older "core" neighbourhoods and newer "suburban" areas, whereas in other cities, this geographical divide is not as apparent; distinct policy issues dominate local politics in different cities. We explain these differences with reference to three important factors: the amalgamations of Halifax, Toronto, Ottawa, and Montreal, which transformed local political dynamics in those cities; variation in the structure of local representative and participatory institutions; and the different growth trajectories of our cities, which shape local government resources and influence which political issues are most salient.

In the final section of our concluding chapter, we offer some thoughts on how students of local politics in Canadian cities can contribute to theory development in urban politics and political science more generally. We argue that multilevel democracy and multilevel governance, political economy, spatial cleavages in politics, and participatory democracy are all fields to which the study of city politics in Canada can make significant contributions. Much has happened in the four decades since Magnusson and Sancton's pioneering volume. It is our hope that the present volume can provide a guide to the landscape of city politics in Canada today and help to inspire a new generation of scholarship on the challenges and promise of local politics in our big cities.

NOTES

1   These cities are identical to the first edition, with one exception: we have replaced Edmonton with Calgary in this edition of the book. We do so because of the increased significance of Calgary, which is now Canada's third-largest municipality, and because, unfortunately, there are currently no close academic observers of Edmonton's civic politics.
2   The 1993 secession from Winnipeg of Headingley, with a population of 4,400, is a small exception.

3   It is important to note that these trends are not unique to central municipalities. Visible minority populations have increased in the suburban municipalities of large urban areas as well. In fact, among the ten most populous municipalities in Canada, the one with the largest proportion of visible minority residents is Brampton, a suburb of Toronto where 81 per cent of the total population of 656,480 identified as visible minority in 2021 (Statistics Canada 2023).

4   Calgary is a partial exception. Its manufacturing base was always very small, and unlike other cities, it has a substantial primary industry employment base as a hub of producer services for the oil and gas industry.

5   Canadian municipalities maintain two budgets. The operating budget funds current expenditures, mostly for wages and benefits, debt servicing costs, and the procurement of services, and must be balanced every year. The capital budget funds the maintenance and expansion of works and facilities, including infrastructure, and is funded by long-term borrowing, intergovernmental transfers, and development levies.

6   The user fees category is smaller in Montreal because transit fare revenues are remitted to the city by the regional transit agency.

REFERENCES

Assunçao-Denis, Marie-Ève, and Ray Tomalty. 2019. "Increasing Cycling for Transportation in Canadian Communities: Understanding What Works." *Transportation Research Part A: Policy and Practice* 123: 288–304. https://doi.org/10.1016/j.tra.2018.11.010.

Bélanger, Éric, Cameron D. Anderson, and R. Michael McGregor, eds. 2022. *Voting in Quebec Municipal Elections: A Tale of Two Cities.* Toronto: University of Toronto Press. https://doi.org/10.3138/9781487540081.

Bourne, Larry S. 1967. Private Redevelopment of the Central City; Spatial Processes of Structural Change in the City of Toronto. University of Chicago, Department of Geography. Research paper no. 112. Chicago: University of Chicago.

Bradford, Neil. 2007. *Whither the Federal Urban Agenda? A New Deal in Transition.* Ottawa: CPRN. https://oaresource.library.carleton.ca/cprn/46924_en.pdf.

Bradford, Neil. 2018. "A National Urban Policy for Canada? The Implicit Federal Agenda." *IRPP Insight* 24. Montreal: Institute for Research on Public Policy. https://irpp.org/research-studies/national-urban-policy-canada-implicit-federal-agenda/.

Bradford, Neil. 2020. "Policy in Place: Revisiting Canada's Tri-Level Agreements." *IMFG Papers on Municipal Finance and Governance* 50. http://hdl.handle.net/1807/102474.

Brenner, Neil. 2009. "Is There a Politics of "Urban" Development? Reflections on the U.S. Case." In *The City in American Political Development*, edited by Richardson Dilworth, 121–40. New York: Routledge.

Breux, Sandra, and Jérôme Couture, eds. 2018. *Accountability and Responsiveness at the Municipal Level: Views from Canada.* Montreal: McGill-Queen's University Press. https://doi.org/10.1515/9780773553743.

Breux, Sandra, Jérôme Couture, and Royce Koop. 2019. "Influences on the Number and Gender of Candidates in Canadian Local Elections." *Canadian Journal of Political Science/Revue canadienne de science politique* 52 (1): 163–81. https://doi.org/10.1017/S0008423918000483.

Brownstone, Meyer, and Thomas J. Plunkett. 1983. *Metropolitan Winnipeg: Politics and Reform of Local Government.* Berkeley: University of California Press.

Charter City Toronto. 2019. "Proposal Overview." Accessed May 24, 2025. https://www
    .chartercitytoronto.ca/proposal-overview.html.
Darchen, Sébastien, and Diane-Gabrielle Tremblay. 2013. "The Local Governance of Culture-led
    Regeneration Projects: A Comparison between Montreal and Toronto." *Urban Research & Practice* 6
    (2): 140–57. https://doi.org/10.1080/17535069.2013.808433.
Dilworth, Richardson, ed. 2009. *The City in American Political Development*. New York: Routledge.
    https://doi.org/10.4324/9780203881101.
Dilworth, Richardson. 2020. Comparative Case Study Methods in Urban Political Development. *Social
    Sciences* 9 (10). https://doi.org/10.3390/socsci9100183.
Doberstein, Carey. 2011. "Institutional Creation and Death: Urban Development Agreements in
    Canada." *Journal of Urban Affairs* 33 (5): 529–48. https://doi.org/10.1111/j.1467-9906.2011.00566.x.
Doering, Jan, Daniel Silver, and Zack Taylor. 2021. "The Spatial Articulation of Urban Political
    Cleavages." *Urban Affairs Review* 57 (4): 911–51. https://doi.org/10.1177/1078087420940789.
Downey, Terrence J., and Robert J. Williams. 1998. "Provincial Agendas, Local Responses: The 'Common
    Sense' Restructuring of Ontario's Municipal Governments." *Canadian Public Administration* (2):
    210–38. https://doi.org/10.1111/j.1754-7121.1998.tb01537.x.
Foran, Max. 2009. *Expansive Discourses: Urban Sprawl in Calgary, 1945–1978*. Edmonton: Athabasca
    University Press. https://doi.org/10.15215/aupress/9781897425138.01.
Good, Kristin. 2009. *Municipalities and Multiculturalism: The Politics of Immigration in Toronto and Vancouver*.
    Toronto: University of Toronto Press. https://doi.org/10.3138/9781442690417.
Good, Kristin. 2019. "The Fallacy of the "Creatures of Provinces" Doctrine: Recognizing and Protecting
    Municipalities' Constitutional Status." *IMFG Papers on Municipal Finance and Governance* 46. http:
    //hdl.handle.net/1807/98264.
Good, Kristin, Enid Slack, Zack Taylor, and Patricia Burke Wood. 2020. *Charting a New Path: Does Toronto
    Need More Autonomy? IMFG Forum* 10. Toronto: Institute on Municipal Finance and Governance,
    Munk School of Global Affairs and Public Policy, University of Toronto. http://hdl.handle.net
    /1807/100602.
Gordon, David L.A. 2022. "The Canadian Dream? Growth Trends in Canada's Suburban and Urban
    Neighbourhoods." In *Suburbia in the 21st Century: From Dreamscape to Nightmare?*, edited by Paul
    Maginn and Katrin B. Anacker, 95–111. London, UK: Routledge. https://doi.org/10.4324
    /9781315644165-8.
Grant, Jill, Alan Walks, and Howard Ramos. 2020. *Changing Neighbourhoods: Social and Spatial Polarization
    in Canadian Cities*. Vancouver, BC: UBC Press. https://doi.org/10.59962/9780774862042.
Horak, Martin. 2012. "Conclusion: Understanding Multilevel Governance in Canada's Cities." In *Sites of
    Governance: Multilevel Governance and Policy Making in Canada's Big Cities*, edited by Martin Horak and
    Robert Young, 339–70. Montreal: McGill-Queens University Press. https://doi.org/10.1515
    /9780773586918-013.
Horak, Martin. 2013. "State Rescaling in Practice: Urban Governance Reform in Toronto." *Urban
    Research & Practice* 6 (3): 311–28. https://doi.org/10.1080/17535069.2013.846005.
Horak, Martin. 2021. "Building Rapid Transit in Canada: The Problem of Governance." *Anuario de
    Derecho Municipal* 14: 243–58. https://doi.org/10.37417/ADM/14-2020_09.
Horak, Martin, and Aaron A. Moore. 2015. "Policy Shift without Institutional Change: The Precarious
    Place of Neighborhood Revitalization in Toronto." In *Urban Neighborhoods in a New Era: Revitalization
    Politics in the Postindustrial City*, edited by Clarence N. Stone, Ellen Shiau, Robert P. Stoker, John
    Betancur, Susan E. Clarke, Marilyn Dantico, Martin Horak, Karen Mossberger, Juliet Musso and

Jefferey M. Sellers, 182–208. Chicago: University of Chicago Press. https://doi.org/10.7208
/9780226289151-011.

Horak, Martin, and Robert Young, eds. 2012. *Sites of Governance: Multilevel Governance and Policy Making
in Canada's Big Cities*. Montreal: McGill-Queen's University Press. https://doi.org/10.1515
/9780773586918.

Joy, Meghan, and Ronald K. Vogel. 2015. "Toronto's Governance Crisis: A Global City Under Pressure."
Cities 49: 35–52. https://doi.org/10.1016/j.cities.2015.06.009.

Keil, Roger. 2000. "Governance Restructuring in Los Angeles and Toronto: Amalgamation or Secession."
International Journal of Urban and Regional Research 24 (4): 758–81. https://doi.org/10.1111
/1468-2427.00277.

Kipfer, Stefan, and Roger Keil. 2002. "Toronto Inc? Planning the Competitive City in the New
Toronto." *Antipode* 34 (2): 227–64. https://doi.org/10.1111/1467-8330.00237.

Kitchen, Harry. 2016. "Is 'Charter-City Status' a Solution for Financing City Services in Canada – or Is
That a Myth?" *SPP Research Papers* 9 (2). https://doi.org/10.55016/ojs/sppp.v9i1.42566.

Latendresse, Anne. 2005. "Municipal Reform and Public Participation in Montreal's Urban Affairs: Break
of Continuity?" In *Metropolitan Democracies: Transformations of the State and Urban Policy in Canada,
France and Great Britain*, edited by Bernard Jouve, 117–32. Milton Park: Routledge. https://doi.org
/10.4324/9781351153089-8.

Leo, Christopher, and Wilson Brown. 2000. "Slow Growth and Urban Development Policy." *Journal of
Urban Affairs* 22 (2): 193–213. https://doi.org/10.1111/0735-2166.00050.

Ley, David. 1994. "Gentrification and the Politics of the New Middle Class." *Environment and Planning D:
Society and Space* 12 (1): 53–74. https://doi.org/10.1068/d120053.

Lucas, Jack. 2016. *Fields of Authority: Special Purpose Governance in Ontario, 1815–2015*. Toronto: University
of Toronto Press. https://doi.org/10.3138/9781487510367.

Lucas, Jack. 2017a. "Patterns of Urban Governance: A Sequence Analysis of Long-Term Institutional
Change in Six Canadian Cities." *Journal of Urban Affairs* 39 (1): 68–90. https://doi.org/10.1111
/juaf.12291.

Lucas, Jack. 2017b. "Urban Governance and the American Political Development Approach." *Urban
Affairs Review* 53 (2): 338–61. https://doi.org/10.1177/1078087415620054.

Lucas, Jack. 2021. "The Size and Sources of Municipal Incumbency Advantage in Canada." *Urban Affairs
Review* 57 (2): 373–401. https://doi.org/10.1177/1078087419879234.

Lucas, Jack. 2022. "Do 'Non-Partisan' Municipal Politicians Match the Partisanship of Their
Constituents?" *Urban Affairs Review* 58 (1): 103–28. https://doi.org/10.1177/1078087420958074.

Lucas, Jack. 2023. "The Ideological Structure of Municipal Non-Ideology." *Urban Affairs Review* 59 (1):
275–93. https://doi.org/10.1177/10780874211038321.

Lucas, Jack, and R. Michael McGregor, eds. 2021. *Big City Elections in Canada*. Toronto: University of
Toronto Press. https://doi.org/10.3138/9781487528577.

Lucas, Jack, R. Michael McGregor, and Aengus Bridgman. 2023. "Spatial Voting in Non-Partisan Cities:
A Case Study." *Electoral Studies* 82:102599. https://doi.org/10.1016/j.electstud.2023.102599.

Lucas, Jack, R. Michael McGregor, and Kim-Lee Tuxhorn. 2021. "Closest to the People? Incumbency
Advantage and the Personal Vote in Non-Partisan Elections." *Political Research Quarterly* 75 (1):
188–202. https://doi.org/10.1177/1065912921990751.

Lucas, Jack, and Alison Smith. 2019. "Multilevel Policy from the Municipal Perspective: A Pan-Canadian
Survey." *Canadian Public Administration* 62 (2): 270–93. https://doi.org/10.1111/capa.12316.

Lynch, Mona, Marisa Omori, Aaron Roussell, and Matthew Valasik. 2013. "Policing the 'Progressive' City: The Racialized Geography of Drug Law Enforcement." *Theoretical Criminology* 17 (3): 335–57. https://doi.org/10.1177/1362480613476986.

Magnusson, Warren, and Andrew Sancton, eds. 1983. *City Politics in Canada.* Toronto: University of Toronto Press. https://doi.org/10.3138/9781487575908.

McGregor, Michael, and Zachary Spicer. 2016. "The Canadian Homevoter: Property Values and Municipal Politics in Canada." *Journal of Urban Affairs* 38 (1): 123–39. https://doi.org/10.1111/juaf.12178.

McGregor, R. Michael, Aaron A. Moore, and Laura B. Stephenson. 2016. "Political Attitudes and Behaviour in a Non-Partisan Environment: Toronto 2014." *Canadian Journal of Political Science* 49 (2): 311–33. https://doi.org/10.1017/S0008423916000573.

McGregor, R. Michael, Aaron A. Moore, and Laura B. Stephenson. 2021. *Electing a Mega-Mayor: Toronto 2014.* Toronto: University of Toronto Press. https://doi.org/10.3138/9781487509651.

Moore, Aaron A. 2013. *Planning Politics in Toronto: The Ontario Municipal Board and Urban Development.* Toronto: University of Toronto Press. https://doi.org/10.3138/9781442699458.

Moore, Aaron A., and Alexandra Caporale. 2023. "The Efficacy of Statutory Public Hearings for Planning: A Comparison of Four Canadian Jurisdictions." *Journal of Urban Affairs* 47 (3): 739–59. https://doi.org/10.1080/07352166.2023.2195664.

Peters, Evelyn J. 2012. *Urban Aboriginal Policy Making in Canadian Municipalities.* Montreal & Kingston: McGill-Queen's University Press. https://doi.org/10.1515/9780773587441.

Podmore, Julie. 2015. "From Contestation to Incorporation: LGBT Activism and Urban Politics in Montréal." In *Queer Mobilizations: Social Movement Activism and Canadian Public Policy*, edited by Manon Tremblay, 187–207. Vancouver, BC: UBC Press. https://doi.org/10.59962/9780774829090-011.

Pourali, Mehrdokht, Craig Townsend, Angela Kross, Alex Guindon, and Jochen A.G. Jaeger. 2022. "Urban Sprawl in Canada: Values in all 33 Census Metropolitan Areas and Corresponding 469 Census Subdivisions between 1991 and 2011." *Data in Brief* 41:107941. https://doi.org/10.1016/j.dib.2022.107941.

Sancton, Andrew. 2000. *Merger Mania: The Assault on Local Government.* Montreal: McGill-Queen's University Press. https://doi.org/10.1515/9780773568914.

Sancton, Andrew. 2006. "Why Municipal Amalgamations? Halifax, Toronto, Montreal." In *Municipal-Federal-Provincial Relations in Canada – Canada: State of the Federation 2004*, edited by Robert Young and Christian Leuprecht, 119–37. Kingston: Institute of Intergovernmental Relations.

Sancton, Andrew. 2008. *The Limits of Boundaries: Why City-Regions Cannot Be Self-governing.* Montreal: McGill-Queen's University Press. https://doi.org/10.1515/9780773574977.

Sancton, Andrew. 2015. *Canadian Local Government: An Urban Perspective.* 2nd ed. Don Mills, ON: Oxford University Press.

Sancton, Andrew. 2016. "The False Panacea of City Charters? A Political Perspective on the Case of Toronto." *SPP Research Papers* 9 (3). https://doi.org/10.55016/ojs/sppp.v9i1.42564.

Siegel, David. 2015. *Leaders in the Shadows: The Leadership Qualities of Municipal Chief Administrative Officers.* Toronto: University of Toronto Press. https://utppublishing.com/doi/book/10.3138/9781442626652.

Siemiatycki, Myer. 2011. "Governing Immigrant City: Immigrant Political Representation in Toronto." *American Behavioral Scientist* 55 (9): 1214–34. https://doi.org/10.1177/0002764211407840.

Siemiatycki, Myer. 2010. "Municipal Voting Rights for Non-Canadian Citizens." Mowat Centre, School of Public Policy and Governance, University of Toronto. Accessed May 24, 2025. https://mowatcentre.munkschool.utoronto.ca/municipal-voting-rights-for-non-canadian-citizens/.

Siemiatycki, Myer, and Anver Saloojee. 2002. "Ethnoracial Political Representation in Toronto: Patterns and Problems." *Journal of International Migration and Integration* 3 (2): 241–73. https://doi.org/10.1007/s12134-002-1013-8.

Silver, Daniel, Zack Taylor, and Fernando Calderón-Figueroa. 2020. "Populism in the City: The Case of Ford Nation." *International Journal of Politics, Culture, and Society* 33 (1): 1–21. https://doi.org/10.1007/s10767-018-9310-1.

Smith, Alison. 2022. *Multiple Barriers: The Multilevel Governance of Homelessness in Canada.* Toronto: University of Toronto Press. https://doi.org/10.3138/9781487548742.

Spicer, Zachary. 2011. "The Rise and Fall of The Ministry of State For Urban Affairs: A Re-Evaluation." *Canadian Political Science Review* 5 (2): 117–26. https://doi.org/10.24124/c677/2011149.

Statistics Canada. 1981a. "Census Profile for Canada, Provinces and Territories, Census Divisions and Census Subdivisions, 1981 Census - Part A - ARCHIVED, 1981 Census - Part A - ARCHIVED, 1981 Census of Population, Statistics Canada Catalogue no. 97-570-X1981004." Accessed November 28, 2023. https://www150.statcan.gc.ca/n1/en/catalogue/97-570-X1981004.

Statistics Canada. 1981b. "Census Profile for Census Metropolitan Areas and Census Agglomerations, 1981 Census - Part A - ARCHIVED, 1981 Census of Population, Statistics Canada Catalogue no. 97-570-X1981002." Accessed November 28, 2023. https://www150.statcan.gc.ca/n1/en/catalogue/97-570-X1981002.

Statistics Canada. 2022. "Canada's Large Urban Centres Continue to Grow and Spread." Accessed July 21, 2023. https://www150.statcan.gc.ca/n1/daily-quotidien/220209/dq220209b-eng.htm.

Statistics Canada. 2023a. "Table 17-10-0136-01. Components of Population Change by Census Metropolitan Area and Census Agglomeration, 2016 Boundaries." Accessed February 22, 2023. https://doi.org/10.25318/1710013601-eng.

Statistics Canada. 2023b. Census Profile, 2021 Census of Population, Brampton. Statistics Canada Catalogue no. 98-316-X2021001. Ottawa. Accessed December 12, 2024. https://www12.statcan.gc.ca/census-recensement/2021/dp-pd/prof/index.cfm?Lang=E.

Stone, Clarence N. 2008. "Urban regimes and the capacity to govern: a political economy approach." In *Power in the City: Clarence Stone and the Politics of Inequality*, edited by M. Orr and V.C. Johnson, 76–107. Lawrence, KS: University of Kansas Press.

Suttor, Greg. 2016. *Still Renovating: A History of Canadian Social Housing Policy*. Montreal: McGill-Queen's University Press. https://doi.org/10.1515/9780773548572.

Taylor, Zack. 2019. *Shaping the Metropolis: Institutions and Urbanization in the United States and Canada.* Montreal: McGill-Queen's University Press. https://doi.org/10.1515/9780773558427.

Taylor, Zack. 2022. "Regionalism from Above: Intergovernmental Relations in Canadian Metropolitan Governance." *Commonwealth Journal of Local Governance* 26: 139–59. https://doi.org/10.5130/cjlg.vi26.8141.

Taylor, Zack. 2024. "From City Autonomy to the Metagovernance of Place." In *Cities and the Constitution: Giving Local Governments in Canada the Power They Need*, edited by Nathalie Des Rosiers, Richard Alpert and Alexandra Flynn, 19–42. Montreal: McGill-Queen's University Press. https://doi.org/10.1515/9780228022084-006.

Taylor, Zack, and Neil Bradford. 2020. "Governing Canadian Cities." In *Canadian Cities in Transition*, edited by Markus Moos, Tara Vinodrai and Ryan Walker, 33–50. Toronto: Oxford University Press.

Taylor, Zack, Marcy Burchfield, and Anna Kramer. 2014. "Alberta Cities at the Crossroads: Urban Development Challenges and Opportunities in Historical and Comparative Perspective." *SPP Research Papers* 7 (12). https://doi.org/10.11575/sppp.v7i0.42465.

Taylor, Zack, and Alec Dobson. 2020. "Power and Purpose: Canadian Municipal Law in Transition." *IMFG Papers on Municipal Finance and Governance* 47. http://hdl.handle.net/1807/99226.

Taylor, Zack, and Gabriel Eidelman. 2010. "Canadian Political Science and the City: A Limited Engagement." *Canadian Journal of Political Science* 43 (4): 961–81. https://doi.org/10.1017/S0008423910000715.

Tremblay, Manon, and Anne Mévellec. 2013. "Truly More Accessible to Women than the Legislature? Women in Municipal Politics." In *Stalled: The Representation of Women in Canadian Governments*, edited by Linda Trimble, Jane Arscott and Manon Tremblay, 19–35. Vancouver, BC: UBC Press. https://doi.org/10.59962/9780774825221-005.

Urbaniak, Tom. 2009. *Her Worship: Hazel McCallion and the Development of Mississauga.* Toronto: University of Toronto Press. https://doi.org/10.3138/9781442688223.

Vojnovic, Igor. 2000. "The Transitional Impacts of Municipal Consolidations." *Journal of Urban Affairs* 22 (4): 385–417. https://doi.org/10.1111/0735-2166.00063.

White, Stephen E. 2023. "Immigrant Voter Turnout and Time: Does Period of Arrival Matter More Than Length of Stay?" *International Migration* 61: 118–32. https://doi.org/10.1111/imig.13150.

Young, Douglas. 2017. "Redefining 'Renewal' in Toronto's High-Rise Suburbs." *Alternate Routes: A Journal of Critical Social Research* 28: 219–32.

2

Montreal

Laurence Bherer, Sandra Breux, and Sophie L. Van Neste

INTRODUCTION

The most significant episode in the institutional political history of Montreal (*Tio'tià:ke*) since the 1980s was undoubtedly the amalgamation of the 27 municipalities on the Island of Montreal. This wave of mergers led to a renewed and considerably enlarged City of Montreal on 1 January 2002, while also affecting other municipalities in the Montreal metropolitan region. Beyond that, the scheme imposed by the Quebec government also comprised a far-reaching reform that completely reorganized the region's institutional landscape. Aside from establishing new large cities such as Longueuil and Montreal, it introduced two new levels of government: a metropolitan authority and sub-municipal boroughs. In this way, the reform achieved both centralization and decentralization at once: centralization around a large city, and an unprecedented degree of decentralization at different, new levels of municipal government.

At the time, the imposition of the mergers and subsequent institutional compromises was the subject of considerable criticism. Montreal's new governance structure has become quite complex, and inequalities in the representation of the populations of the various territories remain. However, twenty years after the reform, we can see that local players have, to varying degrees, taken ownership of the different levels. This far-reaching institutional change has completely transformed local political dynamics, opening up new political arenas.

On the one hand, elected officials and local players have had to learn to think in a "metropolitan" way, with the Communauté métropolitaine de Montréal (CMM) taking time to establish itself as a political space that transcends the traditional fragmentation of interests between a downtown and its suburban outer rings. After the failure to adopt a first metropolitan plan in 2006, the members of the CMM succeeded in reaching a consensus to adopt a

Figure 2.1: Municipal boundaries in Montreal before the 2002 merger

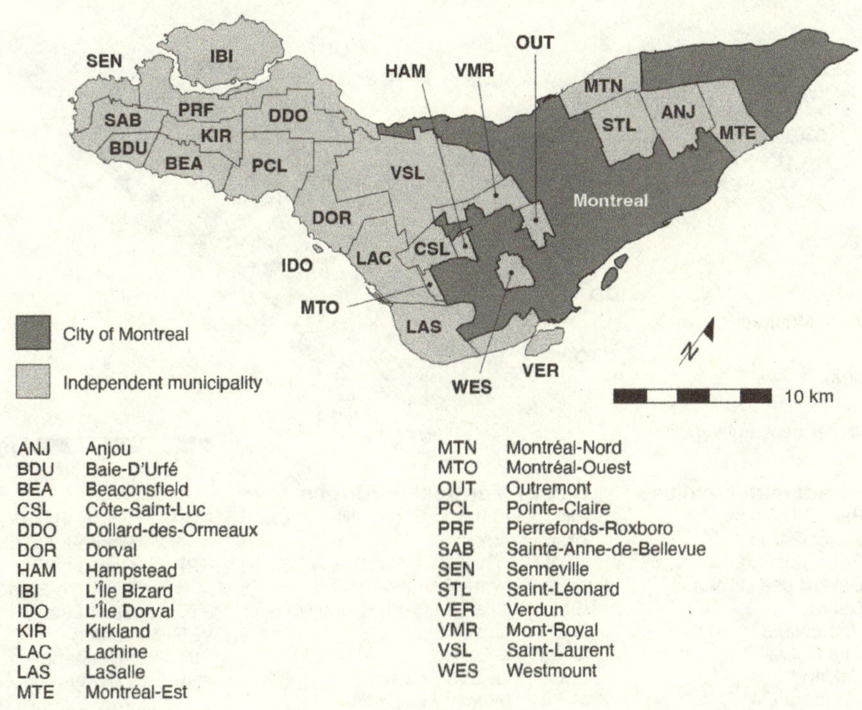

ANJ	Anjou	MTN	Montréal-Nord
BDU	Baie-D'Urfé	MTO	Montréal-Ouest
BEA	Beaconsfield	OUT	Outremont
CSL	Côte-Saint-Luc	PCL	Pointe-Claire
DDO	Dollard-des-Ormeaux	PRF	Pierrefonds-Roxboro
DOR	Dorval	SAB	Sainte-Anne-de-Bellevue
HAM	Hampstead	SEN	Senneville
IBI	L'Île Bizard	STL	Saint-Léonard
IDO	L'Île Dorval	VER	Verdun
KIR	Kirkland	VMR	Mont-Royal
LAC	Lachine	VSL	Saint-Laurent
LAS	LaSalle	WES	Westmount
MTE	Montréal-Est		

Metropolitan Land Use and Development Plan in 2012. This plan notably calls for densification and a transit-oriented development approach – elements that remain structuring objectives of metropolitan governance to this day.

On the other hand, the decentralization into Montreal's new boroughs has spawned the emergence of new intra-local dynamics, leading some to say that the merger has resulted less in "one island, one city," as Mayor Pierre Bourque had promoted in the late 1990s, than in "one island, cities" (Collin and Bherer 2018). Community groups and citizens' associations, already very present in Montreal democracy, have appropriated the boroughs to invest in local issues. The new political culture created by the reform has also led to the emergence of a new, enduring local political party, Projet Montréal, which bears some resemblance to the Rassemblement des citoyens et citoyennes de Montréal (RCM) a few decades earlier. Founded in 2004 in the wake of the mergers, Projet Montréal used the boroughs to achieve its first political gains, demonstrating to Montrealers its ability to lead. Nevertheless, there is considerable variation among the boroughs. Some centrally located boroughs feature an active political life, with decisions and local mobilizations that are widely publicized, while others, located in the so-called old suburbs, have undergone little political renewal.

Figure 2.2: Municipal boundaries in Montreal after the 2006 demerger

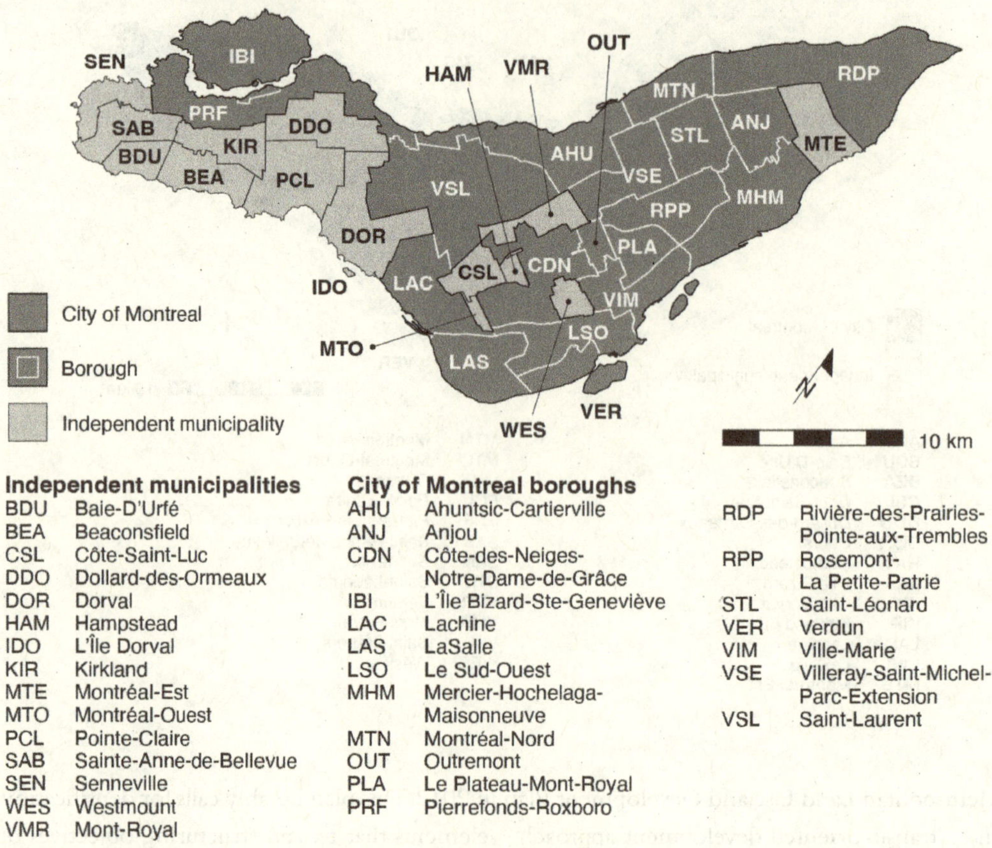

Independent municipalities

BDU Baie-D'Urfé
BEA Beaconsfield
CSL Côte-Saint-Luc
DDO Dollard-des-Ormeaux
DOR Dorval
HAM Hampstead
IDO L'Île Dorval
KIR Kirkland
MTE Montréal-Est
MTO Montréal-Ouest
PCL Pointe-Claire
SAB Sainte-Anne-de-Bellevue
SEN Senneville
WES Westmount
VMR Mont-Royal

City of Montreal boroughs

AHU Ahuntsic-Cartierville
ANJ Anjou
CDN Côte-des-Neiges-
     Notre-Dame-de-Grâce
IBI L'Île Bizard-Ste-Geneviève
LAC Lachine
LAS LaSalle
LSO Le Sud-Ouest
MHM Mercier-Hochelaga-
     Maisonneuve
MTN Montréal-Nord
OUT Outremont
PLA Le Plateau-Mont-Royal
PRF Pierrefonds-Roxboro

RDP Rivière-des-Prairies-
     Pointe-aux-Trembles
RPP Rosemont-
     La Petite-Patrie
STL Saint-Léonard
VER Verdun
VIM Ville-Marie
VSE Villeray-Saint-Michel-
     Parc-Extension
VSL Saint-Laurent

Notes: Although a number of former municipalities "demerged" from the amalgamated Montreal in 2006, the post-demerger City of Montreal is much larger than it was prior to the 2002 merger. Borough boundaries are indicated with white lines. While some boroughs correspond to municipalities that had merged with Montreal, others are subdivisions of Montreal as it was before the merger.

In this chapter, we discuss the importance and consequences of the mergers as a way to highlight the political dynamics of Montreal over the last twenty years. We begin by describing the politics of municipal mergers and detailing the institutional, demographic, and budgetary consequences of this restructuring. We then focus on how Montrealers have appropriated the city's new institutional architecture, in particular with regard to the city's representative and participatory institutions. Finally, we examine patterns of social mobilization, already firmly rooted prior to the mergers, and their evolution in light of the municipal reorganization. In this way, we will show that the mergers contributed to the creation of new political dynamics on a variety of scales.

INSTITUTIONAL, SOCIAL, AND ECONOMIC CHANGE: THE MERGERS

From Mergers to Demergers

Montreal is located on an island at the confluence of two important waterways: the St. Lawrence River and the Ottawa River, which splits into the Rivière des Prairies and the Milles-îles River as it flows through the metropolitan area. Montreal's unique geography has played a significant role in shaping the city's early economic development and its urban and metropolitan politics, and it also figures in the recent institutional developments discussed in this chapter. In the nineteenth and twentieth centuries, Montreal rapidly began annexing surrounding municipalities and expanding its public works and infrastructure, culminating in the recent consolidation project aimed at creating "one island, one city." While suburban forms are not rare on the island, the rivers have separated the city from what is generally thought of as its suburbs on the south and north shores (to the north lies another island that contains Laval, the second largest city in the region and third largest in the province). Together, this forms a metropolitan region connected across rivers by some twenty-eight bridges and tunnels. The rivers and the mountain (Mount Royal) have long been iconic elements of Montreal's landscape and prized focal points for its urban development.

Montreal's multilevel reform at the dawn of the 2000s is a complex story with many twists and turns. In addition to the mandatory merger of all municipalities on the Island of Montreal in 2002, Bill 170 also reformed the territorial organization of the province's three metropolitan regions: Montreal, Quebec, and Outaouais. This led to the creation of two new institutional levels: (1) sub-municipal bodies – the boroughs (the boundaries of which by and large match those of the former suburban towns, but also of neighbourhoods within the former City of Montreal), with powers of planning, management of local services, and local urban planning; and (2) a supramunicipal body at the city-region level – the CMM, with a mandate for consultation, planning, management of regional facilities, and economic development.

Bill 170 represented a radical departure from the Quebec government's previous strategy for municipal mergers. The focus shifted from an approach based on the goodwill of municipalities to an imposed amalgamation for several dozen municipalities, including the City of Montreal. This restructuring of the municipal system generated considerable discontent among part of the population and many elected municipal officials, so much so that municipal mergers became the main issue during the 2003 provincial election campaign. It could be said that the Quebec Liberal Party, led by Jean Charest, won the election due to its promise to hold referendums on undoing the imposed municipal amalgamations (Bherer 2006). Once in power, and before holding the referendums, the Charest government gave the affected municipalities time to propose a reform of their institutions, which would convince voters to join the new municipality. Montreal's mayor at the time, Gérald Tremblay, proposed giving more powers to the 27 boroughs. The message was that the boroughs could become a real locus of power, and that this could compensate for the loss of their former home, or parent, municipality. The Tremblay reform provided for boroughs to be responsible for collecting special taxes, taking legal action and exercising significant local powers, including managing borough personnel, modifying the urban plan and urban planning regulations, and scheduling referendums. Bill 170 suggested that the borough president be appointed by the committee of local elected officials. Mayor Tremblay, on the other hand, proposed that the holder of this position be directly elected by the population, and thus become a borough mayor.

While this decentralizing reform was implemented, it was not desired by voters in all of the boroughs. In the referendums promised by the Charest government, held in June 2004, eight boroughs on the Island of Montreal met the criteria for re-founding the municipality. Those criteria were that a minimum of 50 per cent of the *votes cast*, representing at least 35 per cent of *registered voters*, would have to approve that option. Voters living on the territory of 15 previous municipalities voted to demerge, in other words, to restore their local governments. Generally, these were municipalities with higher-income levels and high property values and a strong English-speaking population (nearly 50 per cent) located in the west of the island (with the exception of Montreal East, which is predominantly French-speaking) (Drouilly and Gagnon 2004). These results confirm the strong sense of identity that a part of the English-speaking community may have felt towards municipal institutions (Sancton 2004).

However, the demerger could hardly qualify as a true return to the situation that had prevailed in Montreal for the previous two decades in terms of supramunicipal coordination. From 1970 to 2001, the Communauté urbaine de Montréal (CUM), which had brought together the 31 municipalities of the Island of Montreal, Île Bizard, and Île de Dorval, was based on a double-majority principle: a majority of representatives from the central city and a majority of mayors from the so-called suburban municipalities. The CUM executive committee was made up of equal numbers of representatives from the suburban municipalities and the City of Montreal (Collin and Bherer 2018). After the referendums on the demergers, the Charest government created a new institution, the agglomeration council, which included the central city and the 15 reconstituted municipalities. Unlike the CUM, the governance of the agglomeration council is based on the demographic weight of each municipality, ensuring that the new institution is dominated by the City of Montreal. The small populations of the demerged cities (300,000 in total) were dwarfed by the 1.7 million inhabitants of Montreal (Ville de Montréal 2021). Moreover, the civil service that implements the decisions of the agglomeration council is de facto that of the City of Montreal.

Almost 50 per cent of municipal budgets on the Island of Montreal were determined by the agglomeration council (Meloche and Vaillancourt 2022), because a significant proportion of the powers usually vested in municipalities were delegated to this new supramunicipal body. The demerged municipalities had thus regained their legal identity, but their powers were scarcely greater than those of the Montreal boroughs, which gained in power and autonomy, thanks to the Tremblay reform implemented before the referendum. These institutional changes permanently transformed the demographic, fiscal, and, above all, political portrait of Montreal and led to the strong decentralization of the city as we know it today.

Demographic and Economic Trends

From the early 2000s to the present day, Montreal's population has risen sharply, largely due to the expansion of its territory as a result of the municipal mergers. Before the 2002 amalgamation, the City of Montreal's population was 1 million. Afterward, the city was home to 1.8 million residents. This figure fell to 1.6 million once the demergers were implemented, representing 87 per cent of the population of the Island of Montreal (Montréal en statistiques 2022).

Since then, the population has continued to grow, and by 2022, the Montreal agglomeration had a population of 2 million (Montréal en statistiques 2022). Between 2016 and 2021, the City of Montreal's population grew by 3.4 per cent. In terms of boroughs, Ville-Marie, which corresponds to Montreal's downtown, saw the greatest demographic increase (17.7 per cent), followed by Sud-Ouest (8.2 per cent), LaSalle (7 per cent), and Montréal-Nord (5 per cent). On the Island of Montreal as a whole, "[o]ver the same period, population growth in the other cities of the agglomeration was significantly lower, at 1.7%" (Montréal en statistiques 2022). Moreover, the population of the Montreal census metropolitan area represents more than half the population of Quebec. The most obvious demographic growth, however, has been in the Montreal suburbs (Montréal en statistiques 2022).

Since the early 1980s, the francization of the economy has continued, with the rise of French-speaking business elites and the development of French-language services (such as advertising or media) that have helped to enhance the value of the French language (Polèse 2018). In just one generation, the wage gap between francophones and anglophones disappeared, representing the so-called Quebec Inc. phenomenon, with several large companies now controlled by francophones (e.g., Cascades, Québécor, Bombardier). The merger of the Chambre de commerce du Montréal métropolitain and the Montreal Board of Trade to form the Board of Trade of Metropolitan Montreal in 1992 could also be seen as an end of the linguistic separation that had prevailed until then.

However, this rapid transformation came at a cost: the mass departure of companies to Toronto in the late 1970s, combined with Toronto's strong demographic growth, had accelerated the city's loss of status as Canada's financial metropolis. This marked the start of two decades of economic slowdown and unemployment. Montreal's unemployment rate was higher than that of other Canadian cities and would not fall below 10 per cent until the late 1990s. This economic decline marked the discourse of economic players throughout this period (Hamel and Poitras 2004), leading to commissions of inquiry and the drafting of expert reports, the best known of which is the (Laurent) Picard report (1986). While economic difficulties were generally associated with the departure of corporate head offices to Toronto, analyses also point to other factors, such as the fact that Montreal was no longer the leading hub of maritime, rail, and air transport due to technological changes and political decisions (Polèse 2018. Observers also pointed to the outdated nature of municipal governance structures, seen as fostering sterile local competition among municipalities even as economic globalization necessitated a more coordinated response. This discourse paved the way for the municipal mergers presented in the previous section (Hamel and Poitras 2004).

The Montreal economy has gradually transformed from an industrial to a service economy. While manufacturing jobs still accounted for 37.6 per cent of total employment in 1951, this figure had dropped to 11.5 per cent by 2011 (Polèse 2018: 198). New sectors developed and became the economic engines of the 2000s, including the technology and video game industries, the pharmaceutical sector, and the aerospace industry. Figures compiled by Mario Polèse in 2006 show that, over the previous 25 years, jobs in the "information technology, engineering and scientific consulting, and management and public affairs consulting" categories had grown the most. Due to the diversification of its economy, Montreal was relatively unaffected by the 2008 crisis, and during the 2010s even had unemployment rates comparable to Toronto's (Polèse 2018). However, Montreal has seen a significant increase in income disparities between neighbourhoods since the 1980s, ranking third after Toronto and Calgary in this respect (Breau et al. 2023).

At the borough and neighbourhood level, the transformation of the Montreal economy has led to the creation of organizations to manage deindustrialization and other changes. Community Economic Development Corporations (CDECs) were set up to establish partnerships between social economy organizations, private organizations, and public authorities and to launch structuring projects for neighbourhoods, such as the conversion of former industrial buildings or places of worship, the promotion of public spaces, or the enhancement of community facilities (Klein et al. 2012).

The diversification of Montreal's economy can also be seen in the location of activities. Since the 1950s and the construction of a network of expressways in subsequent decades, there has been a suburbanization of certain types of jobs, alongside a strengthening of the downtown core as the main employment hub. Starting with the 1980s, Montreal became more polycentric, with four additional employment hubs located on the Island of Montreal, as well as medium-sized employment poles on the north and south shores. These hubs are generally specialized around specific economic activities, for example, downtown features financial and insurance services, hospitals, and universities; the Saint-Laurent/Dorval area features medium- and high-tech manufacturing, transportation, and warehousing; and Laval features services, retail, and leisure (Shearmur 2018). However, the COVID-19 pandemic and the new norm of telecommuting have weakened downtown Montreal, where many offices are now empty several days a week, creating concern among many Montreal stakeholders and elected officials. Nevertheless, thanks to the construction of many new residential buildings since the first decade of the 2000s, the downtown area now has a sizeable residential population that exhibits a certain dynamism.

NEW POLITICAL DYNAMICS

From mergers to demergers and beyond, municipal reorganization has restructured the city's electoral system, giving Montreal a unique political structure in Quebec and Canada. Each of Montreal's 19 boroughs is headed by a mayor and a borough council, which is made up of borough councillors and city councillors. Montreal is the only city in Quebec to have directly elected borough mayors; the other large cities divided into boroughs choose a president from among the councillors elected in that borough. In Montreal, borough councillors sit only on the borough council, while city councillors sit on both the city council and the borough council. This restructuring has greatly increased the number of elected municipal officials, from 51 before the municipal mergers to a total of 103 (46 city councillors, 18 borough mayors, 38 borough councillors and 1 city mayor). The following sections show in concrete terms how this new institutional architecture influences local dynamics, focusing on three aspects: voting, partisan forces, and participatory democracy and community action.

Voting and Electoral Participation

Although the borough system was designed to allow the old towns to retain a certain distinctiveness, it also creates electoral complexity for voters and candidates. On the one hand, electors vote for different numbers of positions depending on the borough in which they live. At the very least, they can vote for three positions (city mayor, borough mayor, city councillor). The maximum number of votes is five (city mayor, borough mayor, city councillor, and as many as two borough councillors). In addition, candidates for mayor are generally better known than candidates for councillor. This is also true of candidates for borough mayor, who are generally better known than candidates for (borough or city) councillor.[1] In such a context, candidates may have some difficulty making themselves known to the electorate and disseminating their ideas. As evidence, a 2017 survey of over 3,000 Quebec voters revealed that "just under a third of voters do not know the position of mayoral candidates on at least one issue" (Dubois and Gélineau 2021:30).

Moreover, voter turnout in Montreal municipal elections, which often makes headlines for being low, is consistently lower than in municipal elections elsewhere in the province. However, this observation must be qualified. First, this phenomenon is not specific to Montreal, and second, in comparison with some other major cities in Quebec, with the exception of 2001, Montreal does not have the lowest voter turnout rates (see Table 2.1). Rather, voter turnout has fluctuated (Latendresse and Frohn 2011).

While several variables explain municipal voter turnout, Philippe R. Dubois and François Gélineau found that voters in larger cities (100,000 inhabitants and over) are significantly more likely than those in smaller cities (under 5,000 inhabitants) to agree with the following state-

Table 2.1: Voter turnout (%) from 2001 to 2021 in five major Quebec cities

	Montreal	Gatineau	Longueuil	Laval	Quebec	Provincial average
2001	49.2	53	45	50	60	Not available
2005	34.9	48	38	31.0	51	44.5
2009	39.4	38.5	38.9	35.7	49.4	44.8
2013	43.3	41.9	34.0	41.1	54.9	47.2
2017	42.5	38.5	33.1	36.3	50.9	44.8
2021	38.3	35.1	34.0	28.8	45.2	38.7

Source: Produced by Jérôme Couture in 2013 and updated with data from the Ministère des Affaires municipales et de l'Habitation (MAMH) and Données Québec.

ment: "Sometimes, municipal politics and council operations seem so complicated that it's hard to understand what's going on" (2021: 32).

The complexity of voting in Montreal is linked to another reality specific to the city, namely the large number of invalid ballots. There are a variety of reasons for this phenomenon, including deliberate ballot spoiling, and voter error leading to the invalidation of a ballot. In general, municipal elections in Quebec's major cities feature a higher percentage of invalid ballots than do elections at the provincial and federal levels. While that percentage is usually the highest in Montreal, a significant decline has been observed over the past 15 years (see Table 2.A at the end of the chapter). Breux and Couture (2014), in the context of the 2009 Montreal election, showed that invalid ballots increased in situations where the elector had to vote for four or five elective positions as well as when the elective position was closest to citizens, such as that of borough councillor. Indeed, this latter situation garnered the highest number of invalid ballots. In that context, the ballot invalidation could also be attributed to two main co-existing realities: the number of choices that the voter has to make, and the importance that the voter attaches to the various elective positions. Might this "hierarchy" of positions be associated with a general lack of information on the various elective positions, or with a presumed lesser importance of borough councillor positions? Élections Montréal has worked hard to disseminate and popularize the city's electoral system. In addition, the Commission de la présidence du conseil (2023) carried out public consultations and submitted a report on electoral participation in Montreal, wherein it made numerous recommendations to improve information for voters in general. While these efforts reflect a genuine desire to create a culture of participation, they cannot ignore the various partisan dynamics that are also likely to influence the decision to go to the polls.

Partisan Politics

Montreal's new institutional architecture must be placed in the city's political and historical context to be properly understood. Mayor Jean Drapeau, renowned for his authoritarianism and major projects (the Montreal Metro rapid transit system, Expo 67, infrastructure for the

Olympic Games, to name but a few), won his eighth term in office in 1982. For several years, however, opposition had coalesced around a new political party, the Rassemblement des citoyens et citoyennes de Montréal (RCM). Led by Jean Doré, the RCM won the 1986 election after Drapeau retired from politics. During his two terms in office, Doré's administration set out to democratize political institutions and embarked on a decentralization project (Hamel 1991). It created nine *arrondissements*, or boroughs, along with specific bodies to enable citizens to obtain information on various aspects of daily life related to municipal responsibilities (notably permit applications – these boroughs were the ancestors of today's boroughs but were really much weaker). Towards the end of the RCM's second mandate, the Doré administration also established borough consultation committees. These consultative bodies, which could make recommendations to the city's executive committee, gave citizens a voice. The Bureau de consultation de Montréal and the city council's standing committees were added to these bodies.

The RCM failed to win a third term, however. In 1994, Pierre Bourque, head of the new Vision Montréal party, was elected mayor. Once in office, Bourque abandoned the RCM's previous consultation initiatives. Re-elected in 1998, he then focused on making Montreal "one island, one city" during his second term.

The new, unified city's first election in November 2001 manifested a new political dynamic that extended beyond the opposition between the pro- and anti-merger factions. On one side was a new party, the Union des citoyens et citoyennes de l'île de Montréal (UCIM), led by Gérald Tremblay and composed mainly of elected officials from the former suburban municipalities, many of whom were also dissidents from the merger. The UCIM also included some members of the RCM: the RCM having been unable to recover from its 1994 defeat, its members chose to merge with the UCIM. On the other side was Bourque's Vision Montréal party. While these two parties dominated the political scene in the early 2000s, this dynamic underwent a change.

Two things happened following the merger. First, borough parties emerged, which presented candidates only in one borough and developed a political platform targeting only that territory. One example is Équipe Anjou, a party created in 2005 by borough mayor Luis Miranda, who had been mayor of this suburban municipality prior to the merger. In every election in which the party ran (it did not run in 2009), it won every seat, reflecting Mayor Miranda's stranglehold on local politics. Other borough parties tended to be short-lived, however. Some reflected local dynamics and maintained a presence spanning several elections, such as Oser Outremont (2005 and 2009 elections). Most are groupings of dissidents from the dominant parties, who focus their political activities within their own boroughs. Except for Équipe Anjou, borough parties have had limited electoral success, even if they represent new ways of organizing Montreal politics.

Second, the predominance of the UCIM (which became Union Montréal, after the de-mergers) and Vision Montréal eroded between 2005 and 2009. A new player, the Projet Montréal party, created in 2004 by Richard Bergeron and a handful of activists, gradually increased its visibility. It is worth remembering that the majority of the former independent suburban municipalities – now boroughs – strongly supported Gérald Tremblay's party from 2001 to

2009.[2] The situation is different in the downtown boroughs, where Union Montréal lost seats to Projet Montréal in particular. Despite the arrival as its head of Louise Harel, the former Parti Québécois Minister of Municipal Affairs who implemented the municipal mergers, Vision Montréal failed to make significant electoral gains, placing second to Union Montréal in the 2009 elections.

Corruption scandals in the municipal sector weakened Union Montréal considerably, ultimately prompting Mayor Tremblay to resign in November 2012. A 2008 journalistic investigation uncovered systemic corruption in the province, revolving primarily around the misuse of political and election campaign finance rules and the awarding of public contracts. The Commission d'enquête sur l'octroi et la gestion des contrats publics dans l'industrie de la construction, commonly known as the Charbonneau Commission and held from 2011 to 2015, identified several corrupt practices, some of which were specific to Montreal.[3] Although Union Montréal was the main party singled out by the inquiry in Montreal, the commission uncovered a large-scale system that also tainted Vision Montréal.

In the wake of this major crisis, numerous elected officials joined other parties or left politics altogether, leaving the political space open to other political hopefuls. This was notably the case for the 21 Union Montréal elected officials who crossed over to the new party created by Denis Coderre on the eve of the 2013 elections, known simply as Équipe Denis Coderre. Coderre, a former federal Liberal minister, presented himself as "the new sheriff in town" who could restore the reputation of municipal politics in Montreal. He served as mayor from 2013 to 2017 and tried, unsuccessfully, to win back the mayoralty in 2021. The context was also favourable to the emergence of new partisan formations, notably Vrai Changement pour Montréal–Équipe Mélanie Joly (who is federal Minister of Foreign Affairs at the time of writing of this chapter) and Coalition Montréal–Équipe Marcel Côté. This proliferation of players in 2013 concealed Projet Montréal's gradual breakthrough and adaptation of its strategy to the institutional structure of the city and, above all, its boroughs.

Projet Montréal's first years were devoted to consolidating its organization and publicizing its ecologically progressive platform. In 2005, the party had only one elected member in the Plateau-Mont-Royal borough. In 2009, Projet Montréal won the borough mayor's office (with Luc Ferrandez, an elected official who would later become highly publicized and controversial), as well as all other borough positions. The mayor of Rosemont-La Petite Patrie, François Croteau, initially elected in 2009 under the Vision Montréal banner, switched to Projet Montréal in 2011. Projet Montréal's initial success at the borough level was crucial in establishing its credibility (Sanger 2021). In different ways, but in keeping with the party's ideals, the Plateau-Mont-Royal and Rosemont–La Petite-Patrie teams made the most of the autonomy granted by the new powers devolved to the boroughs to pursue the party's goal of building dense, humane, and lively neighbourhoods. In both boroughs, Projet Montréal altered local urban planning bylaws and practices to enhance public space, reduce car traffic, and increase pedestrian safety. Examples include installing sidewalk projections to make street intersections safer, developing a temporary network of public squares, making changes to the local traffic plan to eliminate through-traffic and direct cars towards major boulevards, and reviewing snow-clearing practices to give pedestrians a more prominent place. Many of the pilot projects, such as the traffic-calming measures and squares, have been gradually replicated by other boroughs. Rosemont–La Petite-Patrie also set up a regulatory framework conducive to citizen greening initiatives (Bherer and Dufour 2021). According to the party, the strategy was to "change the perception of urban space, street by street" (Sanger 2021).

Despite ongoing controversies surrounding Luc Ferrandez as Plateau-Mont-Royal's mayor, the 2013 election returned him to power, alongside Rosemont–La-Petite-Patrie's François Croteau. Most importantly, Projet Montréal became the official opposition at city hall with 28 elected members, having succeeded in getting 18 new city councillors elected against Denis Coderre's victorious team. After Richard Bergeron left the party and its leadership in 2016 to join Mayor Coderre's executive committee, Valérie Plante was elected first leader of Projet Montréal and then became the first woman to be elected mayor of Montreal in 2017. Her victory was largely attributed to the fact that voters perceived Plante as being closer to the political centre than her opponent, the more conservative Denis Coderre, and to the fact that Projet Montréal was also able to attract voters further to the left of the spectrum (Breux et al. 2022). Re-elected with control of the mayoralty and a majority on council in 2021, Projet Montréal also won 10 borough mayoralties. From 2009 to 2021, the party has stood out from the other contenders for its ability to systematically field 103 candidates, something that other parties have not always managed to do. In anticipation of the 2025 municipal elections, Valérie Plante announced in October 2024 that she would step down as head of Projet Montréal at the end of her term. The challenge will be to see whether the democratic vitality and mobilization of Projet Montréal activists will continue when the new leader is appointed.

Projet Montréal is clearly distinct from other recent Montreal parties. Unlike "équipes" established as vehicles for mayoral candidates, it belongs to a select group of structured municipal parties, such as the historic examples of the Rassemblement des citoyennes et citoyennes de Montréal and, in Quebec City, the Regroupement populaire de Québec. Since 2009, the party has distinguished itself by its focus on democratic transparency and participation. At that time, the party pledged to "post all information relating to the donations it receives on its website" (Latendresse and Frohn 2011: 109) – a promise which the party has, in fact, lived up to ever since. In general, the party's platform stands out for its commitments to active mobility, the built environment, and the environment. An analysis of the profiles of elected officials since 2009 – when Projet Montréal's breakthrough became more visible – shows that the individual commitments of the party's elected officials in these areas is constant and clearly differentiated from those of the other parties' elected officials (Breux and Van Neste 2022). This commitment echoes the various debates taking place on the Montreal scene.

Diversified Participatory and Community Approaches

The post-2000 municipal restructuring has institutionalized and expanded practices of participatory democracy across the agglomeration (albeit with marked differences between boroughs) and from one municipality to the next. To understand the effect of the mergers, we will first examine the participatory democracy model that existed before (Bherer and Collin 2018). The RCM was elected in 1986 on the promise of breaking with the authoritarian style that had held sway over the city for some three decades and of democratizing municipal political life. The party delivered on this promise by creating the Bureau de consultation de Montréal (BCM) in 1988, with a mandate to organize regular public hearings. The experiment was short-lived, however, as in 1994 the new mayor, Pierre Bourque, abolished the BCM and reverted to a centralizing leadership style (Trépanier and Alain 2008).

This approach was at odds with the diversity and vitality of Montreal's civil society. Repeated complaints from community groups and civic associations about the mayor's authoritarian style led the then-Minister of Municipal Affairs, Louise Harel, to establish a commission of inquiry into the need for public consultation. The report of the Tremblay Commission (named after its chair, Gérald Tremblay, who would become the first mayor of the new city after the mergers), tabled in 1999, recommended the creation of an autonomous, impartial public agency with a mandate to organize public hearings for all major development projects and major city policies. The Office de consultation publique de Montréal (OCPM) was created in 2002 following the amalgamation, and it gradually became a key democratic institution recognized by Montreal civil society. Between 2002 and 2021, the OCPM organized 157 public consultations. Additional citywide public consultation bodies have likewise been introduced since, such as the Bureau de l'expérience citoyenne (Office of citizen experience). Having changed its name several times over the years, this Bureau to this day initiates public consultations, organizes the participatory budget (see below), and operates the Réalisons Montréal platform, which centralizes many of the city's participatory mechanisms, including those of the boroughs. City council committees also hold public consultations on various issues.

Managing the transition following the mergers and decentralization to the boroughs also helped create new democratic spaces (Bherer and Breux 2012). The new city organized a major summit in 2002 that generated several initiatives, including the creation of a Chantier sur la démocratie, a consultative body made up of President Dimitri Roussopoulos, one of Montreal's leading democratic activists, and various members of Montreal's civil society. This project led to the adoption of a right to initiate a public consultation. Implemented in 2009, this right enables citizens to request that a public consultation be held if they collect at least 15,000 signatures citywide, or a smaller number not exceeding 5,000 for a borough-wide debate (with the number determined according to the borough's population). Since 2010, nine consultation initiatives have been approved, while 16 have been rejected by the city or abandoned by the applicants.

Several have had a lasting impact on public debate in Montreal. These include the public consultation on urban agriculture in 2011 and the one on racism and systemic discrimination in 2019. Their influence on decision making is, likewise, not negligible. Several recommendations resulting from these consultations were subsequently adopted. Examples include the ban on advertising flyers and the creation of the position of Anti-Racism Commissioner in Montreal, whose primary role includes ensuring that the recommendations arising from the *Consultation publique sur le racisme et la discrimination systémique* are implemented.

In addition, each borough has its own specific participatory dynamic, the boroughs having been in charge of local urban planning and development rules since the second wave of decentralization in 2004. Some boroughs have experimented with a great variety of mechanisms, while others have opted for a more minimalist interpretation of public consultation and urban planning requirements. This means that access to institutions and municipal life varies according to the borough in which a resident lives.

Projet Montréal's rise to power, first at the borough level and then citywide, has opened up new opportunities to experiment with participatory innovations. First, participatory budgeting processes were introduced in two stages. A short participatory budget experiment in the Plateau borough from 2006 to 2008 (Patsias et al. 2013) has given way, since 2020, to the creation of participatory budgets in the boroughs of Ahuntsic and Hochelaga-Maisonneuve and across the greater city of Montreal. Second, the city and certain boroughs have implemented a policy of supporting citizen initiatives, whereby citizens decide to produce a collective good in their community on their own, with the help of public authorities. These are generally small-scale interventions, such as greening tree patches and alleyways, setting up community fridges, coordinating garage sales, or setting up bike repair shops on a vacant lot. The Rosemont–La Petite-Patrie borough has been a forerunner in this field, gradually putting in place several forms of support for citizen initiatives, particularly in terms of greening public spaces. Among these are providing permaculture training, issuing small grants, donating land and seeds, offering networking opportunities, easing rules and regulations, occupying spaces or premises free of charge or at a symbolic cost, and so on. The borough has also encouraged the self-management of new neighbourhood parks, where citizens are responsible for their animation.

Municipal restructuring and the special powers conferred on the boroughs have also provided additional opportunities for community action and social mobilization. Community organizations and sub-municipal consultation mechanisms have always played an important role in local politics in Montreal. In fact, the first neighbourhood tables date back to the 1970s, on the one hand under the influence of a global movement for healthy neighbourhoods, and on the other in the wake of the Quebec community movement (Sénécal et al. 2008; IMSDSL 2015). The place of Montreal community organizations in local politics had thus already been observed and recognized prior to the merger. Indeed,

Geneviève Cloutier and Muriel Sacco argue that "for a long time, community groups, through the *tables de concertation*, were more influential players than district councillors in the governance of neighbourhoods by having more financial resources at their disposal" (2012: 17 [our translation]).

Montreal's "neighbourhood table" experience became a model for local intersectoral cooperation (Bacqué 2005; Sénécal et al. 2008), funded first by the City of Montreal, then in collaboration with the Centraide Foundation and the Direction de santé publique. Following the amalgamation, funding was released to extend the model to the former suburban municipalities (IMSDSL 2015). Accordingly, the number of neighbourhood tables has risen from 20 to 32 between 1997 and today. Neighbourhood tables bring together community organizations in a neighbourhood with public partners (boroughs, health centres, schools, institutions) on an intersectoral basis, with the aim of fighting poverty and improving living conditions. Although neighbourhood tables existed before the mergers and the decentralization of powers to the boroughs, since that time it is often in strong partnership with the boroughs that the neighbourhood tables deploy their projects and influences (Sénécal et al. 2008). A case in point is the integrated urban revitalization program. Launched in 2002, following the Sommet de Montréal, this program established direct links between the neighbourhood tables and the boroughs to facilitate the consultation and planning of cross-cutting measures for the urban revitalization of disadvantaged neighbourhoods.

Community organizations also have an institutionalized place in particular issue sectors, where they contribute as much to the co-production of municipal programs and services for the most vulnerable as to broader collective mobilization. First, citizens' committees had emerged in the 1960s and 1970s to defend living environments against technocratic urban planning and infrastructure interventions as well as authoritarian municipal administration (Godbout and Collin 1977; Hamel 1991). Housing committees emerged in the 1970s to defend tenants' rights and investment in social housing. Indeed, housing is cited as the theme generating the most conflictual action in Montreal in the 1990s (Fontan et al. 2013). Montreal community organizations are also active in the field of community economic development, inspired by the American experience of community development corporations (Bacqué 2005).

In the 1980s and 1990s, community action in Montreal moved towards a position of partnership with the state. Pierre Hamel describes how citizen committees in Montreal gradually professionalized and moved from a confrontational mode to a more pragmatic one where they would, in collaboration with the state, participate in providing services for the most disadvantaged, improving living conditions and fighting poverty. While some see this as co-opting the movement, others see it as a form of innovation in relations between the state, civil society, and citizens. Inspired by the experiences of working-class Montreal neighbourhoods, such as Pointe-Saint-Charles, and a philosophy of social intervention linked to the Catholic past, the community movement would like the most vulnerable segments of the population to be part of the solutions, and for community groups to support these people in developing projects that are more adapted and conducive to empowerment than top-down public intervention (Dufour and Guay 2019). Community organizations were particularly solicited during the pandemic, a crisis that highlighted both the importance of these networks and their lack of resources to meet social needs.

SOCIAL MOVEMENTS AND URBAN COALITIONS

In addition to the linguistic cleavage and struggles for local participatory democracy that marked the periods of merger and demerger in the first decade of the 2000s, the most important points of division and public debate in recent years in Montreal have concerned the themes of urban planning, housing and transportation, and mobility (Fontan et al. 2013), alongside issues of diversity. On the first themes, the borough level has facilitated alliances with community players and citizens' movements, as well as faster, but also conflict-ridden, progress in the neighbourhoods led by Projet Montréal. These issues also feed into debates and policy at the metropolitan and provincial levels.

While issues around language and diversity are commonly discussed in Montreal, they are less divisive for Montrealers than what might be expected by outsiders. On this point, the divide between Montreal and the rest of Quebec seems wide. In their book, Amiraux et al. (2017) speak of the everyday experience of diversity in Montreal, with certain discomforts, tensions, and demands, yet in spaces of sociability that contrast drastically with media and political representations outside Montreal. Local associations also celebrate and participate in the social and cultural dynamism of different communities and the improvement of the living environment, with, for example, strong networks in the notably Afrodescendant communities of Little Burgundy and Saint-Michel (Kapo, 2020; Boudreau and Rondeau 2021).

Community actors and academic literature have revealed the presence of systemic racism, racial profiling, and everyday discrimination in, for example, housing, employment, and education. Certain communities, including Black communities and notably the Haitian community, have been particularly victimized by harassment and profiling, insofar as the media tends to cover the topics of "street gangs" and "difficult" neighbourhoods in ways that reinforce logics of stigmatization and police brutality. These dynamics, commonplace in many other cities, are discussed in the case of Montreal by Maxime Aurélien and Ted Rutland (2023). These issues are also reflected in a certain cleavage within municipal politics in the approach to urban security and the role of the police. During the 2021 election campaign, for example, Valérie Plante emphasized community policing and social workers, in stark contrast to her opponent Denis Coderre, who sought to increase the number of regular police officers. Antiracism and a reduction in the police department's budget were also the main themes of Balarama Holness's mayoral election campaign.

Although always present, antiracism struggles have taken on a more prominent role in the Montreal landscape over the past five years (Zoghlami 2023). Nevertheless, racism was until recently a taboo in Quebec and Montreal, in a climate characterized by interethnic cohabitation and a history of division of urban space between French- and English-speaking communities (Germain and Rose 2000; Germain et al. 2014; Boudreau and Rondeau 2021). Certain moments brought together the antiracism movement in Montreal and Quebec. First, activism against police violence and racial profiling were reactivated with the death of a young person in Montreal-North in 2003 as well as other more recent events and the influence of the Black Lives Matter movement (Zoghlami 2023). Second, the death of Joyce Echaquan from the Manawan community, who was subjected to racial discrimination by hospital nursing staff in 2020, sparked mobilization against racism towards Indigenous people. The City of Montreal officially recognized the existence of systemic racism and put in place an action plan to tackle it in 2019, following social mobilization and the initiation of a public consultation on the subject, based on the above-mentioned right of initiative.

Montreal has also been marked by activism, conflicts, and coalitions around mobility and infrastructure. At the turn of the century, expressway projects and local traffic-calming interventions motivated the formation of new citizen coalitions and street protests in Montreal. The great debates surrounding the construction of an east–west expressway in the 1970s, which was opposed by a citizens' coalition supported by architects and heritage advocates (Poitras 2009), led to a moratorium on new expressway projects. When expressway reconstruction projects (the Turcot Interchange and Notre-Dame Street East) returned to the agenda at the turn of the century, it was in a context of mandatory environmental assessment at the provincial level and public participation at the provincial and municipal levels. Opposition to these projects has left its mark on the Montreal political scene, first through tensions, and then through strong alliances between environmentalists, the above-mentioned community groups, the Agence de santé publique de Montréal and even the City of Montreal against the provincial Ministry of Transport (Van Neste and Sénécal 2015). In the past, major modernist expressway and urban renewal projects were supported or put forward by the City of Montreal (Poitras 2009). In the 2000s, however, an urban coalition was formed around health issues related to urban pollution and accidents affecting pedestrians and cyclists, the impact of expressways on the urban fabric, expropriations and the increased environmental impact of automobile traffic. These infrastructure reconstruction projects arose at the very moment when the City of Montreal was implementing its urban plan (the second in the city's history) and institutionalizing public participation mechanisms. This discrepancy between, on the one hand, Montreal's burgeoning participatory urban planning practices and, on the other, the provincial road infrastructure model disconnected from urban realities and local concerns, led to strong criticism of the so-called provincial Ministry of Highways: "Montreal is not a field of corn through which a highway passes" (BAPE 2009: 25).

The issue of mobility and transportation has a prominent place in Montreal's history. First, it has been a key factor in the formation of alliances leading to broad social mobilization. Second, it is at the heart of Projet Montréal's political platform and its approach to the boroughs. Community organizations, environmental groups, and local citizens' associations, in parallel with their participation in the movement opposing expressway projects, have also been developing initiatives for car alternatives in Montreal neighbourhoods. These groups are pushing for traffic calming on local streets and the promotion of pedestrian and cycling facilities, with significant new funding from philanthropists for the protection of the most vulnerable segments of the population. No fewer than 165 groups focused on promoting active mobility and traffic calming were counted between 2006 and 2011 in Montreal (Van Neste and Martin 2018). Emerging during the tumultuous period of changes in Montreal's governance, they established new links among environmental and urban ecology non-governmental organizations (NGOs), neighbourhood tables and community organizations, street-level citizen groups, and boroughs. They share a common discourse centred on the idea of a dynamic local community that supports walking and cycling, while making specific demands to the boroughs that are responsible for local amenities (Van Neste and Martin 2018). The boroughs in the central districts have in turn taken up the challenge, and have invested heavily in transforming local streets. In a few cases, such as Plateau–Mont-Royal, the redevelopment of two or three streets was the subject of controversy in the mainstream provincial media, where there was talk of an unprecedented "war on the car." Based on media coverage alone, these initiatives could be interpreted as anti-democratic. However, they stem from citizen activism and alliances among community, ecological, and public health organizations, demonstrating a broader movement than just the Plateau–Mont-Royal mayor's initiative. In other boroughs, analogous changes have been carried out with less fanfare, but always in response to pressure from citizens and civil society. Nevertheless, media coverage portrayed the debate as a divide between the central districts and the rest of Greater Montreal (Van Neste and Martin 2018). It is true that the central districts led by Projet Montréal have moved faster than the outlying sectors of the agglomeration. Studies also show a conflict between the interests of cyclists on the one hand and those of car drivers on the other, who deplore a lack of parking spaces in Montreal. Indeed, the term *bikelash* has even been coined to refer to the widespread discontent experienced by this latter group of people. In addition, studies show an unequal distribution of investments, notably gaps in lower-income and outlying districts, such as Montreal North and Montreal East, which are disadvantaged in terms of public transport (Breau et al. 2023; Rodrigue et al. 2023).

Transportation is also an area of tension between the municipal and provincial levels, and one in which Projet Montréal has played an important role. Indeed, the mobilizations around the Turcot Interchange and Notre-Dame Street East marked a major tension over the relative influence, or even veto, that the city could exert over major infrastructure projects. This battle is also linked to the expertise of players specializing in urban planning and public health, who tried to change the transportation planning paradigm at the provincial level. For example, following the mobilizations surrounding the redevelopment of the Turcot Interchange, the provincial Ministry of Transport allowed the City of Montreal to develop a counter-proposal for the interchange's redevelopment. This counter-proposal was drafted by Richard Bergeron, leader of Projet Montréal, whom Mayor Tremblay had included on the executive committee for this purpose in 2010. Similar tensions between the municipal and provincial levels have recently resurfaced in debates surrounding the development of the city's rapid transit system, the Réseau express métropolitain de transport collectif (REM). The unilateral efforts of the provincial public pension fund, the Caisse de dépôt et placement du Québec, and its failure to recognize the concerns and expertise of municipalities on the Island of Montreal, provoked major opposition that ultimately resulted in the abandonment and modified governance of the project for one leg of the route, which was taken over by the City of Montreal with multi-party partnerships. This time, the city, through Mayor Valérie Plante, who presented herself as the "mobility mayor," succeeded in its gamble of having a real place at the table.

In addition to transportation infrastructure, on which Montreal's leadership has been repeatedly tested, energy infrastructure has also been the subject of a tug-of-war, in this case between the Montreal region and federal authorities. In the mobilization against oil pipelines, the mayor of the City of Montreal has played a leading role. Mayor Coderre was portrayed in the media as one of the decisive players in the National Energy Board's 2014 about-face on the Enbridge pipeline in Quebec, and in the halting of proceedings for Trans Canada's Energy East pipeline in 2016 – even more decisive than the significant opposition from civil society and Indigenous communities. Coderre relied on a monitoring committee that included active participation from the suburbs affected by the proposed pipeline route, and expertise developed at the CMM. In these debates, he always presented himself as president of the CMM, and not as mayor of Montreal, to justify the legitimacy of his speeches and the breadth of municipal opposition, frequently telling the media that he represented 3.9 million people. At the height of the debate, in 2016, 61 per cent of newspaper articles (national and local) on oil pipelines in the Montreal region addressed the CMM's position (Van Neste and Couture-Guillet 2022). Leakage risks for drinking water intakes were among the most important concerns. However, the social movement went beyond the municipalities and included local citizen mobilization, a broad alliance of Indigenous communities and a strong student presence.

As a result, the CMM gained not only greater visibility with the region's population but also a legitimacy it had previously lacked with the various suburban municipalities. Created in tandem with the municipal amalgamations, the CMM faced great difficulty during its first ten years of existence in adopting a metropolitan development plan that would attract the support of the suburban rings to the north and south of the Island of Montreal (Boudreau and Collin 2009). Lagging behind other Canadian city-regions (Filion and Kramer 2012), it was only in

2012 that such a plan was adopted in which the 82 municipalities agreed on targets for residential density and an urban structure based on transit-oriented development principles. At the time, the metropolitan question was interpreted by analysts as the domain of experts and elites (Bherer and Sénécal 2009), and the consensus around the 2012 plan focused above all on a negotiated compromise between the City of Montreal and suburban municipalities on density and access to public transit (Roy-Baillargeon 2017). While groups pushed for the metropolitan plan to include natural environmental protections and a green belt, the adopted plan and regulations disappointed them.

At the city level, action to tackle climate change has become an even greater priority with the election of Valérie Plante and Projet Montréal. The city had already embarked on a process with sustainable development plans (2005) and climate change mitigation and adaptation plans (2013, 2015). Under Projet Montréal, the city presents itself as an international leader, notably in the C40 global network of cities united to tackle climate change. Montreal joined Vancouver in following the C40 network's timetable for adopting a climate plan by 2020, with the aim of achieving carbon neutrality by 2030. Some of its boldest experiments, however, have been at the borough level, namely in matters concerning mobility, the ecological transition, or participatory greening. It was Plateau-Mont-Royal borough mayor Luc Ferrandez, who, as the city's head of large parks, in 2018 halted a large-scale real estate development project on wetlands in the west end of the island to allow for what is now being promoted as one of the largest nature parks in Canada. Despite this bold gesture, Ferrandez resigned from his position the following year, explaining, "[by remaining in office as an elected official], I feel like I'm fooling citizens into believing that we're collectively taking all the necessary steps to slow the rate of destruction of our planet" (Ferrandez 2019).

CONCLUSION

Overall, Montreal has changed a great deal since the 1980s. The most visible aspect of this change has been the complete transformation of its institutional architecture, which has profoundly affected the way in which Montrealers engage in local politics, get involved as citizens, and build local mobilizations. The so-called municipal merger reform in fact did much more than simply amalgamate numerous municipalities into a single one. It also created of two new levels of government in Montreal: the Communauté métropolitaine de Montréal and the 19 boroughs. The reform has been particularly beneficial for local social groups, which have used the boroughs to make themselves heard – a situation that contrasts with the former City of Montreal, where everything was centralized around the city council. The advent of the boroughs also benefited Projet Montréal, which was able to politically control certain boroughs and thus prove itself as a serious political alternative, culminating in the election of the party's first mayor, Valérie Plante, in 2017.

Projet Montréal's rapid progress is impressive and provides a rare example of a municipal political party with a dynamic partisan life that has managed to endure with each passing election. Projet Montréal has become a formidable adversary, to the point where opposition parties – often coalitions revolving around mayoral candidates more so than conventional parties (Couture et al. 2018) – have difficulty making themselves heard. However, as in any political power dynamic, Projet Montréal will have to be vigilant if it seeks to maintain its position of power. A first, fairly common scenario when one party holds considerable sway over the political spectrum is the rise of internal adversaries. Party unity then threatens to be undermined by tensions between party factions regarding policy choices. In a second scenario, these tensions could then be appeased or mitigated through the diversity of the city's institutional venues. For example, Projet Montréal councillors who disagree with central decisions may simply withdraw to their own boroughs, thus resuming the party's original strategy of fully exploiting borough powers. In a third scenario, the contemporary issues of mobility, antiracism, and climate change, which are so central to Projet Montréal, could become the very ones that will lead to the party's demise, given their divisive nature among the electorate. Moreover, the dynamism of partisan life and local activism contrast with voter turnout, which remained more or less the same before and after the reform that led to municipal mergers and the multilevel system of local and metropolitan governance.

Decentralization has also benefited local struggles and concerted action. Prior to the mergers, Montreal had already experimented with local collaborative governance through the creation of neighbourhood tables. The creation of a decentralized political space, along with the creation of the boroughs, enabled this movement to continue, with local networks becoming more structured. Montreal is a textbook example of decentralization and distribution of political spaces for representation and participation. This aspect should not, however, obscure the fact that the influence of municipal mergers is still clearly visible within the municipal administration, particularly within the boroughs, where the working cultures, often unique, inherited from the former administrations may still be prevalent and contribute to making information sharing less fluid. We still know very little about how Montreal's decentralized public administration works. Is the room to manoeuvre experienced by elected representatives at the borough level also found in the public service? From the viewpoint of public servants, what effects have mergers and decentralization had on career opportunities and the division of labour? Are administrative procedures difficult to coordinate?

Finally, the unequal distribution of policy and governance innovations across boroughs highlights the differences between them and the inequalities associated with decentralization. One example is the uneven distribution of bike paths and bike-sharing stations, which are more prevalent in central boroughs. Democratic practices and access to elected officials are also highly varied, which ultimately creates several regimes of "municipal citizenship" in Montreal.

APPENDIX

Table 2.A: Percentage of invalid ballots in Montreal by type of elective office, compared with other major cities in Quebec

		2009	2013	2017	2021
Montreal	City Hall	3.44	2.68	2.46	1.84
	Borough halls	3.94	4.02	3.22	2.64
	City councillors	4.12	2.67	3.33	2.70
	Borough councillors	4.60	4.69	3.83	3.62
Gatineau	City Hall	1.61	1.56	1.76	0.75
	City councillors	2.05	2.08	1.89	1.72
Laval	City Hall	2.47	4.96	1.90	1.63
	City councillors	3.20	4.56	2.37	1.78
Longueuil*	City Hall	2.93	3.75	2.17	1.14
	City councillors	2.65	2.78	2.29	1.59
Quebec	City Hall	1.61	1.77	1.37	0.83
	City councillors	1.65	1.78	1.70	1.29

* It should be noted that Ville de Longueuil has two borough councillor positions. In 2013, the rate of invalid votes for these two positions was 2.87, whereas it was 4.21 in 2021. Data for 2009 and 2017 are not available

Source: Adapted from Breux and Couture (2014) and updated with data from the Ministère des Affaires municipales et de l'Habitation (MAMH) and Données Québec.

NOTES

1 Within the Ville-Marie borough, the city mayor is de facto the borough mayor. The mayor also appoints two elected officials to the borough. Voters from Ville-Marie borough have only two votes.

2 "This political factor converges with other factors, notably socioeconomic and linguistic. Indeed, [...] voters with the highest average incomes mainly support Union Montréal. What's more, we also know that a majority of anglophones and allophones (although again, this does not exclude all francophones) support Union Montréal" (Latendresse and Frohn 2011: 115; our translation).

3 This commission was chaired by Justice France Charbonneau. First, Union Montréal had set up a nominee system that enabled companies to ask their employees to donate money to a party. The company would then show the party the amounts it had collected and, in exchange, demand privileged access to public contracts. This practice undermines the principle of popular financing, which underpins Quebec's political financing system (only individuals can donate, and only up to a set maximum amount), because the money is actually donated by the company, not by an individual. The second method is false invoicing: a company asks an accomplice firm to send it false invoices, which are paid in cash. A few days later, the accommodating firm sends back part of the sum, taking care to keep a commission for itself. The main company thus accumulates a kitty that it can use to pay the rebates Union Montréal demands in return for public contracts.

REFERENCES

Amiraux, Valérie, Julie-Anne Boudreau, and Annick Germain. 2017. "Vivre Ensemble ? Les Modes de Cohabitation à Montréal, Entre Conflits et Convivialités," Atelier 10. In *Vivre Ensemble à Montréal: Épreuves et Convivialités*, edited by Annick Germain, Valérie Amiraux, and Julie-Anne Boudreau, 8–17. Montreal: Centre de recherche interdisciplinaire sur la diversité et la démocratie (CRIDAQ), coll. "Forme."

Aurélien, Maxime, and Ted Rutland. 2023. *Il Fallait se Défendre. L'histoire du Premier Gang de Rue Haïtien à Montréal*. Montreal: Mémoire d'encrier.

Bacqué, Marie-Hélène. 2005. "Action Collective, Institutionnalisation et Contre-pouvoir: Action Associative et Communautaire à Paris et à Montréal." *Espaces et Sociétés* 4 (123): 69–84. https://doi.org/10.3917/esp.123.0069.

Bherer, Laurence. 2006. "Les Valeurs Libérales, la Démocratie Locale et les Défusions Municipales: Des Citoyens sans Ville ou des Villes sans Citoyens." In *Le Parti Libéral: Enquête sur les Réalisations du Gouvernement Charest*, edited by Éric Bélanger, Louis Imbeau, and François Pétry, 338–60. Quebec: Presses de l'Université Laval.

Bherer, Laurence, and Sandra Breux. 2012. "The Diversity of Public Participation Tools: Complementing or Competing With One Another?" *Canadian Journal of Political Science* 45 (2): 379–403. https://doi.org/10.1017/S0008423912000376.

Bherer, Laurence, and Jean-Pierre Collin. 2018. "Urban Issues and Political Mobilization: From Subsidiarity to Institutionalized Governance." In *Montreal: The History of a North American City*, edited by Dany Fougères and Roderick MacLeod, 379–415. Montreal: McGill-Queen's University Press. https://doi.org/10.2307/j.ctt2111gbs.51.

Bherer, Laurence, and Pascale Dufour. 2021. "Participation Informelle: Où et Comment S'engage-t-on à Montréal Aujourd'hui ?" In *Montréal en Chantier: les Défis d'une Métropole pour le XXIe Siècle*, edited by Jonathan Durand Folco, 171–88. Montreal: Écosociété.

Bherer, Laurence, and Gilles Sénécal. 2009. "La Métropole: La Ville Continue ou le Territoire Sans Légitimité ?" In *Le Métropolisation et Ses Territoires*, edited by Laurence Bherer and Gilles Sénécal, xiii–xxiii. Quebec: Presses de l'Université du Québec.

Boudreau, Julie-Anne, and Jean-Pierre Collin. 2009. "L'espace Métropolitain comme Espace Délibératif?" In *La Métropolisation et Ses Territoires*, edited by Laurence Bherer and Gilles Sénécal, 269–86. Quebec: Presses de l'Université du Québec.

Boudreau, Julie-Anne, and Joëlle Rondeau. 2021. *Les Mondes Urbains de la Jeunesse: L'action Politique Esthétique à Montréal*. Quebec: Presses de l'Université Laval. https://doi.org/10.2307/j.ctv23khnnt.

Breau, Sébastien, Megan Wylie, Kevin Manaugh, and Samantha Carr. 2023, "Inclusive Growth, Public Transit Infrastructure Investments and Neighbourhood Trajectories of Inequality in Montreal." *Environment and Planning A: Economy and Space*, Online First. https://doi.org/10.1177/0308518X231162091.

Breux, Sandra, and Couture, Jérôme. 2014. "Explaining Invalid Votes in a Non-Politicized Context and in a Low-Turnout Environment: The Case of One Municipal Canadian Election." Paper presented at the 64th Political Studies Association Annual International Conference, Manchester, United Kingdom.

Breux, Sandra, Jérôme Couture, and Anne Mévellec. 2022. "Does the Left-Right Axis Matter in Municipal Elections?" In *Voting in Quebec Municipal Elections, a Tale of Two Cities*, edited by Éric Bélanger, Cameron D. Anderson, and Michael McGregor, 94–116. Toronto: University of Toronto Press. https://doi.org/10.3138/9781487540081-008.

Breux, Sandra, and Sophie L. Van Neste. 2022. "Differentiated Profiles of Elected Officials in Montreal: A Specific Party Identity Around Mobility, Urban Planning, and Environment?" *Frontiers in Political Science* 4 (March): 745777. https://doi.org/10.3389/fpos.2022.745777.

Bureau d'audiences Publiques sur L'environnement (BAPE). 2009. *Projet de Reconstruction du Complexe Turcot à Montréal, Montréal-Ouest et Westmount: Rapport d'enquête et d'audience Publique.* Bureau d'audiences publiques sur l'environnement. https://www.bape.gouv.qc.ca/fr/dossiers/reconstruction-complexe-turcot-montreal-montreal-ouest-westmount/.

Cloutier, Geneviève, and Muriel Sacco. 2012. "Les Mouvements Sociaux Urbains dans les Politiques Socio-Urbaines: Le Cas du Quartier Sainte-Marie à Montréal." *L'information Géographique* 76 (1): 58–73. https://doi.org/10.3917/lig.761.0058.

Collin, Jean-Pierre, and Laurence Bherer. 2018. "One Island, Several Cities: Montreal and its Immediate Suburbs." In *Montreal: The History of a North American City*, edited by Dany Fougères and Roderick MacLeod, 132–63. Montreal: McGill-Queen's University Press. https://doi.org/10.2307/j.ctt2111gbs.44.

Commission de la Présidence du Conseil. 2023. *La Participation aux Élections Municipales à Montréal.* Montreal: Ville de Montréal. http://ville.montreal.qc.ca/pls/portal/docs/page/commissions_perm_v2_fr/media/documents/rapport_participationelectorale_v2_20230210.pdf.

Couture, Jérôme, Sandra Breux, and Laurence Bherer. 2018. "Political Accountability and Responsiveness: What Is the Role of Municipal Political Parties?" In *Accountability and Responsiveness at the Municipal Level: Views from Canada*, edited by Sandra Breux and Jérôme Couture, 76–104. Montreal: McGill-Queen's University Press. https://doi.org/10.1515/9780773553743-006.

Drouilly, Pierre, and Alain-G. Gagnon. 2004. "Tout ça, Pour ça: La Stratégie de l'Autruche." *Le Devoir*, July 2: A7.

Dubois, Philippe R., and François Gélineau. 2021. *Les Motifs de la Participation Électorale aux Élections Municipales Québécoises: Le Cas de 2017.* Quebec: Chaire de recherche sur la démocratie et les institutions parlementaires, Université Laval. https://hal.archives-ouvertes.fr/hal-03172581.

Dufour, Pascale, and Lorraine Guay. 2019. *Qui Sommes-Nous pour Être Découragées ? Conversation Militante.* Montreal: Écosociété.

Ferrandez, Luc. 2019. Facebook, May 14. https://www.facebook.com/permalink.php?story_fbid=10205489320908286&id=1709066378.

Filion, Pierre, and Anna Kramer. 2012. "Transformative Metropolitan Development Models in Large Canadian Urban Areas: The Predominance of Nodes." *Urban Studies* 49 (10): 2237–64. https://doi.org/10.1177/0042098011423565.

Fontan, Jean-Marc, Pierre Hamel, and Richard Morin. 2013. *Ville et Conflits. Action Collective, Justice Sociale et Enjeux Environnementaux.* Quebec: Presses de l'Université Laval, coll. "Études urbaines."

Germain, Annick, Xavier Leloup, and Martha Radice. 2014. "La Cohabitation Interethnique dans Quatre Quartiers de Classes Moyennes à Montréal: Deux Petites Leçons Tirées des Discours sur la Diversité." *Diversité Urbaine* 14 (1): 5–24. https://doi.org/10.7202/1027812ar.

Germain, Annick, and Damaris Rose. 2000. *Montréal. The Quest for a Metropolis.* Chichester, UK: John Wiley & Sons.

Godbout, Jacques, and Jean-Pierre Collin. 1977. *Les Organismes Populaires en Milieu Urbain: Contre-Pouvoir ou Nouvelle Pratique Professionnelle ?* Montreal: INRS-Urbanisation (Institut national de la recherche scientifique).

Hamel, Pierre. 1991. *Action Collective et Démocratie Locale. Les Mouvements Urbains Montréalais.* Montreal: Les Presses de l'Université de Montréal.

Hamel, Pierre, and Claire Poitras. 2004. "Déclin et Relance Économique d'une Agglomération Métropolitaine. Le Discours et les Représentations des Élites Économiques à Montréal." *Recherches Sociographiques* 45 (3): 457–92. https://doi.org/10.7202/011466ar.

Initiative montréalaise de soutien au développement social local (IMSDSL). 2015. "Cadre de Référence de l'Initiative Montréalaise de Soutien au Développement Social Local – Des Quartiers où Il Fait Bon Vivre !" http://www.tablesdequartiermontreal.org/wp-content/uploads/2017/01/11-CADRE_REFERENCE_INITIATIVE_MONTREALAISE_15_JUIN_2015.pdf.

Kapo, Leslie Touré. 2020. "Les Aventures Ordinaires des Jeunes Montréalais.e.s Racialisé.e.s." Doctoral thesis in urban studies, INRS (Institut national de la recherche scientifique).

Klein, Juan-Luis, Dario Enríquez, Ping Huang, and Reina Victoria Vega. 2012. "Le Développement Économique Communautaire et la Cohésion Sociale à Montréal: Un Rôle de Médiation et d'Intermédiation." *Économie et Solidarités* 42 (1): 9–35. https://doi.org/10.7202/1029008ar.

Latendresse, Anne, and Winnie Frohn. 2011. "Fracture Urbaine, Legs Institutionnel et Nouvelles Façons de Faire: Trois Mots Clés d'une Lecture des Élections Municipales à Montréal." In *Les Élections Municipales au Québec: Enjeux et Perspectives*, edited by Sandra Breux and Laurence Bherer, 85–122. Quebec: Presses de l'Université Laval. https://doi.org/10.1515/9782763792736-004.

Meloche, Jean-Philippe, and François Vaillancourt. 2022. *Le Partage du Financement des Services de l'Agglomération de Montréal en 2020: État des Lieux, Analyse et Éléments de Comparaison*. Cahier scientifique du CIRANO (Centre interuniversitaire de recherche en analyse des organisations), 2022S-22.

Montréal en statistiques. 2022. Population et Démographie. Montreal: http://ville.montreal.qc.ca/pls/portal/docs/page/mtl_stats_fr/media/documents/population%20et%20d%c9mographie_population%20totale%20en%202021.pdf.

Patsias, Caroline, Anne Latendresse, and Laurence Bherer. 2013. "Participatory Democracy, Decentralization and Local Governance: The Montreal Participatory Budget in the Light of "Empowered Participatory Governance." *International Journal of Urban and Regional Research* 37 (6): 2214–30. https://doi.org/10.1111/j.1468-2427.2012.01171.x.

Picard, Laurent. 1986. *Rapport du Comité Consultatif au Comité Ministériel sur le Développement de la Région de Montréal*. Ottawa: Ministre des Approvisionnements et Services Canada.

Poitras, Claire. 2009. "Repenser les Projets Autoroutiers: Deux Visions de la Métropole Contemporaine." In *La Métropolisation et Ses Territoires*, edited by Gilles Sénécal and Laurence Bherer, 107–24. Quebec: Presses de l'Université du Québec. https://doi.org/10.2307/j.ctv18pgmj4.8.

Polèse, Mario. 2018. "Montreal's Economy since 1930." In *Montreal: The History of a North American City*, edited by Dany Fougères and Roderick MacLeod, 164–205. Montreal: McGill-Queen's University Press. https://doi.org/10.2307/j.ctt2111gbs.45.

Rodrigue, Lancelot, Aryana Soliz, Kevin Maunaugh, and Ahmed El-Geneigy. 2023. "Situating Divergent Perceptions of a Rapid-cycling Network in Montréal, Canada." *Active Travel Studies* 3 (2). https://doi.org/10.16997/ats.1355.

Roy-Baillargeon, Olivier. 2017. *Le TOD Contre la Ville Durable ? Utiliser le Transport Collectif pour Perpétuer le Suburbanisme Dispersé dans le Grand Montréal*. Environnement urbain /Urban Environment, June. https://doi.org/10.7202/1050577ar.

Sancton, Andrew. 2004. "Les Villes Anglophones au Québec: Does It Matter That They Have Almost Disappeared? *Recherches Sociographiques* 45 (3): 441–56. https://doi.org/10.7202/011465ar.

Sanger, Daniel. 2021. *Saving the City: The Challenge of Transforming a Modern Metropolis*. Montreal: Véhicule Press.

Sénécal, Gilles, Geneviève Cloutier, and Patrick Herjean. 2008. "Le Quartier comme Espace Transactionnel: l'Expérience des Tables de Concertation de Quartier à Montréal." *Cahiers de Géographie du Québec* 52 (146): 191–214. https://doi.org/10.7202/019588ar.

Shearmur, Richard. 2018. "Montreal, 1950–2010: The Spatial Economy Transformed." In *Montreal: The History of a North American City*, edited by Dany Fougères and Roderick MacLeod, 206–39. Montreal: McGill-Queen's University Press. https://doi.org/10.2307/j.ctt2111gbs.46.

Trépanier, Marie-Odile, and Martin Alain. 2008. "Planification Territoriale, Pratiques Démocratiques et Arrondissements dans la Nouvelle Ville de Montréal." In *Renouveler l'Aménagement et l'Urbanisme: Planification Territoriale, Débat Public et Développement Durable*, edited by Mario Gauthier, Michel Gariépy, and Marie-Odile Trépanier, 221–46. Montreal: Les Presses de l'Université de Montréal. https://doi.org/10.1515/9782760625273-010.

Van Neste, Sophie L., and Annabelle Couture-Guillet. 2022. "The Urban Politicization of Fossil Fuel Infrastructure: Mediatization and Resistance in Energy Landscapes." *Environment and Planning C: Politics and Space* 40 (8): 1801–18. https://doi.org/10.1177/23996544221111657.

Van Neste, Sophie L., and Deborah G. Martin. 2018. "Place-Framing against Automobility in Montreal." *Transactions of the Institute of British Geographers* 43 (1): 47–60. https://doi.org/10.1111/tran.12198.

Van Neste, Sophie L., and Gilles Sénécal. 2015. "Claiming Rights to Mobility through the Right to Inhabitance: Discursive Articulations from Civic Actors in Montreal." *International Journal of Urban and Regional Research* 39 (2): 218–33. https://doi.org/10.1111/1468-2427.12215.

Ville de Montréal. 2021. "Population Totale en 2016 et en 2021 – Taux de Croissance – Densité." Annuaire statistique de l'agglomération de Montréal. Montreal. http://ville.montreal.qc.ca/pls/portal/docs/page/mtl_stats_fr/media/documents/population%20totale_taux%20de%20croissance_densit%c9_2016-2021.pdf.

Zoghlami, Khaoula. 2023. "Peut-Être si C'est Toi, Ça Ne Va Pas Fonctionner": l'Antiracisme au Québec Face aux Limites de la Représentation Stratégique." In *Québec en Mouvements. Continuité et Renouvellement des Pratiques Militantes*, edited by Pascale Dufour, Laurence Bherer, and Geneviève Pagé. Montreal: Les Presses de l'Université de Montréal.

3

Toronto

Martin Horak and Zack Taylor

INTRODUCTION

In the first edition of *City Politics in Canada*, Warren Magnusson wrote that Toronto is "a pressure-cooker for political action and for new definitions of the urban issues that face Canadians," but that the structure of local government constrains political responses to these issues (Magnusson 1983b, 94). Forty years later, the Toronto Magnusson wrote about no longer exists, yet his analysis still rings true.

From 1953 to 1998, Toronto was one of six lower-tier municipalities that made up the two-tier Metro system of local government (Figure 3.1). In 1998, the provincial government amalgamated Metro and its constituent municipalities into a single City of Toronto, which, by 2022, was home to three million people. This city is in turn at the core of the Greater Toronto Area, a sprawling, rapidly growing conurbation of nearly seven million people. Sustained growth over the last 40 years has gone hand in hand with economic and social transformation. The industrial and commercial Toronto of the 1970s and 1980s has metamorphosed into a leading post-industrial global city, a place that is at once highly diverse and deeply socially divided.

Despite Toronto's structural, economic, and social transformation, its local politics today remains institutionally constrained. The post-amalgamation City of Toronto has a larger population than six of Canada's provinces. It is deeply involved in a wide range of social, economic, and environmental policy fields. Yet it is at once too big and too small: too big to reflect and respond to the interests of its many constituent communities, yet too small to govern the city-region as a whole. Since amalgamation, the city has faced two persistent institutional challenges. The first is a *challenge of legal standing and fiscal capacity*, in which the city has been unable to move beyond its weak legal foundation and property-based revenue system. The second is a condition whereby the city's enormous population size and centralized political institutions combine

Figure 3.1: The constituent municipalities of Metro Toronto prior to amalgamation

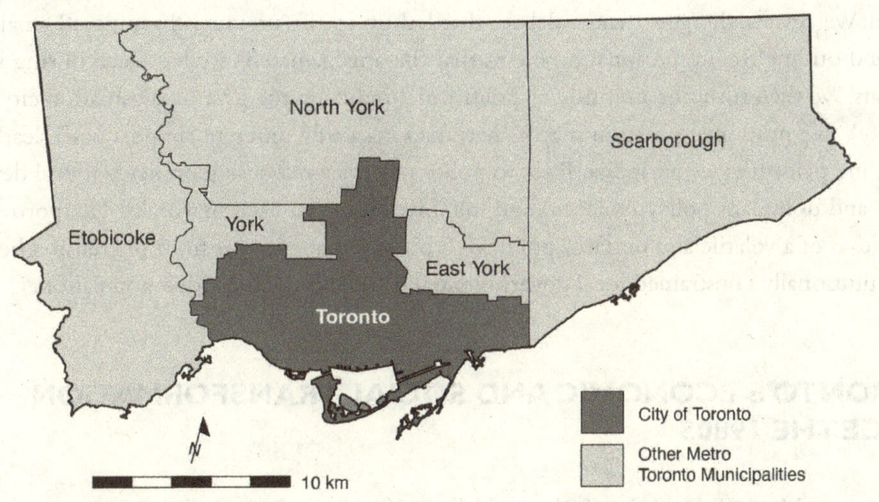

with a weak executive to produce government that is distant from residents, yet fractious and indecisive at the apex. We call this the *problem of institutional intermediation*. These challenges are exacerbated by the rapid pace of change, as the city's population has grown and diversified and physical redevelopment has transformed urban space.

Post-amalgamation Toronto is not only ethno-culturally diverse and socially unequal, it also has a deep social and political divide between residents in the old core and those in the post-war suburbs. This divide predates amalgamation, but amalgamation forced residents with different political cultures and preferences together into the same political system. In this context, the institutional challenges Toronto faces have impeded its ability to function as a meaningful venue for democratic representation and an effective vehicle for policymaking and implementation. Although the city has a substantial base of self-consciously left-progressive residents, its limited powers and high dependence on property-based revenues have constrained its ability to develop and implement policies that address social inequality. Its weak legal status has left it open to unilateral provincial action on issues ranging from financing to executive structure. Its huge size and lack of decentralized governing bodies have limited its responsiveness to local issues and preferences in different parts of the city, as well as to emerging sectoral demands from a diverse and unequal citizenry. Instead, the city has periodically lurched from one governing style and set of priorities to another, as different socio-spatial coalitions have captured power in local elections. Meanwhile, the weak executive structure Toronto possessed until 2022 has inhibited decisive political leadership on big-picture policy concerns.

Our chapter presents a historical and developmental account of Toronto's politics over the last forty years. We begin with an overview of Toronto's remarkable demographic, economic, and spatial transformation since the 1980s. We then discuss the city's pre-amalgamation politics between 1983 and 1998, paying attention to both tiers of local government in the Metro system. We review the governance debate that led up to the provincially imposed amalgamation and outline the institutional problems that the amalgamated city has faced during its first 25 years. We then turn our attention to politics and policy in the post-amalgamation city. Since the city's five post-amalgamation mayors have had markedly different support bases, leadership styles, and priorities, we periodize Toronto politics by mayor, focusing on key political developments and dominant policy concerns and initiatives issues in each mayoralty. The portrait that emerges is of a volatile and unstable politics – a politics that emerges from the tension between an institutionally constrained local government and a diverse and divided population.

TORONTO'S ECONOMIC AND SOCIAL TRANSFORMATION SINCE THE 1980S

A visitor to Metro Toronto in 1981 would have found a bustling industrial city of over two million people, its downtown a mix of older commercial buildings and new office towers, with manufacturing enterprises and warehouses to the east and west. Surrounding this core were neighbourhoods of red brick houses, interspersed with concrete private rental high-rises constructed over the previous 25 years. Some of these neighbourhoods were gentrifying but remained mixed by class and ethnicity. Ethnic quarters, such as Little Italy, Chinatown, and Greektown, adjoined those inhabited by mostly British-descended middle- and working-class residents, knit together by streetcar and subway lines.

Should our visitor have ventured beyond the core City of Toronto, they would travel, most likely by car, into the interwar neighbourhoods of the boroughs of York and East York and the automobile-oriented post-war landscapes of the outer boroughs – Etobicoke, North York, and Scarborough (Relph 2014; White 2016). Metro and lower-tier municipal planners had imposed strict land-use segregation as these municipalities grew rapidly in the post-war period. While vast expanses of industrial and commercial lands were designated, the jobs they contained were largely inaccessible, except by car, to the residents of the single-detached house neighbourhoods, punctuated by clusters of high-rise rental apartments, that surrounded them. Forty years later, the main features of these urban landscapes remain, yet much has changed as powerful forces have reshaped the city's economy, society, and physical form.

Population Growth and Demographic Transformation

Toronto has been one of the fastest growing cities in Canada for decades (see Figure 1.1 in Chapter 1). As of 2022, the total population of the City of Toronto had surpassed three million and the number of dwellings – houses, apartment units, and so on – had grown from 776,380 in

Figure 3.2: Percentage of immigrant and of non-official language mother tongue, 1981–2021

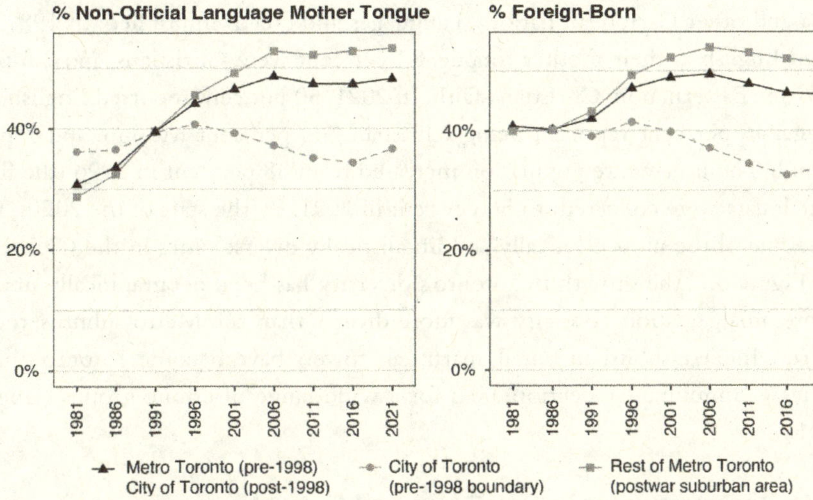

Source: Census of Canada, various years.

1981 to 1,112,930.[1] That the number of residents has increased by 50 per cent in forty years in a territory that, at the outset, had very little remaining "greenfield" land is remarkable – especially when compared to its "rustbelt" peer cities in the American Midwest and Northeast, which have experienced absolute population decline and property abandonment.

Yet even as the territory governed by today's City of Toronto has experienced rapid population growth and ongoing urban development, the metropolitan area beyond the city's limits, commonly referred to by its "905" telephone area code, has grown even faster. In 1981, the surrounding jurisdictions of Halton, Peel, York, and Durham had a combined population of 1,280,315; by 2021, this had more than tripled to 3,917,985. As a result, the relative weight of today's Toronto within the broader metropolitan area has decreased over time.

While in a boom, the City of Toronto has continued to attract new residents from other parts of Ontario and the rest of Canada, virtually all population growth since 1981 has occurred through international immigration. Toronto has consistently experienced *negative* net intraprovincial migration since at least 2006; each year, more non-immigrants leave Toronto for elsewhere in Ontario than arrive. Without immigration, Toronto's population would be declining. This is evidenced by the growing proportion of Toronto residents who were born outside Canada (see Figure 3.2).

Toronto had long been a city of immigrants, home to substantial Italian, Portuguese, Greek, and East and Central European populations. Indeed, 41 per cent of Torontonians were foreign-born in 1981, only slightly less than 46 per cent in 2021. Yet starting in the 1980s, the face of Toronto rapidly changed as British and European immigration to Canada declined in favour of flows from South Asia, first Hong Kong and then mainland China, the Philippines, and Jamaica and other Caribbean nations. Languages and creeds multiplied: In 1981, 68 per cent claimed English as their mother tongue, 82 per cent were Christians, and 3.5 per cent belonged to an "Eastern, non-Christian" faith. In 2021, 50 per cent reported English as their first language, 46 per cent reported being Christian, 9.6 per cent Muslim, and 7 per cent Hindu or Sikh. The non-white population increased from 38 per cent in 1996 (the first year in which such data were collected) to 56 per cent in 2021. By the start of the 2020s, Toronto had become one of the most ethnically and linguistically diverse cities in the Global North. As seen in Figure 3.2, the growth in Toronto's diversity has been geographically uneven. In 1981, the pre-amalgamation core city was more diverse than the Metro suburbs; today, the reverse is true. Indeed, suburban rental apartment towers have become Toronto's principal "arrival cities" – immigrant reception areas for a wide range of ethnic groups (Goel 2023; Hiebert 2015).

Economic Restructuring in the Core and Suburbs

The immigration-driven transformation of Toronto paralleled a wrenching economic transformation. The Toronto of 1981 had a diversified economy. The exodus of capital and headquarters functions from Montreal to Toronto in the late 1970s, spurred by the rise of Quebec separatism, had consolidated Toronto's position as Canada's dominant financial centre (Esman 1987, 400). In the downtown core, a new generation of glass and steel skyscrapers was being built to house offices in financial and other producer services. To the east and west of downtown lay nineteenth-century industrial zones that contained a range of manufacturing activities. Outside the City of Toronto, Metro had established vast industrial parks integrated with highway and rail networks. By 1981, the suburban boroughs contained 53 per cent of all jobs within Metro, with 31 per cent in manufacturing, and many manufacturing workers lived in the post-war suburban zone.

Manufacturing employment declined rapidly starting in the 1980s, a trend accelerated by the signing of the Canada–US Free Trade Agreement (1988) and its successor, the North American Free Trade Agreement (1993). Toronto was hit hard by the global recession of the early 1990s, but out of the ashes of the old economy, a new service-based knowledge economy emerged as Toronto moved into the new millennium. Between 1983 and 2021, total employment in Toronto grew from about 1.1 million to 1.45 million, but manufacturing employment declined by half (a loss of almost 175,000 jobs), while FIRE (finance, insurance, and real estate) sector employment increased by almost 60 per cent, or about 50,000 jobs (City of Toronto 2014, 2021b). As shown in Figure 3.3, these processes had geographically uneven impacts, with manufacturing declining further in the core, which has increasingly specialized in FIRE and other producer services sectors.

Figure 3.3: Employment by sector, core versus suburbs, 1981–2021

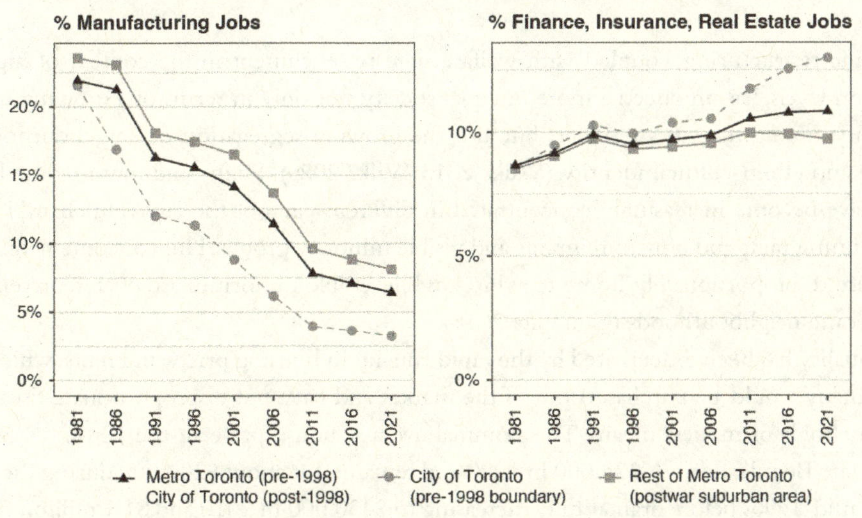

Jobs are located by place of work, including working from home, not by place of residence. The dip in core-area employment in 2021 is due to the COVID-19 pandemic. Source: Census of Canada, various years.

Physical Landscape Change

These social and economic transformations have spurred change in the city's built environment. Almost 250,000 more people now live within the boundaries of the former City of Toronto than did so in 1981, and 400,000 more people now live in the post-war suburbs. A demographic shift to smaller households, together with steadily rising real estate prices since the 1990s, has fuelled demand for smaller residential units. New housing has almost entirely taken the form of high-rise condominium apartments. Most of these have been built on brownfields on the waterfront and other areas, although the replacement of low-density with high-density housing also occurs (Dingman 2018). At the same time, many former manufacturing zones have been converted to car-oriented clusters of "big-box" retail stores (Hernandez and Simmons 2006).

Infill and redevelopment have considerably altered the composition of the housing stock. Belying the tourist's image of Toronto as a city of houses, single-detached dwellings today comprise only 23 per cent of the housing stock; the share made up by high-rise apartments has grown from one-third in 1981 to almost half. The number of high-rise apartment units has more than doubled, from 260,000 to over 540,000, with 57% of them in the suburbs. Especially in core neighbourhoods, the historical residential fabric has also been transformed by more or less subtle forms of gentrification, including teardown-and-rebuilds and extensive renovations of existing single-family housing (Buckley and Brauen 2022; Walks and Maaranen 2008).

Rising Inequality and Segregation

Economic restructuring, coupled with welfare state retrenchment in the context of high immigration levels, has produced a more unequal society, not only in terms of a growing income inequality among the city's residents but also the growing segregation of neighbourhoods by income and ethno-cultural identity (Walks 2013; Walks 2020). High- and low-income households have become increasingly concentrated in different parts of the city (Hulchanski 2010), as have immigrants and non-immigrants and visible minority groups. High-income neighbourhoods are disproportionately home to whites, while visible minorities are over-represented in low-income neighbourhoods (Contenta 2018).

Inequality has been exacerbated by the rapid run-up in housing prices and rents, which have priced many would-be purchasers out of the market and have put disproportionate fiscal stress on households of modest means. The nominal average house price in the Toronto Regional Real Estate Board's area was $75,000 in 1980 and remained at about $200,000 during the recessionary mid-1990s before dramatically increasing to $430,000 in 2010 and $1.1 million in 2024 (Toronto and Region Real Estate Board 2023). As of 2022, the average rent for a one-bedroom apartment had topped $2,000 (CBC News 2022).

The City of Toronto is not the place it was forty years ago. Immigration has made it more populous and diverse. While some manufacturing industries remain, deindustrialization and post-industrialization have created an economy dominated by the service sector and polarized between high- and low-wage jobs. Neighbourhoods are increasingly segregated by ethnicity and class. Old industrial and commercial lands have been radically transformed by residential infill and redevelopment, while owner-led gentrification has both preserved and subtly altered the fabric of many long-established residential areas. As we will see, these changes have layered new political divides and demands over old ones, underpinning the fractious and unstable politics that have characterized Toronto since amalgamation. Before we turn to these politics, however, let us look back at what came before. We look at the forces that produced the institutional arrangements that shape and constrain politics in Toronto today.

THE 1980S AND 1990S: FROM EVOLUTION TO AMALGAMATION

Until 1998, what is now the City of Toronto was governed by a two-tier local government structure. The upper-tier Municipality of Metropolitan Toronto led urban structure planning; financed and operated infrastructure, including major roads, transit, and water and sewer systems; and delivered policing and social services. The six lower-tier municipalities (Toronto, York, East York, North York, Etobicoke, and Scarborough) were responsible for local planning and zoning, building regulation, parks, and property services such as waste collection and fire services. Created by the province in 1953, the Metro model was widely regarded as an innovative response to the challenge of metropolitan governance. In the 1950s and 1960s Metro planned and financed the infrastructure needed to support the development of post-war suburbs within its boundaries (Magnusson 1983b, 109). By the 1980s, however, Metro was built out and the urbanization frontier had moved far beyond its boundaries. While Metro's population grew by about 11 per cent between 1981 and 1996, to 2.4 million, the population of the four neighbouring regional municipalities more than doubled to 1.9 million.

The transition to a mature, slower-growth city set the stage for the fiscal challenges and policy battles that defined politics in Metropolitan Toronto in the 1980s and 1990s. At the upper tier, a Metro government – whose primary mission had once been to finance capital projects such as roads, sewers, and subway lines – now spent 75% of its budget on transit operations, policing, and provincially mandated social services and faced a fiscal squeeze due to slow local tax revenue growth combined with provincial fiscal downloading. In the lower-tier suburban boroughs of Etobicoke, Scarborough, and North York, boosterish politicians doubled down on a low-tax, development-friendly approach to governing. Meanwhile, in the old central city of Toronto, a council divided between "moderate" and "radical" reformers grappled with reconciling disparate objectives, promoting development, preserving neighbourhoods, and supporting social and quality-of-life programs. The challenges of governing a mature metropolis with a two-tier system, combined with growing concerns that Metro was losing businesses and residents to surrounding suburbs, sparked a vigorous debate in the 1990s on how local government in the whole city-region might be restructured – a debate that would be unexpectedly cut short in 1997 by the creation of an amalgamated City of Toronto.

Metro Politics: Leadership Drift and Electoral Reform

In the 1980s, the upper-tier Metro government faced serious problems of political leadership. Most of the big-ticket services that Metro delivered were either tightly regulated by the provincial government (social services, social housing) or governed by arm's length commissions (police, public transit), so Metro Council's degree of political control was limited. In addition, Metro councillors were indirectly elected as part-time delegates from lower-tier municipal councils, so they were strongly incentivized to prioritize lower-tier concerns (Mellon 1993, 42; Feldman 1995, 210). Any big-picture leadership on Metro Council thus had to come from its chair, which councillors chose from among their own ranks. But former Etobicoke mayor Dennis Flynn, who succeeded Paul Godfrey as Metro Chair in 1984, "had neither Godfrey's charisma nor a clear agenda of what he wanted to accomplish" (Feldman 1995, 212).

Responding to the sense of drift on Metro Council, the Liberal provincial government appointed a task force to deal with the issue in 1986. Drawing on recommendations made a decade earlier in the Robarts Report (Robarts 1977), this task force proposed a 34-member Metro Council consisting of 28 directly elected councillors, plus the mayors of the six lower-tier municipalities. The chair would be chosen by the council from among the 28 directly elected councillors, since both the province and Metro itself were opposed to direct election of the chair on the grounds that "enormous power and influence could be exercised by a Chairman elected at-large" (Metro response to task force report, cited in Feldman 1995, 210).

Direct elections for Metro Council took place in 1988. The new council chose Alan Tonks – the former mayor of the lower-tier municipality of York – as chair, a position that he held until Metro was abolished in 1997. Faced with more active and engaged directly elected councillors, "Tonks tend[ed] to spend a considerable amount of time quietly building consensus" (Feldman 1995, 212). This behind-the-scenes "broker" style of leadership, so common among Canadian mayors (Sancton 2021), helped to assure Tonks' political longevity. At the same time, Metro Council as a body became more assertive in the era of direct elections, taking strong positions and sometimes raising tension with other levels of government. The new era began with the Metro government's contentious 1988 decision to build Metro Hall, an expensive new complex of buildings for Metro's council and bureaucracy, but the controversy did not stop there.

A particularly contentious issue was property value assessment. Property taxes in Metropolitan Toronto were based on assessments of the value of individual properties that were often decades old. Assessments were most outdated in the core City of Toronto, which led to the over-valuation of property in post-war suburbs such as North York and Etobicoke, whose residents complained that they paid an unfair proportion of Metro taxes as a result. In 1989, the new directly elected Metro Council proposed to implement market value assessment to address the issue (Horak 1998, 12). The move was vociferously opposed by Toronto, which went so far as to put a non-binding question about separation from Metro on its 1994 election ballot over the issue (Tomalty 1997). Toronto's opposition succeeded in staving off reassessment until 1997, when the Conservative provincial government imposed it province-wide (Bird et al. 2012).

Metro Council also began to stake out more assertive positions on other issues such as regional planning in the 1990s. The 1992 Metro Strategic Plan, a project that Alan Tonks strongly championed, called for Metro to have a "leading role" in shaping development at the city-regional scale, a turn of phrase that sparked strong objections from lower-tier municipalities (Feldman 1995, 217). By 1995, Metro was calling for the extension of its own boundaries to cover the whole Toronto region (Ontario 1996, 164). Writing in 1995, Feldman concluded that "it is only now, nearly 40 years after its formation that the Corporation of the Municipality of Metropolitan Toronto … has come into its own" (Feldman 1995, 206). Ironically, the Metro experiment would abruptly end only two years later.

Suburban Boosterism and Downtown Progressivism: Politics at the Lower Tier

As Magnusson observed, the existence of Metro with its social and human service functions entrenched lower-tier municipalities in their "traditional role" as managers of urban growth (Magnusson 1983b, 127). Metro's various lower-tier municipalities approached this role in different ways. Patterns of local politics in the post-war suburban boroughs of North York, Scarborough, and Etobicoke had been forged during the 1950s and 1960s. Rapid urbanization spurred a boosterist politics in which mayors and councils focused on keeping property taxes low and facilitating large-scale property development. As suburban growth slowed in the 1980s and 1990s, boosterism nonetheless remained dominant, facilitated by a Metro that largely absorbed the responsibility and cost of dealing with emerging social challenges (Horak 1998). Perhaps the best-known representative of suburban boosterism was Mel Lastman, who served as mayor of North York between 1972 and 1997. A tough-talking businessman who had founded Bad Boy, a discount furniture and appliance chain, Lastman governed with the support of tight circle of loyal councillors and private-sector developer allies (Stanwick 2000). Like long-serving mayors in other growing suburbs – such as Hazel McCallion in Mississauga – Lastman traded on an outsized public persona, which belied a style of politics that relied on a politically passive public and left little room for public engagement in decision making.

By contrast, an active and engaged resident base was at the heart of politics in the old core City of Toronto, which was still strongly shaped by the legacy of "reform." Coalescing around opposition to 1960s-style urban redevelopment, reformers had captured power in Toronto in 1972 and, under the leadership of mayor David Crombie (1972–78), had entrenched resident participation in planning and development as a cornerstone of local politics (Magnusson 1983b, 118–22). However, the reform movement was, from the outset, divided between middle-class homeowners concerned with preserving the character of historic neighbourhoods, and "radical" working-class activists concerned with issues such as social housing, rent control, police reform, and subsidized daycare (Magnusson 1983b, 126; Caulfield 1974, 12–13). Furthermore, the self-conscious progressivism of reform politics rested on paradoxical foundations, because the innovative services and programs rolled out by reform councils were largely bankrolled by steady property tax assessment growth in Toronto's Central Business District.

In 1980, Art Eggleton, supported by Toronto's business community, narrowly defeated incumbent "radical" reform mayor John Sewell (1978–80). Although he presided over councils that were strongly divided – primarily on approaches to development – Eggleton was re-elected three times and served until 1991, making him Toronto's longest-serving mayor. In a context where Toronto's growth was slowing and business elites talked anxiously of the need to enhance the city's competitiveness, Eggleton found a ready constituency for pro-development politics. He championed a new 1982 Official Plan that loosened development restrictions implemented in the 1970s and pushed to create the Toronto Economic Development Corporation (TEDCO) in 1986 as a vehicle for redeveloping the derelict port lands (Todd 1996). At the same time, he maintained restrictive development guidelines in established middle-class neighbourhoods and supported social and environmental policy initiatives dear to progressive reformers. Like his Metro counterpart Alan Tonks, Eggleton was above all a broker, adept at finding the middle ground on controversial issues (Ron Kanter, personal interview, February 6, 2023).

After Eggleton stepped down in 1991, the divisions on Toronto Council became more diffi-cult to manage. In the summer of 1989, Toronto's overheated real estate market crashed; this was followed by a deep recession that lasted until 1992. The downturn raised the stakes in the local debate between politicians, who argued that Toronto needed to become more entrepreneurial to secure jobs and investment in the global economy, and progressive reformers, who argued that now more than ever Toronto needed to support struggling and marginalized residents. In the 1991 elections, alderman June Rowlands, a former chair of the Toronto Police Commission with links to the provincial Progressive Conservative Party, defeated New Democratic Party (NDP)-endorsed candidate Jack Layton to become Toronto's first female mayor. Three years later, Rowlands was defeated by Barbara Hall, a moderate progressive who had support across ideological lines, but nonetheless failed to maintain a working coalition on council. A sense of systemic crisis gripped Toronto City Hall, feeding an emerging debate about large-scale restruc-turing of Toronto's local government.

The Regional Restructuring Debate and Toronto Amalgamation

By the early 1990s, Metro was a mature city with big-city costs, and its non-residential prop-erty taxes were substantially higher than those in surrounding municipalities. Business leaders expressed concern that businesses and wealthier residents were relocating to neighbouring mu-nicipalities, in some cases actively lured by local tax incentive programs (Isin and Wolfson 1999, 57). Worries about the hollowing out of the Metro core only deepened in the wake of the early 1990s recession, from which Metro was slower to recover than the outer suburbs. Meanwhile, burgeoning suburban sprawl and traffic congestion in the rapidly growing 905 belt highlighted the need for region-wide coordination of planning and development (Taylor 2019, 127–8). Political conflict between the upper and lower tiers in Metro further contributed to the sense that the whole Greater Toronto Area (GTA) was ripe for major structural reform.

In 1995, Bob Rae's provincial NDP government appointed a GTA Task Force, headed by United Way Chair Anne Golden, to consider options for structural reform. The task force, strongly influenced by the rhetoric of global competitiveness that dominated discussions of metropolitan reform at the time, aimed to position the GTA as "the economic engine that drives not only Ontario but the whole of Canada" (Ontario 1996, 33). After consid-ering several options, the task force recommended that Metro and the four surrounding upper-tier municipalities (Halton, Durham, Peel, and York regions) should be replaced by "a single Greater Toronto regional government with a more limited number of functions" (On-tario 1996, 166). However, by the time the task force issued its report in January 1996, the Rae government had been replaced with a Progressive Conservative provincial government under Mike Harris.

The Harris Conservatives had won in a landslide in the fall of 1995, running on a radical platform that promised a 30 per cent cut to the provincial income tax rate *and* balanced provincial budgets (Ibbitson 1997, ch. 3). For most of 1996, the issue of GTA governance sat on the back burner as the Conservatives focused on reversing the Rae government's increases to social assistance funding and radically cutting numerous provincial programs to fulfil their election promise. However, in December 1996, they introduced legislation that would amalgamate the municipalities within Metro into one "Megacity" of Toronto, while leaving the two-tier systems in the GTA's outer suburbs untouched. This move took most observers by surprise. The Conservative government, which had been convinced by bureaucrats in the Ministry of Municipal Affairs that consolidating small and rural local governments in Ontario would fit with its cost-cutting agenda, had been pursuing amalgamations across much of the province since late 1995 (Sancton 2000). Amalgamation of Metro was not an option that had been seriously considered in the GTA restructuring debate, however, and until late 1996, the Conservatives had seemed more inclined to abolish Metro than the lower-tier municipalities. The government justified amalgamation by claiming that it would save money by reducing waste and duplication in Metro, and that it would help Toronto's competitiveness on the global stage. However, it produced little systematic analysis to support these claims (Horak 1998, 16).

Initial local reaction to the announcement was mixed, with politicians in Metro's lower-tier municipalities calling it a threat to local democracy, while Metro and the business community expressed cautious support (Isin and Wolfson 1999, 63). The city's major daily newspaper, the *Toronto Star*, had advocated Metro amalgamation for some time, and it greeted the announcement enthusiastically (Horak 1998). However, such local support largely evaporated in January 1997, when the province announced its Local Services Realignment (LSR) reforms. These involved a major reorganization of provincial–municipal funding responsibilities. The province would upload education funding, while partly or fully downloading fiscal responsibility for social assistance, social housing, old age homes, and public transit to municipalities. The government claimed that at a province-wide scale, the swap was "revenue neutral," but most of the downloaded functions such as transit and social services were in heavy demand in Metro. Toronto's transit system was already suffering from years of underinvestment in infrastructure maintenance, while the city's social service caseloads had risen significantly during the 1990s recession. The LSR reforms thus threatened to load significant new costs onto Toronto's local government (Isin and Wolfson 1999, 65).

At this point, many politicians and civic leaders in Toronto came to see amalgamation and LSR as "a two-pronged attack on the city by a provincial government rooted in a suburban and small-town support base" (Horak 2008, 14), and local opposition spread. Reform activists from the old core City of Toronto founded a civic group called Citizens for Local Democracy, which organized marches, media events, and demonstrations. The lower-tier municipalities held non-binding plebiscites in March 1997 that delivered a 76 per cent vote against amalgamation (Horak 1998). Such opposition notwithstanding, the provincial government moved ahead with the amalgamation plan, replacing the two-tier Metro system with the single-tier City of Toronto.

THE NEW CITY OF TORONTO: CONFRONTING INSTITUTIONAL CHALLENGES

The amalgamated City of Toronto came into being on 1 January 1998. It was, in all senses, enormous. At its creation in 1998, Toronto was home to 2.4 million people and had an operating budget of $5.6 billion. By 2024, the population had surpassed 3 million, the municipal operating budget was over $17.1 billion, and the city had 42,299 employees (City of Toronto 2024a,b). As of 2013, it was the sixth largest unit of government (at any level) in Canada in terms of spending (Horak 2013). As we will see, Toronto's huge size has sometimes enhanced its visibility and clout on the intergovernmental stage (Horak 2008, 17). However, the amalgamated city has faced two enduring institutional challenges. The first is the challenge of legal standing and fiscal capacity. Toronto has chafed under the restrictions of Ontario's general-purpose municipal legislation and confronted a significant imbalance between revenues and expenditures. In addition, the city has faced the problem of institutional intermediation – a complex web of institutionally embedded representation and leadership problems that have impeded the city's ability to respond to its diverse population and set policy priorities. In this section, we review the evolving political response to these challenges since 1998, setting the stage for our later discussion of politics and policy in post-amalgamation Toronto.

The Challenge of Legal Standing and Fiscal Capacity

Amalgamation and LSR demonstrated the provincial government's virtually unlimited legal control over Toronto's local government. Not only did the province summarily abolish the Metro system in the face of strong opposition, but by shifting funding responsibility for transit and social housing to municipalities, it loaded anywhere between $163 million and $275 million in new annual costs onto the city's property tax base (Isin and Wolfson 1999, 67; Horak 2008, 14). Local outrage over these changes continued to fester, especially among progressive civic leaders and activists from the urban core. In the early years after amalgamation, as Mayor Lastman repeatedly begged the province for bailouts to plug the annual fiscal deficit and bureaucrats struggled to blend seven municipal administrations into one, a coalition of civic leaders – including famed urbanist Jane Jacobs and philanthropist Alan Broadbent – published the Greater Toronto Charter, which called for the province to establish a Greater Toronto authority with enhanced legal and fiscal powers (Keil and Young 2003; Jacobs and Rowe 2000). But there was little appetite on the ground for a regional solution to Toronto's governance challenges. GTA-wide tax pooling for social services, which the province had introduced in 1998 to offset the costs of LSR for Toronto, was widely resented by 905-area municipalities. The Greater Toronto Services Board, a GTA-wide coordinating body established by the provincial government in 1999, only lasted for two years (Horak 2008, 21–23).

While the prospect of regionalism faded, a push by the City of Toronto for a new relationship with other levels of government proved more fruitful. Starting in 2000, council advocated for stand-alone legislation that would exempt Toronto from the province-wide Municipal Act and give it additional powers and sources of revenue. Meanwhile, a coalition of big-city mayors from across Canada began calling for long-term urban funding commitments from the federal government (Horak 2008, 24). This agenda was aided by a temporary political conjuncture. After years of antagonism between Toronto and Queen's Park, long-time downtown progressive councillor David Miller won office, only days after the provincial Liberal Party won a sweeping majority government with strong electoral support in Toronto on a centre-left platform that emphasized cooperative relations with municipalities. Before the end of the 2003, Paul Martin succeeded Jean Chrétien as Liberal prime minister, and, after the June 2004 federal election, John Godfrey was appointed to cabinet as Minister of State for Infrastructure and Communities with a mandate to implement a "New Deal for Cities and Communities" (Bradford 2007).

In 2005, Martin's minority Liberal government struck a budget deal with the NDP, headed by former progressive Toronto councillor Jack Layton. The deal delivered two cents of the federal gas tax to Canadian municipalities for infrastructure. These and other intergovernmental transfers for infrastructure began to flow to Toronto from both levels of government (see Figure 3.4). Between 2003 and 2009, the inflation-adjusted value of intergovernmental transfers to Toronto went up by 70 per cent, significantly easing Toronto's post-amalgamation fiscal problems. In addition, the provincial Liberal government incrementally "uploaded" selected cost-shared expenditures after 2008, thereby reducing municipal spending on land ambulance, court services, and social services (Reid 2022). Provincial–municipal alignment was challenged by the 2008 recession, however, which forced the province to retrench promised spending, and federal engagement was curtailed with the 2006 election of a Conservative government with less interest in municipal entanglements.

In 2004, the McGuinty government established a joint task force with Toronto to develop a stand-alone City of Toronto Act. The Act came into force following the 2006 municipal election. It gave Toronto some new powers, such as the power to establish corporations, extend commercial opening hours, to restrict rental housing conversions, and to establish design guidelines (Ontario 2006). More significantly, the Act gave the city access to several new sources of revenue, including excise taxes on tobacco, alcohol, and entertainment; a property transfer tax; road tolls; taxes on vehicle registration and parking; and an outdoor advertising tax. These new revenue powers had significant potential, but political considerations have limited their use. During his second term, Miller did introduce the Municipal Land Transfer Tax and the Personal Vehicle Registration Tax, both of which piggybacked onto existing provincial levies (see Figure 3.4). The vehicle registration tax raised modest revenue during its three years of existence. Much hated by drivers, it was

Figure 3.4: Inflation-adjusted operating revenue per occupied dwelling unit by source, 1998–2023 (1998 dollars)

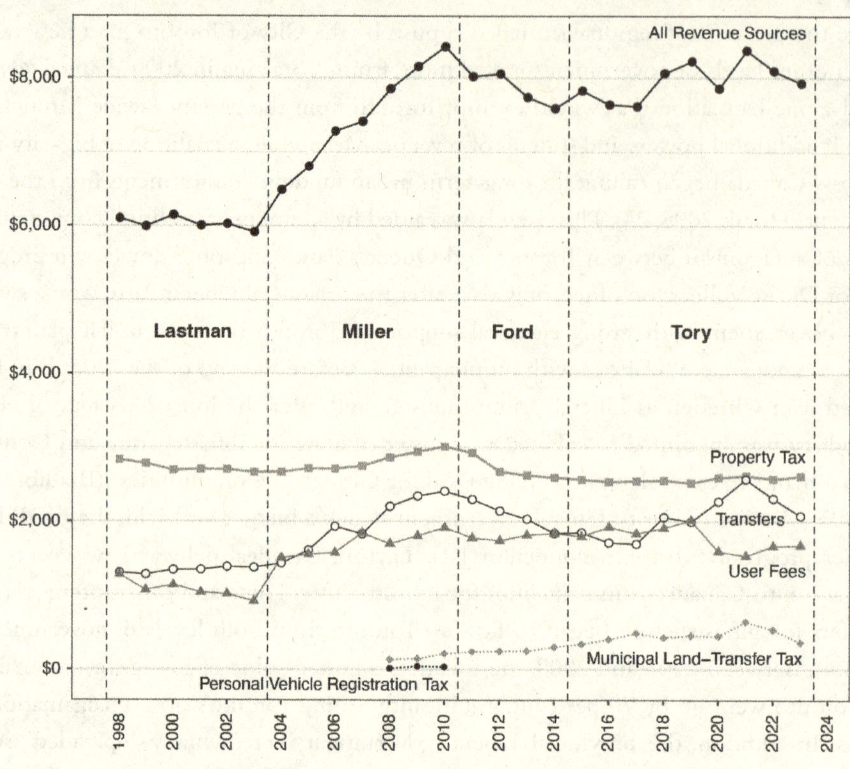

Notes: Toronto's inflation-adjusted revenue per dwelling increased dramatically between 2003 and 2010, while David Miller was mayor, but property tax per dwelling remained constant. The difference was made up by increased reliance on user fees, grants from other levels of government, and the Municipal Land Transfer Tax. Note that the property tax category in this calculation includes non-residential property tax revenues and payments in lieu of taxes by other governments; residential property tax revenue could not be isolated. Sources: Dwelling counts interpolated from quinquennial Census counts, 1996–2021; revenue by source from city budgets (1998–2003) and annual financial reports (2008, 2012, 2015, 2019, 2021, and 2023); consumer price index from Statistics Canada, Table 18-10-0004-01 Consumer Price Index, monthly, not seasonally adjusted (Toronto region) (January values rebased to 1998), https://doi.org/10.25318/1810000401-eng.

repealed in 2010 when Rob Ford was elected as mayor, riding a wave of disaffection in Toronto's post-war suburbs. The land transfer tax has become a major source of revenue for the city. In 2024, it generated $880 million, accounting for over 5 per cent of all operating revenues (City of Toronto 2024b). By contrast, the parking lot, alcohol, and entertainment taxes remain unimplemented, and a 2016 proposal by mayor John Tory to impose road tolling on the Gardiner Expressway and Don Valley Parkway, city-owned highways leading into the city's core, was vetoed by a provincial Liberal government facing an upcoming election (Rieti 2017).

The city's fiscal situation has thus improved significantly since the late 1990s (see Figure 3.4). However, political opposition and the desire to remain competitive with neighbouring municipalities have constrained the growth and diversification of local revenues. Crucially, property taxes remain by far the strongest single revenue source, and the city's most significant new revenue source – the land transfer tax – is likewise property-based. Toronto's fiscal sustainability thus depends heavily on property values and real estate activity, structurally placing the city in a situation where it is in the interests of political leaders and bureaucrats alike to respond, first and foremost, to the interests of property owners and to pursue "developmental" policies as an indirect revenue generation strategy. As we will see, when combined with the inaccessibility of local politics to citizens, this fiscal dependence on property-based revenues has greatly limited the willingness of politicians to allocate resources to social concerns.

Because the city depends heavily on a limited number of property-related revenue sources, it is highly vulnerable to fiscal shocks. This became all too apparent in 2020, when a steep drop in transit fares and other user fee revenues during the COVID-19 pandemic put acute stress on Toronto's finances. User fee revenues have been slow to recover and, together with rolled-forward COVID-era expenses, these fiscal challenges contributed to a 2023 budget shortfall of $1.08 billion, leaving the city, once again, reliant on other levels of government for emergency funding (City of Toronto 2023). Furthermore, a substantial capital spending backlog, principally for transit and social housing, has persisted since amalgamation. Earlier federal and provincial retrenchment created a situation where the municipality shouldered the lion's share of the fiscal burden of maintaining the state of good repair of existing facilities, and by 2014, the city faced a $5.2 billion backlog of unfunded capital spending for transit and social housing alone (Pennachetti 2014, 62–63).

As for the city's legal standing, the provisions in the City of Toronto Act in no way changed the fundamental legal subordination of Toronto to the provincial government. This became very clear in 2018, when Doug Ford's newly elected provincial Conservative government suddenly cut the number of council seats in Toronto in half, a mere two months before the municipal election. The move sparked a renewed outcry, with activists and some scholars calling for Toronto and other municipalities to be better protected against unilateral provincial action (Charter City Toronto 2019; Good 2019). A court challenge to the province's actions was quashed by the Supreme Court in 2021, however. For all its enormous size, economic clout, and social diversity, Toronto legally remains a standard Canadian municipality.

The Problem of Institutional Intermediation in Toronto

As we saw earlier, Toronto is an increasingly socio-culturally diverse and unequal city. In addition, amalgamation internalized a deep divide between the core and the post-war suburbs. This divide is multifaceted, with differences in social structure and political culture layered on top of each other. Since the 1990s, Toronto's post-war suburbs have become the primary landing spot for new immigrants. They are now less white, more culturally diverse, and less wealthy than the largely gentrified core of the city (see pages 64–5). At the same time, pre-amalgamation differences between core and inner suburbs, in terms of levels of resident engagement in local politics, have persisted. Voter turnout in the inner suburbs has consistently been several percentage points lower than in the city core (Siemiatycki and Marshall 2014), and neighbourhood resident associations are heavily clustered in the core (Flynn 2019). As we will discuss, the core and suburbs also have different voting preferences in mayoral elections, making the core–suburb divide a highly salient political cleavage in Toronto.

The difficulties of governing this diverse, unequal, and geographically divided city have been exacerbated by a persistent crisis of institutional intermediation (Silver et al. 2020), which stems from the combination of the city's huge size, its political centralization, and its weak executive. Although the city has a non-partisan and ward-based system of representation, which is generally thought to encourage responsiveness to neighbourhood-level communities (Koop and Kraemer 2016), the sheer size of the city means that wards have become very large, and councillors are distant from the needs and preferences of local communities, to say nothing of individual residents.

Post-amalgamation Toronto initially had a 56-member council with 28 two-member wards. Prior to the second elections for the new city in 2000, the province reduced the number of councillors from 56 to 44 and introduced single-member wards with an average of about 60,000 residents each. As we saw earlier, the geography of population change in Toronto has been uneven, with most growth occurring in compact nodes of high-rise development downtown and along major transportation arteries. As a result, by 2014, ward populations varied widely, prompting the city to undertake a comprehensive ward boundary review. The review recommended an increase to 47 wards for the 2018 municipal election. But in July 2018, just before the start of the municipal campaign period, premier Doug Ford suddenly cut the number of wards in Toronto to 25, aligning their boundaries with those of federal and provincial ridings (Benzie 2018). Ford justified the move on the grounds that it would speed up the work of council and save money, and followed through despite widespread local protest. As a result, the average ward now has 112,000 residents and includes multiple neighbourhoods and social groups.

The sense that councillors do not represent Toronto's diversity is underscored by a serious gap in descriptive representation on council. Since amalgamation, the proportion of visible minority councillors has averaged a mere 10 per cent in a city where about half of the population identifies as visible minority (Siemiatycki 2011). Visible minority representation on council went up to about 20 per cent in the 2022 election, in part due to the efforts of Progress Toronto, an advocacy group founded in 2018 that helped to recruit and support the campaigns of young progressive council candidates from visible minority backgrounds (Ahmed 2022). This increase in visible minority representation was widely seen as a breakthrough (Kopun 2022), underlining just how deeply underrepresented visible minority residents have been on council.

In Canada and beyond, large urban municipalities often give voice to local communities by devolving some powers to sub-municipal elected bodies, such as Montreal's 19 borough councils (see the Montreal chapter in this volume). But in Toronto, political power is centralized at the citywide level. Initially, post-amalgamation Toronto had six "community councils" that were based on the boundaries of the former lower-tier municipalities and made up of local area councillors. However, to minimize the prospects of conflict with city council and promote political integration after a forced merger, community councils received only consultative powers. Their main job was (and is) to hear resident input on local issues and make non-binding recommendations to city council (Spicer 2016, 135, 139). In 2003, the six community councils were consolidated into four to even out their size, and in 2007, under the new City of Toronto Act, they received delegated legislative authority for a few local zoning and property-related matters. They still exist today, but their authority is limited, they have virtually no dedicated resources, and each covers a huge geographical area. As such, they primarily function as gatekeepers that filter which issues make it to city council, rather than as meaningful decentralized decision-making bodies. In so doing, they provide opportunities for ratepayer groups and councillors to oppose densification and the placement of homeless shelters, group homes, and other social facilities in neighbourhoods.

Even though political power in Toronto is highly centralized, until recently the city has had a weak executive. Initially, post-amalgamation Toronto had a classic Canadian "weak" mayor with no substantial executive powers at all. A large, non-partisan council with a weak executive led to slow and messy decision making, featuring heavy doses of logrolling and issue-by-issue coalition building (Horak 2013). Over time, local and provincial initiatives have strengthened executive authority. In 2005, the city conducted an internal governance review that recommended greatly strengthening the mayor's powers. After a backlash from community groups, the city watered the proposal down, and following the 2006 municipal elections, the mayor got the power to appoint a deputy mayor, committee heads, and an executive committee (Horak 2008, 31).

A more dramatic change occurred in 2022, when – at the encouragement of mayor John Tory – the Ford government rolled out "strong mayor" powers for Toronto and Ottawa. After the October 2022 local elections, the mayors of these two cities received wide-ranging powers, including the power to appoint the chief administrative officer, senior managers, and the heads of boards and commissions; the power to propose the budget and veto council amendments to it, subject to a two-thirds council override; and – perhaps most controversially – the power to veto council decisions that conflict with "provincial priorities," and to propose by-laws that advance these priorities, which could be passed with one-third-plus-one support on council (Taylor et al. 2023). The provincial government explicitly defined "provincial priorities" as building more new housing and the infrastructure necessary to support it, positioning the mayor as a cudgel that beats down not-in-my-backyard (NIMBY) opposition to housing development on behalf of the province. However, many of the new powers are not yoked to the provincial housing agenda, and with the resignation in February 2023 of mayor John Tory – who had become a reliable ally of the provincial government – it is an open question how Toronto's new strong mayor system of executive leadership may be used in the future.

In short, after 25 years the City of Toronto has not shaken the institutional challenges that it faced at birth. It is legally weak and subject to provincially dictated change. It depends heavily on revenues from a narrow and volatile set of property-related taxes and user fees. And its politics is highly centralized and distant from the kaleidoscope of neighbourhoods and ethno-racial groups that make up the city. Together, these crises have produced a mayoral and council politics in Toronto that is geographically divided and volatile, yet fundamentally limited in terms of what gets on the policy agenda. Often, as we will see, it is not council but Toronto's vast and professionalized public service that has taken the lead in engaging the public on specific policy issues and responding to emerging concerns.

POLITICS AND POLICY IN POST-AMALGAMATION TORONTO

Conflicts between the divergent political agendas found in different parts of Metro Toronto persisted after the consolidation of Metro, the City of Toronto, and the suburban boroughs into the new single-tier municipality, shaping the new city's politics and policy. While Metro permitted the six lower-tier municipalities to evolve as separate political worlds, making different policy choices within their jurisdiction, amalgamation brought these into direct confrontation. As the new administration set about harmonizing tax rates, service standards, contracting practices, and work rules for unionized workers, so too did opportunities for the different municipalities' long-established governance norms and policy preferences to clash. These divergences persist a quarter century later, unresolved by periodic institutional fixes.

Our discussion of politics and policy since amalgamation begins chronologically, periodized by mayor. We recognize that the ability of Toronto's "weak" mayors to dictate the city's policy direction is limited, at least until the province's imposition of "strong mayor" powers in late 2022. Nevertheless, the sharp contrast in style and substance between Toronto's mayors reflects the divided politics of the city, and the temperament and agenda of each mayor have shaped Toronto's governance and politics. Our focus is on four policy areas: planning and development practices, housing policies and initiatives, transit and transportation, and the local–provincial relationship. We then review the persistent geographic cleavages that electorally divide Torontonians, and we conclude with a discussion of public engagement and policy innovation. In a context where electoral politics takes place at a large scale and politicians are structurally incentivized to privilege property development and property services, administrators play a leading role in engaging with (and responding to) grassroots activism and emerging interests.

The Booster: Mel Lastman, 1998–2003

Despite the seismic shift in local government structure, incumbency effects were strong in the first post-amalgamation elections, and the new Toronto council contained many familiar faces, including 18 of 28 previous Metro councillors, and numerous mayors and councillors from the former lower-tier municipalities. The leading candidates for the new mayoralty were the outgoing mayors of the two largest municipalities: Toronto's Barbara Hall and North York's Mel Lastman. While Hall was a progressive inclined towards incrementalism and consensus building, Lastman cultivated an image as a fast-talking booster, his gaffe-prone tendencies only amplifying his perceived authenticity. As one journalist wrote, the Hall–Lastman contest was a study in contrasts: they "are entirely different types of politicians and the divisions between them on style and substance are stark" (Campbell 1997). Lastman won 52–46 per cent by racking up high margins in North York. The surprise was that Hall did so well, and Lastman so poorly, given that only one-quarter of the megacity's electors lived in the former City of Toronto.

Lastman easily won re-election in 2000 with over 80 per cent support. His only rival was the late environmental activist Tooker Gomberg, a serial candidate in Edmonton (once successfully) and Montreal. While dismissed as a gadfly, Gomberg was endorsed by well-known environmentalist David Suzuki and urbanist Jane Jacobs (Rankin 2000), and promoted causes that later became mainstream: expanding the city's authority over local issues and expanding cycling infrastructure. Lastman showed little interest in policy matters in his second term, focusing instead on bringing mega-events to the city – an ill-fated bid for the 2008 summer Olympic Games and, more successfully, the 2002 Catholic World Youth Day. Deputy mayor Case Ootes managed much of council's day-to-day agenda. Exhausted and citing ill health, Lastman announced he would not run in the 2003 election.

Winning elections was one thing for Lastman; governing the new city was another. In his first three-year term, the city lurched from one crisis to the next as administrators drawn from across the former municipalities slowly developed working relationships and constructed institutions. Political direction to administrators was uneven at best. The large council was unwieldy, and the new committee structure failed to promote efficient deliberation. Siegel (2015, 86–115) reports that the megacity's first chief manager, Mike Garrett, never developed a productive relationship with the mayor and council, and that many managers who had come over from the dissolved municipalities competed rather than collaborated with one another. At Lastman's instigation, council fired Garrett in 2001. Soon after, a far-reaching scandal came to light involving improper computer system procurement practices. A judicial inquiry found evidence of unethical and possibly illegal behaviour by members of council and several senior administrators, ultimately recommending overhauling codes of conduct and conflict-of-interest rules, creating lobbyist and gifts registries, and an integrity commissioner (Bellamy 2005; Fernando 2007).

The administrative side of amalgamation would take the better part of the next decade to digest. Policy issues that predated amalgamation remained stalled as a result. As housing values and prices rebounded after their collapse in the early 1990s, and the federal and provincial governments retrenched housing and social programs (Suttor 2016, ch. 6), visible homelessness and other public signs of distress increased. The appearance of "squeegee kids" on the city's roads sparked something of a moral panic among the city's elites (Parnaby 2003). The four-stop Sheppard line "stubway" in North York – a product of Lastman's lobbying of the provincial government in the early 1990s and the first subway expansion since 1980 – opened in 2002

(Adel and Bow 2017). However, the transit system languished as the province eliminated its operating subsidy, fares increased, and ridership stagnated (Doucet and Doucet 2022, 10–11). Still recovering from a fatal subway crash in 1995, the Toronto Transit Commission (TTC) focused its shrinking capital budget on "state of good repair" rather than service improvements (Horak 2008). Moreover, the ability to absorb the levelling-up effects of harmonizing service levels and unionized worker compensation was hamstrung by Lastman's promise, supported by council, to freeze property tax rates during his first term of office. Two major municipal worker strikes occurred in 2000 and 2002 (Fanelli 2016, 31–33), and Lastman was forced to beg the province for emergency loans and grants to plug operating budget gaps (Boudreau et al. 2009, 77). As inflation increased in Lastman's second term, council was forced to increase residential property taxes to compensate, but real operating spending per household remained about the same (see Figure 3.4).

One important policy change that occurred during Lastman's mayoralty was the adoption of a new, citywide official plan. Lastman himself had little to do with the plan, which emerged out of a compromise between city planners and council. Under this plan, three-quarters of the city's land area was designated as "stable neighbourhoods" and protected greenspace, where little development was expected to occur. Growth was to flow to transit-linked centres and corridors, including the downtown core, and industrial and waterfront brownfields (Boudreau et al. 2009, ch. 6). This represented a political trade-off much like the one that had existed in the pre-amalgamation "reform" City of Toronto: The interests of those who owned single-family homes were protected, while the welcome mat was rolled out for private developers – especially developers of high-density residential towers – in other parts of the city.

The Progressive Technocrat: David Miller, 2003–2010

The 2003 election featured five major candidates. The early front-runner was Barbara Hall. Also in the race were councillor David Miller, a former NDP candidate and labour lawyer who had emerged as a clean-government champion during the computer leasing scandal investigation, and John Tory, a well-known corporate CEO, lawyer, and Conservative Party insider. Rounding out the pack were York borough councillor and populist federal MP John Nunziata (who had been kicked out of the Liberal caucus) and Tom Jakobek, a former high-profile councillor who had been implicated in the computer leasing scandal. Hall's centrist, policy-light campaign was outflanked on the left and right by Miller and Tory, respectively (Lu 2003). Although Tory attracted support with calls to hire more police and cut taxes, Miller prevailed with 43 per cent of the vote.

Miller's campaign was animated by a simple metaphor: the broom. While many Torontonians, especially in the suburbs, likely cared little about his promises to stop the construction of a bridge to the Island Airport and redevelop the waterfront, they could get behind the message of a fresh start, clean air and streets, and clean government after the disarray of Lastman's second term. In 2006, he easily trounced Jane Pitfield, a veteran business-friendly councillor from the prosperous East York neighbourhood of Leaside. By 2009, however, Miller's shine had worn off. Real operating spending had significantly increased (even if the residential property tax burden remained flat – see Figure 3.4), enabling his conservative opponents to characterize him as a "tax-and-spend" New Democrat. His attempts to restrain labour costs provoked a bruising, thirty-nine-day strike. By the end of his second term, his public-sector union base had turned its back on him, and he decided not to run for a third term.

While Lastman had been largely devoid of policy ideas, Miller championed a series of ambitious fiscal reforms and policy initiatives. On the fiscal front, Miller had significant success. As discussed earlier, the new City of Toronto Act passed by the province in 2006 afforded the city new legal and fiscal powers, and sympathetic Liberal federal and provincial governments enhanced fiscal transfers and uploads. Miller and his budget chiefs, first David Soknacki, then Shelley Carroll, used the new fiscal powers to increase the city's overall revenue while diversifying revenue sources. The land transfer and vehicle registration taxes accomplished both objectives while shielding residential homeowners, the most reliable voters, from property tax increases. Shifting the cost of solid waste management to user fees also reduced dependence on the property tax while addressing the long-term crisis of what to do with the city's garbage. Incentivizing recycling by pricing garbage reduced the city's demand for landfill capacity. Figure 3.4 shows that when adjusting for inflation and housing growth, the real per-dwelling property tax burden remained constant over time, even as transfers, user fees, and new revenue sources grew.

Miller also had an ambitious substantive policy agenda, which aimed to tackle head-on some of the most important policy challenges the city faced: inadequate public transit capacity, neighbourhood inequality and deprivation, and the need to invest in aging high-rise rental housing. While some of his efforts were successful, others were soon starved of resources or were abandoned after he left office. We will look at three major initiatives in turn.

Transit expansion. Beyond Mel Lastman's Sheppard "stubway," Toronto's higher-order transit system had not been expanded in decades. In 2007, Miller announced Transit City, a 120-km network of seven new surface light rail lines linking suburban locations to one another and to the core (Horak 2013). However, the city had nowhere near the necessary resources to build these lines, so their construction depended on intergovernmental transfers. The province incorporated several lines into its transit planning blueprint and negotiated cost-sharing arrangements with the city, but the network was scaled back after the province and federal government reduced their spending commitments in response to the post-2008 recession (Horak 2013). Planning and environmental assessments for several of the lines had begun by the time Miller left office in 2010, but Miller's successor Rob Ford campaigned against Transit City and during his mayoralty, the city largely abandoned the plan.

Neighbourhood investment. In response to the report of the Strong Neighbourhoods Task Force (Harding et al. 2005), which had been convened after a rash of gun violence rocked some of the city's neighbourhoods, the city developed a Strong Neighbourhoods Strategy to funnel additional community investments into thirteen "Priority Neighbourhoods.'" Mostly located in the suburbs, these neighbourhoods score high on indices of deprivation and are disproportionately home to racialized immigrant communities. The program was meant to provide for a major infusion of public money into physical and social infrastructure, and the city called for a five-year investment agreement among all three levels of government to fund it. However, the federal government did not participate, and the province provided only limited resources, mostly for policing initiatives. The program was scaled back in the implementation, with most funding going towards physical infrastructure and policing, rather than towards the kinds of employment training and social initiatives that the Strong Neighbourhoods Task Force had called for (Horak and Moore 2015). The strategy has continued in various iterations, but it has primarily become a spatial "lens" for allocating existing city resources, with virtually no dedicated funding of its own (City of Toronto 2017).

Rental housing renewal. Relatedly, the goal of the mayor's signature Tower Renewal initiative was to rehabilitate the almost 1,200 residential towers built in Metro Toronto between the 1940s and the 1980s. These are home to 20 per cent of Toronto's population, and two-thirds are in the private rental or ownership market (Young 2017, 221). The impetus for Tower Renewal was climate change. Recladding the poorly insulated buildings would reduce one of the city's largest sources of greenhouse gas emissions. It was soon attached to other policy agendas, including the economic revitalization and social development of what came to be called "tower neighbourhoods." Tower Renewal has outlasted Miller and continues to be championed by the city, although funds for implementation are scarce and progress remains slow. Like Toronto's efforts to redevelop the Regent Park and Lawrence Heights social housing neighbourhoods (Horak and Moore 2015), it rests on the ability to leverage private investment in a booming real estate market. Enabling new private development in the interstitial spaces between the often widely spaced towers would increase density and create a mixed-tenure, mixed-use urban fabric while generating the funds to pay for the revitalization of the existing towers. This approach has been criticized for characterizing immigrant communities' informal use of interstitial spaces as problematic (Valzania 2022), and for creating new profit-making opportunities for private landlords at the expense of tenants (August 2020).

The Populist: Rob Ford, 2010–2014

Despite a vocal draft movement by civic and business leaders, John Tory, who in 2010 was head of CivicAction, a high-profile good-government advocacy group, decided not to enter the 2010 race. This opened the door to Etobicoke councillor Rob Ford, a back-to-basics conservative with few friends on council, who was known for his erratic behaviour, derogatory comments, and relentless attacks on Miller's fiscal policies (Mahoney 2010). Although the former mayor was not on the ballot, one columnist portrayed the election as "a referendum on David Miller … a vote for Ford is a vote against Miller" and his perceived policy failures (Siddiqui

2010). Decrying downtown elites while championing the silent majority he dubbed "Ford Nation," he advanced a simple populist message: "respect for the taxpayer" meant "stopping the gravy train": cutting property taxes, reducing spending, and privatizing services (Siddiqui 2010; Silver et al. 2020; McGregor et al. 2021, ch. 5).

Ford's principal opponent was George Smitherman, an openly gay provincial Liberal cabinet minister with deep roots in downtown neighbourhood politics who had been Barbara Hall's chief of staff when she was mayor of the former City of Toronto. The third candidate was Miller's deputy mayor, Joe Pantalone, who campaigned on continuity with Miller's policies from his political base in downtown Little Italy. As Ford climbed in the polls, Smitherman abandoned his centrist campaign and tacked to the right, attracting campaign support from traditional business elites uncomfortable with Ford (Rider 2010). Ford won handily, with over 80 per cent of his support coming from the Metro suburbs, while Smitherman and Pantalone performed well only in the former City of Toronto (McGregor et al. 2021, 91–100). Once elected, Ford moved quickly to implement his agenda.

Ending the "war on the car." As he had promised during the campaign, Ford convinced council to redirect funds away from Miller's surface light rail plan and towards underground rail. He deployed two justifications. First, dedicated surface transit lines take up road space used by automobiles; in so doing, they represent a "war on the car" foisted on suburbanites by downtown elites (Markson 2010). Second, Ford argued that, out of fairness, suburban areas "deserved" the subways that downtowners enjoy (Peat 2013). In fact, Transit City's selection of light rail over much more expensive tunnelled subways had been driven by an analysis of forecast ridership in relation to cost; subways were not found to be justified in relatively low-density suburban areas. Nonetheless, Ford's politics trumped transit planning considerations, and much of Transit City was shelved in favour of pursuing the Scarborough subway project, which proposed extending the TTC's major east–west subway line while decommissioning the aging Scarborough Rapid Transit elevated monorail.

Budget cuts and Core Services Review. Miller had sought to address the city's structural fiscal problems by augmenting and diversifying revenues. Ford took the opposite approach, pursuing across-the-board budget cuts, privatization, and elimination of services. He did so by not raising property taxes, even by the rate of inflation (in effect lowering the burden on taxpayers), contracting out garbage pickup in a quarter of the city, and embarking on a "Core Services Review" – a $3 million external audit of the city's expenditures by accounting firm KPMG. The review found very little potential for cuts (KPMG LLP 2011). Members of the public and user groups vigorously protested service cuts to libraries and bus routes, including at a 22-hour executive committee meeting in 2011 (Dale and Rider 2011).

Within eighteen months of taking office, Ford's erratic and abrasive behaviour had alienated most councillors, and he lost the support of his hand-picked executive committee. When his abuse of alcohol and crack cocaine was publicly revealed, igniting an international media firestorm, council voted to transfer most of his authority to his deputy mayor, Norm Kelly. Few new policy initiatives emerged as council limped to the end of the 2010–14 term.

The Incrementalist: John Tory, 2014–2023

Rob Ford remained popular with his core supporters despite his personal woes and isolation on council, and pledged to run in the 2014 election (Towhey and Schneller 2015, 50). Two months before the election, however, Ford revealed that he had been diagnosed with a rare form of cancer and stepped aside in favour of his brother Doug, who had won his former council seat in 2010. Running on a disciplined low-tax/small-government message, and continuing to champion "Ford Nation," Doug Ford polled higher than his brother had. Opposing him was John Tory, the Progressive Conservative Party-connected blue-blood lawyer and executive, who had been persuaded to run by civic and business leaders, and Olivia Chow, a former federal MP for the left-wing NDP and, prior to that, city councillor for a downtown ward. While in 2003 Tory had run as a law-and-order cutter, this time he tacked to the centre. He ran on a promise to bring civility back to council and end the circus atmosphere of the Ford years. He appealed to suburban commuters with a "Smart Track" transit plan that would supplement on-going investments in light rail and subway expansion with twenty-two new stations and more frequent service on commuter rail lines operated by the provincial government. Chow ran on a platform of social investment, especially in immigrant neighbourhoods. Ultimately, Tory won with strong support in higher-income areas, Ford carried suburban areas with concentrations of low-income and immigrant residents, and Chow performed best in low-income and gentrifying areas in the core (Doering et al. 2021; McGregor et al. 2016).

Tory would win again in 2018 and 2022 against centre-left challengers who championed urbanist values: improving the quality of public amenities, expanding transit, and encouraging walking and cycling. In 2018, former chief city planner Jennifer Keesmaat, drafted by NDP-backed progressive forces, matched Chow's citywide 2014 vote share of 23 per cent but received 40,000 fewer votes than Chow had. In 2022, as voter turnout continued to decline in successive elections, international urbanism consultant Gil Peñalosa received an 18 per cent vote share, with half the number of votes received by Keesmaat in 2018. The election maps of all these challengers were broadly similar: higher support in the older, pre-war city and little support in the post-war suburbs. For his part, Tory retained support in higher-income areas while adding suburban neighbourhoods that had supported Ford in 2014 (McGregor and Pruysers 2021). The secret of Tory's repeated electoral success may well be the opposite of Miller's in 2003 and Ford's in 2010. While they had projected clearly defined, if opposed, messages that resonated with voters on different sides of Toronto's core–suburb political cleavage, Tory's pitch to voters centred more on his personality and character than on his policy proposals (McGregor et al. 2021, Ch. 4). Voters rewarded Tory for his image of calm and reasonable centrism.

There was, however, a cost in policy terms to projecting an image as a cautious pragmatist in a deeply divided city. To bridge the geographical and social divides in Toronto, Tory pursued an incoherent policy agenda that produced few significant policy accomplishments. A decade later, his "Smart Track" plan – the cornerstone of his 2014 campaign – proved to be little more than a mirage, with none of it constructed yet (Boisvert 2021). He accommodated downtown councillors (and angered his suburban allies) by supporting arts funding, bike lanes on arterial roads during the COVID-19 pandemic, and – unsuccessfully – tolls on city-owned highways and a major new downtown park. After considerable public controversy, he publicly came out against police "carding," the use of racial profiling in street checks primarily directed at Black youth (CBC News 2015). At the same time, his hand-picked executive committee was dominated by suburban councillors, and he strongly supported an expensive reconstruction of the downtown Gardiner Expressway, which progressives have long advocated removing. He also supported the Scarborough subway project inherited from Rob Ford, which is now years late and well over budget, and is forecast to attract fewer riders than Miller's less costly and more comprehensive surface light rail plan.

Perhaps Tory's most consistent policy position was fiscal restraint. Throughout his mayoralty, he kept property tax increases low by restraining new spending and deferring capital investments and maintenance (although his support for reconstructing the Gardiner Expressway went against the grain of this fiscal conservatism).[2] Unlike Miller, who pursued an ambitious social policy agenda that relied in part on (unstable) funding from other levels of government, Tory did not propose major new spending initiatives. In fact, even in the face of a steadily worsening housing affordability crisis in his second term, Tory was slow to support implementation and funding for the subsidized housing components of an ambitious housing plan that the city rolled out with his support in 2019 (City of Toronto 2019). He gave progressives mostly symbolic victories while largely pursuing a suburb-friendly agenda. Critics across the political spectrum raised the prospect of a city in which the rich get richer, key workers cannot afford to live, and the quality of the public realm, infrastructure, and services erodes (CBC News 2023; Maddeaux 2023).

The Progressive Pragmatist? Olivia Chow, 2023–

Tory's tenure came to a sudden end on 10 February 2023, when he resigned after a media investigation revealed that he had an inappropriate sexual relationship with a staff member less than half his age (Keenan 2023). In an instant, he exploded his carefully cultivated image as the moderate, ethically unimpeachable "only grown-up in the room." The hastily organized 26 June by-election that followed featured 102 candidates, but 7 rose to the top tier: former councillor Ana Bailão, who had been closely allied with Tory; former mayoral candidate, councillor, and NDP MP Olivia Chow; sitting councillors Brad Bradford and Josh Matlow; the Doug Ford-endorsed former police chief Mark Saunders; Liberal Member of Provincial Parliament (MPP) Mitzie Hunter; and right-wing newspaper columnist Anthony Furey. Not since 2003 have so many "serious" candidates run for mayor in an open race.

Olivia Chow, running with support from local activist groups such as Progress Toronto, focused her campaign on housing affordability and homelessness, transit access, tenant rights, the environment and climate change, and enhancing the public realm. Importantly, she ruled out exercising the "strong mayor" powers imposed by the Conservative provincial government at Tory's request. She consistently polled in first place throughout the race, and no clear challenger to her emerged, until John Tory endorsed Bailão a week before election day. Despite the ignominious end to Tory's mayoralty, his endorsement of Bailão turned out to be effective. Chow was pushed over the top by advance votes; had only election-day votes been counted, Bailão would have won. In the end, Chow won with 37 per cent of the vote to Bailão's 33 per cent, improving on her 2014 performance by adding strong support in Scarborough to her core downtown base. Following on the heels of suburban ward victories by several progressive visible minority candidates in 2022, Chow's victory showed that it is possible to bridge the core–suburb divide in Toronto politics.

While many of Chow's supporters expected her to champion progressive causes, her first two years in office signalled creative pragmatism and ideological flexibility when dealing with a Conservative provincial government and a divided council. She quickly established a productive relationship with premier Doug Ford. The premier and mayor negotiated a "new deal" in November 2023 that committed the province to giving the city over $1 billion over three years for transit operations and homelessness programs, transferring ownership of the Gardiner Expressway (and potentially $2 billion in associated maintenance costs over ten years) to the province, and $758 million to purchase new subway cars (Ontario 2023). In addition, Chow resisted calls from her progressive base to oppose the province's decommissioning of the Ontario Science Centre, the commercial redevelopment of Ontario Place, and the potential expansion of the Island Airport. In 2024, council supported her proposed 9.5 per cent property tax increase in her first budget. She also voted to increase the police budget, a motion opposed by five of her progressive council allies (Gray 2024). Whether "ducking fights" in the pursuit of pragmatic incrementalism will allow her to maintain her electoral coalition in 2026 remains to be seen (Keenan 2024).

Enduring Cleavages, Submerged Interests, and Activist Administration

While Toronto's politics since amalgamation has had many idiosyncratic twists and turns, our mayor-centred review suggests that the geographical divide in political culture and preferences between core and suburbs has been a dominant structuring feature of electoral politics and policymaking. In a quantitative analysis of mayoral election results from 1997 to 2018, Doering et al. (2021) find that 56 per cent of variation in neighbourhood-scale vote shares is explained by the place characteristics of neighbourhoods, including car commuting, density, and distance from City Hall. By contrast, despite their obvious presence in Toronto, ethno-cultural diversity and socio-economic inequality have played a much smaller role. They found that indicators of socio-economic status (income, occupation, and educational attainment) explained only 20 per cent of mayoral candidate support, and ethnicity only 7 per cent.

Enduring differences between the political cultures and material interests of the core and suburban areas of the city have dominated Toronto's electoral politics since amalgamation. These map, albeit imperfectly, onto the pre-amalgamation boundary between the City of Toronto and the outer boroughs. Amalgamation did not create this geographic cleavage. Combining the two-tier system into a single city has simply made it more politically salient. By contrast, the many competing interests and preferences of Toronto's radically diverse and socially divided population have been curiously politically submerged. Indeed, even though Toronto is a leading site of progressive activism in Canadian politics more generally, at the local level the electoral and political presence of "progressive" grassroots mobilizations of many kinds – social inclusion, racial justice, environmental sustainability, housing affordability – has been weak. In part, this may stem from the dual character of progressivism in Toronto. The dominant strand of progressivism remains rooted in the pro-neighbourhood "reform" movement of the late 1960s and early 1970s, which was led by middle-class gentrifier professionals in the urban core. As such, their policy agendas were often shaped by their material interests as property owners. However, as the 2022 elections and the 2023 mayoral by-election suggest, new progressive voices are emerging, especially in the post-war suburbs, which more closely reflect the rapidly changing ethnic and socio-economic composition of the city.

Some scholars have interpreted the political marginality of bottom-up progressive activism in post-amalgamation Toronto as an expression of the ascendancy of neoliberal ideology and practice in Toronto (Boudreau et al. 2009; Joy and Vogel 2015). We suggest a more institutionally focused explanation. As we discussed, post-amalgamation Toronto has not moved past the fiscal and legal constraints that existed when Magnusson wrote forty years ago, constraints that yoke local politics to property and the province. What *has* changed is the scale of the municipality, which is now a large, centralized entity that offers few entry points and political opportunities for emerging interests and grassroots initiatives, especially those not focused on property concerns.

In this context, mechanisms to engage citizens and local interest groups operate mostly at the administrative level rather than the political one. Indeed, case study research on policymaking in Toronto across a wide variety of sectors – ranging from social inclusion policy (Fitzgibbon and Mitchell 2021), urban agriculture (Hammelman 2019), and neighbourhood revitalization to immigrant service provision (Hudson et al. 2017), cycling policy (Wilson and Mitra 2020), participatory budgeting (Flynn 2016; Petite 2020), and arts and culture (Silver 2013) – consistently shows that public engagement on specific policy issues is led by administrators, who sometimes act as advocates for local societal interests and activists, either allowing them to pursue their interests "under the radar" or, conversely, relaying their priorities to politicians (Wilson and Mitra 2020; Horak and Moore 2015). However, the same body of studies also shows that bureaucratic advocacy is often insufficient to secure sustained fiscal resources, and innovative policies that reach beyond the city's core mandate of property development and services to property have a tendency to be deprived of funding after a few years, regardless of their substantive success (Wilson and Mitra 2020; Hudson et al. 2017; Hammelman 2019).

CONCLUSION

Toronto is a high-velocity city, with its social and cultural make-up, economy, and physical environment perpetually changing. The search for an appropriate means of governing this flux has continued since the Second World War, unfolding as a series of variably successful institutional fixes. The two-tier Metro fix succeeded in delivering an orderly development of the post-war suburbs, but by the 1980s, it governed a shrinking proportion of the city-region's population and was riven by internal conflict. The amalgamation fix of 1998 did little to address the fundamental governance challenges facing the Toronto city-region. Instead, it created a city that was at once too big and too small, which has faced two persistent and unresolved institutional crises that have shaped its politics: the crisis of legal standing and fiscal capacity, and the crisis of institutional intermediation.

The 2006 City of Toronto Act could not exempt the city from what is in Canada a fundamental constitutional fact: that municipalities are legal creations of provincial legislatures, exercising delegated powers, and therefore vulnerable to unilateral provincial interventions. This has been magnified since former city councillor and mayoral candidate Doug Ford's elevation to premier in 2018. Ford has taken a singular interest in Toronto's governance, wielding provincial authority in the local sphere as if trying to be mayor and premier at the same time (Gee 2023). The city also remains fiscally dependent on other levels of government, especially for capital purposes. While Mayor Miller used the new powers in the City of Toronto Act to expand the city's revenues through the Municipal Land Transfer Tax, this has reinforced the overall dependence on property-based financing and the real estate market, a dependence that in turn reinforces the political and policy focus on property-related concerns at the expense of others.

Politically, amalgamation brought together divergent policy preferences and modes of governing in the old city core and the post-war suburbs that remain unreconciled. Political authority in Toronto is centralized at the citywide level, and the political process offers few meaningful entry points for local residents and grassroots groups. Meanwhile, the city and the provincial government have both pursued efforts to strengthen the mayor's executive powers, while the province has repeatedly reduced the number of councillors (and enlarged their wards) in the name of efficiency, further weakening democratic accountability and community representation.

Toronto's institutional crises have collided with the social realities of a large and increasingly divided city, resulting in a governance gap. This gap has three main elements: representational, substantive, and temporal. The representational element of the governance gap flows from the city's electoral politics. As we have seen, non-partisan mayoral elections on a huge scale have mobilized the power of the enduring core–suburb cleavage, while submerging ethno-racial and class cleavages, as well as a host of more particular sectoral interests. Meanwhile, at the council level, descriptive representation has been strikingly skewed with a predominantly white and male council governing an exceptionally ethno-culturally diverse city. However, the election of several visible minority councillors in 2022, followed by Olivia Chow's victory as mayor, suggests that a new era of more diverse political representation may be emerging.

The substantive element of the governance gap involves a strong bias towards developmental policies and property-based concerns, and a limited responsiveness to social and redistributive concerns. This bias is, of course, not unique to Toronto – it is common at the local level across Canada. However, in a large and socially unequal city like Toronto, this bias further entrenches problems – such as socio-spatial inequality and the unaffordability of housing – that no level of government has been able to effectively respond to, thus contributing to a broader systemic policy gap. Whatever policy responsiveness and innovation exist have largely been the product of action by city bureaucrats, who have often failed to secure adequate budget support for initiatives that reach beyond the city's core mandate to develop and serve property.

Finally, the temporal element of the governance gap involves the difficulty of setting and pursuing long-term strategic policy directions. In issue areas where policymaking requires an extended time horizon, short-term considerations and electorally driven priority shifts have often undermined policy continuity. The most obvious example of this is in the field of transit infrastructure, where the return of intergovernmental funding in the early 2000s has been followed by nearly twenty years of local political wrangling over concrete plans and priorities, with the net effect that much money has been spent and relatively little accomplished.

It is important to temper this rather grim-sounding assessment of Toronto's politics by acknowledging that post-amalgamation Toronto has had numerous policy successes. Many of these successes have been underwritten by the city's continued economic dynamism. For instance, a failed bid for the 2008 Olympics produced an arm's-length waterfront redevelopment authority, funded by all three levels of government, that has leveraged strong private investor interest to regenerate parts of the city's long-derelict port lands. An overwhelmingly white council notwithstanding, Toronto has been widely recognized as a leader in multilingual and multicultural services (Siemiatycki 2011). Moreover, while Toronto's developers continue to churn out residential high-rises, the city is also developing a reputation for investing in high-quality public spaces and cultural facilities. Toronto's vast and capable bureaucracy is often a prime motor of such policy innovation. Programs ranging from multilingual services (Siemiatycki 2011) to neighbourhood investment (Horak and Moore 2015) and cycling infrastructure (Wilson and Mitra 2020) have been developed and championed by administrators, partly compensating for the chronic lack of council leadership on policy issues. But bureaucratic policy innovation is not the same thing as good democratic governance, and the question of how to bridge Toronto's governance gaps – and indeed, to what extent it is even possible to do so within the institutional fix imposed by amalgamation – remains unresolved.

NOTES

1 The 2021 Census reported a population of 2,731,571, however this is almost certainly an undercount due to mobility during the COVID-19 pandemic and persistent gaps in enumerating the immigrant population. Using a different methodology, Statistics Canada estimates the 2022 population at 3,025,467, the increase from the previous year stemming entirely from international immigration (Statistics Canada 2023b).

2 Tory did create a new property tax, the City Building Levy, in 2019. It is a property tax surcharge phased in over several years, with revenues used to fund capital budgets for transit and social housing.

REFERENCES

Adel, Aaron, and James Bow. 2017. "The Sheppard Subway." *Transit Toronto*, Last modified July 20, 2017. https://transittoronto.ca/subway/5110.shtml.

Ahmed, Nairah. 2022. "Progress Toronto and its Candidates Give Municipal Politics a Good Shake." *Canada's National Observer*, Last Modified November 1, 2022. https://www.nationalobserver.com/2022/11/01/news/progress-toronto-its-candidates-gave-municipal-politics-good-shake.

August, Martine. 2020. "The Financialization of Canadian Multi-Family Rental Housing: From Trailer to Tower." *Journal of Urban Affairs* 42 (7): 975–97. https://doi.org/10.1080/07352166.2019.1705846.

Bellamy, Denise E. 2005. *Report. Toronto Computer Leasing Inquiry and Toronto External Contracts Inquiry*. 4 vols. Toronto. https://www.toronto.ca/ext/digital_comm/inquiry/inquiry_site/index.html.

Benzie, Robert. 2018. "Ford to Slash Toronto City Council to 25 Councillors from 47, Sources Say." *Toronto Star*, July 26. https://www.thestar.com/news/queenspark/2018/07/26/ford-to-slash-toronto-city-council-to-25-councillors-from-47-sources-say.html.

Bird, Richard M., Enid Slack, and Almos T. Tassonyi. 2012. *A Tale of Two Taxes: Property Tax Reform in Ontario*. Cambridge, MA: Lincoln Institute of Land Policy.

Boisvert, Nick. 2021. "John Tory's SmartTrack Plan Shrinks Again as City Prepares to Debate Line's Future." *CBC News*, January 27. https://www.cbc.ca/news/canada/toronto/smarttrack-exec-committee-1.5888690.

Boudreau, Julie-Anne, Roger Keil, and Douglas Young. 2009. *Changing Toronto: Governing Urban Neoliberalism*. Toronto: University of Toronto Press. https://doi.org/10.3138/9781442603363.

Bradford, Neil. 2007. *Whither the Federal Urban Agenda? A New Deal in Transition*. Ottawa: CPRN.

Buckley, Michelle, and Glenn Brauen. 2022. "Building Space, Building Value: Residential Space Additions and the Transformation of Low-Rise Housing in Toronto." *Urban Geography*: 1–21. https://doi.org/10.1080/02723638.2022.2100169.

Campbell, Murray. 1997. "How Style Closed Gap in Megacity Mayoral Race." *Globe and Mail*, November 8.

Caulfield, Jon. 1974. *The Tiny Perfect Mayor: David Crombie and Toronto's Reform Aldermen*. Toronto: Lorimer.

CBC News. 2015. "John Tory Calls for End to 'Illegitimate, Disrespectful' Practice of Carding." June 7. https://www.cbc.ca/news/canada/toronto/john-tory-calls-for-end-to-illegitimate-disrespectful-practice-of-carding-1.3103855.

CBC News. 2022. "Average Rents in Canada Soar Above $2K for First Time Ever, New Data Suggests." December 14. https://www.cbc.ca/news/canada/toronto/rental-costs-canada-1.6685602.

CBC News. 2023. "Years of 'Austerity Budgets' Keeping Toronto in State of Disrepair, Critics Say." January 14. https://www.cbc.ca/news/canada/toronto/austerity-budgets-trend-critics-2023-1.6714383.

Charter City Toronto. 2019. "Proposal Overview." https://www.chartercitytoronto.ca/proposal
-overview.html.

City of Toronto. 2014. *Toronto Employment Survey 2013*. Toronto: City of Toronto.

City of Toronto. 2017. "Toronto Strong Neighbourhoods Strategy 2020." https://www.toronto.ca/wp
-content/uploads/2017/11/9112-TSNS2020actionplan-access-FINAL-s.pdf.

City of Toronto. 2019. "HousingTO: 2020–2030 Action Plan." https://www.toronto.ca/wp-content/upl
oads/2020/04/94f0-housing-to-2020-2030-action-plan-housing-secretariat.pdf.

City of Toronto. 2021a. "2021 Consolidated Financial Statements." https://www.toronto.ca/legdocs
/mmis/2022/au/bgrd/backgroundfile-227976.pdf.

City of Toronto. 2021b. *Toronto Employment Survey 2021*. Toronto: City of Toronto.

City of Toronto. 2023. "City of Toronto 2023 Budget Protects Frontline Services and Invests in Housing,
Community Safety, Transit, Parks and Emergency Services." February 15. https://www.toronto.ca
/news/city-of-toronto-2023-budget-protects-frontline-services-and-invests-in-housing-community
-safety-transit-parks-and-emergency-services/.

City of Toronto. 2024a. "Quarterly Workforce Statistics – September 2024." https://www.toronto.ca/city
-government/data-research-maps/workforce-statistics/.

City of Toronto. 2024b. *2024 City of Toronto Budget Summary*. Toronto: City of Toronto. https://www.
toronto.ca/wp-content/uploads/2024/05/9569-2024-City-of-Toronto-Budget-Summary.pdf.

Contenta, Sandro. 2018. "Toronto is Segregated by Race and Income. And the Numbers are
Ugly." *Toronto Star*, September 30. https://www.thestar.com/news/gta/toronto-is
-segregated-by-race-and-income-and-the-numbers-are-ugly/article_9b208c16-2197
-5b0c-95ff-d65b6b315e22.html.

CTV Toronto. 2010. "Ford Declares War on Streetcars in Transit Plan." September 9. https://www
.ctvnews.ca/toronto/article/ford-declares-war-on-streetcars-in-transit-plan/.

Dale, Daniel, and David Rider. 2011. "Ford Unswayed by 22 Hours of Talk, Teen's Tears." *Toronto Star*,
July 30. https://www.thestar.com/news/gta/2011/07/30/ford_unswayed_by_22_hours_of
_talk_teens_tears.html.

Dingman, Shane. 2018. "Toronto Land Zoned for Employment Use Is Shrinking." *Globe and Mail*,
September 25. https://www.theglobeandmail.com/real-estate/article-toronto-land-zoned
-for-employment-use-is-shrinking/.

Doering, Jan, Daniel Silver, and Zack Taylor. 2021. "The Spatial Articulation of Urban Political
Cleavages." *Urban Affairs Review* 57 (4): 911–51. https://doi.org/10.1177/1078087420940789.

Doucet, Brian, and Michael Doucet. 2022. *Streetcars and the Shifting Geographies of Toronto*. Toronto:
University of Toronto Press. https://doi.org/10.3138/9781487510183.

Esman, Milton J. 1987. "Ethnic Politics and Economic Power." *Comparative Politics* 19 (4): 395–418.
https://doi.org/10.2307/421814.

Fanelli, Carlo. 2016. *Megacity Malaise: Neoliberalism, Public Services, and Labour in Toronto*. Halifax, NS:
Fernwood Publishing.

Feldman, Lionel D. 1995. "Metro Toronto: Old Battles – New Challenges." In *The Government of World
Cities: The Future of the Metro Model*, edited by L.J. Sharpe. Chichester, UK: Wiley & Sons.

Fernando, Shanti. 2007. "Ethics and Good Urban Governance in Toronto: The Bellamy Report and
Integrity in Public Service." *Canadian Public Administration* 50 (3): 437–48. https://doi.org/10.1111
/j.1754-7121.2007.tb02136.x.

Fitzgibbon, Joanne, and Carrie L. Mitchell. 2021. "Inclusive Resilience: Examining a Case Study of
Equity-Centred Strategic Planning in Toronto, Canada." *Cities* 108. https://doi.org/10.1016/j.cities
.2020.102997.

Flynn, Alexandra. 2016. "Participatory Budgeting – Not a One-Size-Fits-All Approach." *Public Sector Digest*, February 1. https://www.osgoode.yorku.ca/wp-content/uploads/2016/01/Flynn.pdf.

Flynn, Alexandra. 2019. "Filling the Gaps: The Role of Business Improvement Areas and Neighbourhood Associations in the City of Toronto." *IMFG Papers on Municipal Finance and Governance* 45. http://hdl.handle.net/1807/95673.

Gee, Marcus. 2023. "Doug Ford is Meddling with Toronto again." *Globe and Mail*, February 17. https://www.theglobeandmail.com/canada/toronto/article-doug-ford-toronto-mayor-race/.

Goel, Natasha. 2023. "Residential Segregation and Inequality: Considering Barriers to Choice in Toronto." *Canadian Geographies / Les géographies canadiennes* 67 (3): 380–93. https://doi.org/10.1111/cag.12827.

Good, Kristin. 2019. "The Fallacy of the "Creatures of Provinces" Doctrine: Recognizing and Protecting Municipalities' Constitutional Status." *IMFG Papers on Municipal Finance and Governance* 46. http://hdl.handle.net/1807/98264.

Gray, Jeff. 2024. "Toronto Passes Budget with 9.5-per-cent Tax Hike, Additional Police Funding." *Globe and Mail*, February 14. https://www.theglobeandmail.com/canada/article-toronto-passes-budget-with-95-per-cent-tax-hike-additional-police/.

Hammelman, Colleen. 2019. "Challenges to Supporting Social Justice through Food System Governance: Examples from Two Urban Agriculture Initiatives in Toronto." *Environment and Urbanization* 31 (2): 481–96. https://doi.org/10.1177/0956247819860114.

Harding, Bob, Shirley Hoy, and Frances Lankin. 2005. *Strong Neighbourhoods – A Call to Action. Report of the Strong Neighbourhoods Task Force.* Toronto: City of Toronto and United Way of Greater Toronto. https://www.unitedwaygt.org/wp-content/uploads/2021/10/2005_UnitedWay_Srong _Neighbourhood_TaskForce_FullReport-1.pdf.

Hernandez, Tony, and J.I.M. Simmons. 2006. "Evolving Retail Landscapes: Power Retail in Canada." *Canadian Geographies* 50 (4): 465–86. https://doi.org/10.1111/j.1541-0064.2006.00158.x.

Hiebert, Daniel. 2015. "Ethnocultural Minority Enclaves in Montreal, Toronto and Vancouver." *IRPP Study* 52. https://irpp.org/research-studies/ethnocultural-minority-enclaves-in-montreal-toronto-and-vancouver/.

Horak, Martin. 1998. *The Power of Local Identity: C4LD and the Anti-Amalgamation Mobilization in Toronto, Research Paper.* Toronto: Centre for Urban and Community Studies, University of Toronto. http://hdl.handle.net/1807/94452.

Horak, Martin. 2008. *Governance Reform from Below: Multilevel Politics and Toronto's 'New Deal' Campaign in Toronto, Canada, Human Settlements Global Dialogue Series.* Nairobi, Kenya: United Nations Human Settlements Programme (UN-HABITAT). https://doi.org/10.1093/law:epil/9780199231690/e1715.

Horak, Martin. 2013. "State Rescaling in Practice: Urban Governance Reform in Toronto." *Urban Research & Practice* 6 (3): 311–28. https://doi.org/10.1080/17535069.2013.846005.

Horak, Martin, and Aaron A. Moore. 2015. "Policy Shift without Institutional Change: The Precarious Place of Neighborhood Revitalization in Toronto." In *Urban Neighborhoods in a New Era: Revitalization Politics in the Postindustrial City*, edited by Clarence N. Stone, Ellen Shiau, Robert P. Stoker, John Betancur, Susan E. Clarke, Marilyn Dantico, Martin Horak, Karen Mossberger, Juliet Musso, and Jefferey M. Sellers, 182–208. Chicago: University of Chicago Press.

Hudson, Graham, Idil Atak, and Charity-Ann Hannan. 2017. "(No) Access TO: A Pilot Study on Sanctuary City Policy in Toronto, Canada." *Ryerson Centre for Immigration and Settlement Working Paper Series* 1. https://doi.org/10.32920/ryerson.14637108.v1.

Hulchanski, J. David. 2010. *The Three Cities within Toronto: Income polarization among Toronto's neighbourhoods, 1970–2005.* Toronto: Cities Centre, University of Toronto. http://hdl.handle.net/1807/126137.

Ibbitson, John. 1997. *Promised Land: Inside the Mike Harris Revolution.* Toronto: Prentice Hall Canada.

Isin, Engin, and Joanne Wolfson. 1999. "The Making of the Toronto Megacity: An Introduction." *York University Urban Studies Programme Working Paper* 21.

Jacobs, Jane, and Mary W. Rowe. 2000. *Toronto: Considering Self-Government.* Owen Sound, ON: Ginger Press.

Joy, Meghan, and Ronald K. Vogel. 2015. "Toronto's Governance Crisis: A Global City Under Pressure." *Cities* 49: 35–52. https://doi.org/10.1016/j.cities.2015.06.009.

Keenan, Ed. 2023. "Just Like That, John Tory Has Dismantled the Persona He Spent His Long Career Building." *Toronto Star.* February 10. https://www.thestar.com/opinion/star-columnists /2023/02/10/just-like-that-john-tory-has-dismantled-the-persona-he-spent-his-long-career -building.html.

Keenan, Ed. 2024. "Olivia Chow has been Ducking Fights since She Took Office. It's Time to Talk About Why." *Toronto Star.* December 18. https://www.thestar.com/opinion/star-columnists /olivia-chow-has-been-ducking-fights-since-she-took-office-its-time-to-talk-about/article _a6c4a0f2-bbc7-11ef-99ff-97dfa857cc5f.html.

Keil, Roger, and Douglas Young. 2003. "A Charter for the People? A Research Note on the Debate about Municipal Autonomy in Toronto." *Urban Affairs Review* 39 (1): 87–102. https://doi.org/10.1177 /1078087403253055.

Koop, Royce, and John Kraemer. 2016. "Wards, At-Large Systems and the Focus of Representation in Canadian Cities." *Canadian Journal of Political Science / Revue Canadienne de Science Politique* 49 (3): 433–48. https://doi.org/10.1017/S0008423916000512.

Kopun, Francine. 2022. "'Anything Is Possible in this City': Wave of Racialized Voices Elected to Toronto Council Brings Hope." *Toronto Star*, October 28. https://www.thestar.com/news/gta/2022/10/28 /anything-is-possible-in-this-city-wave-of-racialized-voices-elected-to-toronto-council-brings-hope.html.

KPMG LLP. 2011. "City of Toronto Core Services Review: Standing Committee Summary." July 24. https://www.toronto.ca/legdocs/mmis/2011/ex/bgrd/backgroundfile-39626.pdf.

Lu, Vanessa. 2003. "Mayoral Race Tightens." *Toronto Star*, October 11. https://www.ekospolitics.com /articles/torstar-10-17-2003e.html.

Maddeaux, Sabrina. 2023. "John Tory Helmed Toronto's Decade of Undeniable Decline." *National Post*, February 11. https://nationalpost.com/opinion/john-tory-brought-toronto-a-decad e-of-undeniable-decline.

Magnusson, Warren. 1983. "Toronto." In *City Politics in Canada*, edited by Andrew Sancton and Warren Magnusson, 94–139. Toronto: University of Toronto Press. https://doi.org/10.3138/9781487575908-004.

Mahoney, Jill. 2010. "Rob Ford and a Decade of Controversy." *Globe and Mail*, August 19. https://www .theglobeandmail.com/news/toronto/rob-ford-and-a-decade-of-controversy/article4330595/.

McGregor, R. Michael, Aaron A. Moore, and Laura B. Stephenson. 2016. "Political Attitudes and Behaviour in a Non-Partisan Environment: Toronto 2014." *Canadian Journal of Political Science* 49 (2): 311–33. https://doi.org/10.1017/S0008423916000573.

McGregor, R. Michael, Aaron A. Moore, and Laura B. Stephenson. 2021. *Electing a Mega-Mayor: Toronto 2014.* Toronto: University of Toronto Press. https://doi.org/10.3138/9781487509651.

McGregor, R. Michael, and Scott Pruysers. 2021. "Toronto." In *Big City Elections in Canada*, edited by Jack Lucas and R. Michael McGregor, 169–92. Toronto: University of Toronto Press. https://doi.org /10.3138/9781487528577-011.

Mellon, Hugh. 1993. "Reforming the Electoral System of Metropolitan Toronto: Doing Away with Dual Representation." *Canadian Public Administration* 36 (1): 38–56. https://doi.org/10.1111/j.1754-7121 .1993.tb02165.x.

Ontario. 1996. *Report of the GTA Task Force*. Toronto: GTA Task Force. https://archive.org/details
/31761118494996.

Ontario. 2006. *City of Toronto Act, S.O. 2006, c. 11, Sched. A*.

Ontario. 2023. "Terms of the New Deal Between Ontario and Toronto." November 26. https:
//www.ontario.ca/page/terms-new-deal-between-ontario-and-toronto.

Parnaby, Patrick. 2003. "Disaster through Dirty Windshields Law, Order and Toronto's Squeegee Kids."
The Canadian Journal of Sociology / Cahiers Canadiens de Sociologie 28 (3): 281–307. https://doi.org
/10.2307/3341925.

Peat, Don. 2013. "Subways in Suburbs First, Then Downtown Relief Line: Mayor Rob Ford." *Toronto
Sun*, October 11. https://torontosun.com/2013/10/11/subways-in-suburbs-first-then-downtow
n-relief-line-mayor-rob-ford.

Pennachetti, Joe. 2014. "Toronto: Looking Forward 2014–2018: Investing for the Future (IMFG's 3rd
Annual City Manager's Address)." *Institute on Municipal Finance and Governance, University of Toronto*.
https://imfg.munkschool.utoronto.ca/research/doc/?doc_id=276.

Petite, Wesley. 2020. "The Promise and Limitations of Participatory Budgeting." *Canadian Public
Administration* 63 (3): 522–8. https://doi.org/10.1111/capa.12385.

Rankin, Jim. 2000. "Mayor's Staff Rejects Debate; Was to Include All 26 Candidates." *Toronto Star*,
October 31, B04.

Reid, Craig. 2022. "Backgrounder: The Upload Agreement (Provincial-Municipal Fiscal and Service
Delivery Review)." *Association of Municipalities of Ontario*, January 13. https://www.amo.on.ca
/advocacy/municipal-gov-finance/upload-agreement-provincial-municipal-fiscal-an
d-service-delivery.

Relph, Edward. 2014. *Toronto: Transformations in a City and its Region*. Philadelphia: University of
Pennsylvania Press. https://doi.org/10.9783/9780812209181.

Rider, David. 2010. "John Tory Is Officially Out of the Mayoral Race – Again." *Toronto Star*, August 6.
https://www.thestar.com/news/gta/city-hall/john-tory-is-officially-out-of-the-mayoral-race-again
/article_214f6fc1-986d-57f7-abab-61d1ffdcc53e.html.

Rieti, John. 2017. "Mayor Tory Decries 'Short-Sighted' Road-Toll Rejection by Province." *CBC News*,
January 27. https://www.cbc.ca/news/canada/toronto/john-tory-road-tolls-1.3954882.

Robarts, John C. 1977. *Report of the Royal Commission on Metropolitan Toronto*. Toronto.

Sancton, Andrew. 2000. *Merger Mania: The Assault on Local Government*. Montreal: McGill-Queen's
University Press. https://doi.org/10.1515/9780773568914.

Sancton, Andrew. 2021. *Canadian Local Government: An Urban Perspective*. 3rd ed. Don Mills, ON: Oxford
University Press.

Saunders, Doug. 2011. *Arrival City: The Final Migration and Our Next World*. Toronto: Vintage Canada.

Siddiqui, Haroun. 2010. "Ford Gave Voice to City's Voiceless Citizens." *Toronto Star*, October 17. https:
//www.thestar.com/news/gta/2010/10/17/siddiqui_ford_gave_voice_to_citys_voiceless_citizens.html.

Siegel, David. 2015. *Leaders in the Shadows: The Leadership Qualities of Municipal Chief Administrative
Officers*. Toronto: University of Toronto Press.

Siemiatycki, Myer. 2011. "Governing Immigrant City: Immigrant Political Representation in Toronto."
American Behavioral Scientist 55 (9): 1214–34. https://doi.org/10.1177/0002764211407840.

Siemiatycki, Myer, and Sean Marshall. 2014. *Who Votes in Toronto Municipal Elections?* Toronto: Maytree
Foundation. https://maytree.com/publications/who-votes-in-toronto-municipal-elections/.

Silver, Daniel. 2013. "Local Politics in the Creative City: The Case of Toronto." In *The Politics of Urban
Cultural Policy: Global Perspectives*, edited by Carl Grodach and Daniel Silver. Oxford, UK:
Routledge.

Silver, Daniel, Zack Taylor, and Fernando Calderón-Figueroa. 2020. "Populism in the City: The Case of Ford Nation." *International Journal of Politics, Culture, and Society* 33 (1): 1–21. https://doi.org/10.1007/s10767-018-9310-1.

Spicer, Zachary. 2016. "A Patchwork of Participation: Stewardship, Delegation and the Search for Community Representation in Post-Amalgamation Ontario." *Canadian Journal of Political Science* 49 (1): 129–50. https://doi.org/10.1017/S0008423916000275.

Stanwick, Hannah. 2000. "A Megamayor for All People? Voting Behaviour and Electoral Success in the 1997 Toronto Municipal Election." *Canadian Journal of Political Science* 33 (3): 549–68. https://doi.org/10.1017/S0008423900000196.

Statistics Canada. 2023a. "Table 17-10-0136-01. Components of Population Change by Census Metropolitan Area and Census Agglomeration, 2016 Boundaries." January 11. https://doi.org/10.25318/1710013601-eng.

Statistics Canada. 2023b. "Table 17-10-0139-01 Population Estimates, July 1, by Census Division, 2016 Boundaries." February 22. https://doi.org/10.25318/1710013901.

Suttor, Greg. 2016. *Still Renovating: A History of Canadian Social Housing Policy*. Montreal: McGill-Queen's University Press.

Taylor, Zack. 2019. *Shaping the Metropolis: Institutions and Urbanization in the United States and Canada*. Montreal: McGill-Queen's University Press. https://doi.org/10.1515/9780773558427.

Taylor, Zack, Karen Chapple, Matt Elliott, Alison Smith, and Gabriel Eidelman. 2023. "*Strong(er) Mayors in Ontario – What Difference Will They Make?*" Institute on Municipal Finance and Governance, University of Toronto. https://hdl.handle.net/1807/127410.

Todd, Graham. 1996. "Restructuring the Local State." In *City Lives and City Forms*, edited by Jon Caulfield and Linda Peake, 173–94. Toronto: University of Toronto Press.

Tomalty, Ray. 1997. "Megacity Madness." *New City Magazine* 17 (3): 9–13.

Toronto Regional Real Estate Board. 2025. "TRREB Historical Statistics." Last modified 2023, Accessed May 24, 2025. https://trreb.ca/wp-content/files/market-stats/market-watch/historic.pdf.

Towhey, Mark, and Johanna Schneller. 2015. *Mayor Rob Ford: Uncontrollable. How I Tried to Help the World's Most Notorious Mayor*. New York, NY: Skyhorse Publishing.

Valzania, Giacomo. 2022. "Towers Once in the Park: Uprooting Toronto's Welfare Landscapes." *Geografiska Annaler: Series B, Human Geography* 104 (3): 227–49. https://doi.org/10.1080/04353684.2022.2032256.

Walks, R. Alan. 2013. *Income Inequality and Polarization in Canada's Cities: An Examination and New Form of Measurement*. Toronto: Cities Centre, University of Toronto. http://hdl.handle.net/1807/94383.

Walks, R. Alan. 2020. "Inequality and Neighbourhood Change in the Greater Toronto Region." In *Changing Neighbourhoods: Social and Spatial Polarization in Canadian Cities*, edited by Jill L. Grant, R. Alan Walks, and Howard Ramos, 79–100. Vancouver, BC: UBC Press. https://doi.org/10.59962/9780774862042-008.

Walks, R. Alan, and Richard Maaranen. 2008. *The Timing, Patterning and Forms of Gentrification and Neighbourhood Upgrading in Montreal, Toronto, and Vancouver 1961 to 2001*. Toronto: Cities Centre, University of Toronto. http://hdl.handle.net/1807/94436.

White, Richard. 2016. *Planning Toronto: The Planners, The Plans, Their Legacies, 1940–80*. Vancouver: UBC Press. https://doi.org/10.59962/9780774829373.

Wilson, Adam, and Raktim Mitra. 2020. "Implementing Cycling Infrastructure in a Politicized Space: Lessons from Toronto, Canada." *Journal of Transport Geography* 86. https://doi.org/10.1016/j.jtrangeo.2020.102760.

Young, Douglas. 2017. "Redefining 'Renewal' in Toronto's High-Rise Suburbs." *Alternate Routes: A Journal of Critical Social Research* 28: 219–32.

4

Halifax

Robert G. Finbow

INTRODUCTION

Halifax (*Kjipuktuk*, or the great harbour) was founded in 1749 on traditional Mi'kmaw territory during the British colonial rule, as an outpost against rival French settlers. The so-called Warden of the North (Raddall 1950) was a military outpost whose prosperity ebbed and flowed with global geopolitics, playing notable strategic roles in both world wars. Made into a regional governmental, administrative, educational, and health centre, Halifax has morphed into a more integrated agglomerative economy. In 1983, David M. Cameron and Peter Aucoin (1983) concluded that Halifax's politics was defined by accommodation and incrementalism in the context of slow growth, but noted that if growth were ever to come, the practice of politics would likely change. While traditional patterns of social, race, class, and economic traditions continue to exert influence on the contemporary metropolis, their prediction has come true to a significant extent as growth and diversification have accelerated. Transformation has been substantial, with increases and diversification of the population, and new directions in the economy adjusting to contemporary technological and social pressures. Political institutions have also undergone transformation in the past twenty-five years, beginning with amalgamation, altered municipal district boundaries and integration of rural and urban areas under common governance. Throughout this process, the tensions between modernization, development, community, and equity remain fluid and problematic. The accountability to citizens in democratic governance practices has also been debated with responsiveness to community uncertain in a system with elements of democratic disconnect and deficit.

The chapter opens with a survey of the city's recent social, economic, and institutional evolution. It first presents a review of changing municipal institutions, especially surrounding amalgamation into a larger urban governance unit. It describes the most significant changes in the structure of local or metropolitan institutions, since the instantaneous and unplanned conversion from limited intercity collaboration via a few shared special purpose bodies into a "unicity," providing comprehensive governance across many issue areas. The chapter then reviews social change (demographics, social diversity, inequality, etc.) and economic change (growth rates, change in economic structure, etc.). This chapter also touches upon prominent themes of contestation in municipal politics, including concerns about balancing development and heritage, legacies of colonialism and structural racism, and overarching relationships with the province in a complex multilevel governance structure. As a result, we unravel the story of a city accustomed to a placid pace of gradual change, facing challenges from sudden acceleration in demographic growth and diversification in an urban space where infrastructure and workforce struggle to keep pace. As the province presses for its priorities rather than acknowledging the city as an "accountable and responsible democratic government" with "permissive authority" (Taylor and Dobson 2020, 61) to act autonomously, problems of democratic deficits and accountability potentially worsen.

Amalgamation and Its Aftermath

The city of Halifax, as it is now called, was created in 1996 as the Halifax Regional Municipality (HRM), with the merger of independent municipalities of Halifax and Dartmouth, neighbouring urbanized areas of Bedford and Sackville, and the entire sprawling rural county of Halifax along the Eastern Shore (see Figure 4.1). This makes it one of Canada's largest cities by land area at 5,476 km^2, almost ten times the land area of the City of Toronto, yet sparsely populated outside the urban core. This renders it problematic in governance terms. Reorganization of municipalities had long been advocated to rationalize service provision, planning, and development. Yet the recommendations for a more compact urban unit – spelled out with strong justification based on criteria of efficiency and subsidiarity in the Graham Commission report (Graham 1975) – were successfully resisted by the subunits. Only modest steps towards collaboration were achieved with the creation of Metro (later Halifax) Transit in 1981. The 1 April 1996 merger of urban core and rural hamlets was the result of a decision by a former Dartmouth mayor, John Savage, who as premier enacted a change he had vowed to resist while in municipal office. The change, focused on spending reductions and planning rationalization, was controversial and opposed vigorously by three of four mayors of constituent units, including members of Premier Savage's own party. One mayor referred to the premier as "a liar and an autocrat" (Canadian Press 1994).

A consultant's report proposed a transfer of responsibilities and financial burdens to a unified municipality encompassing the sprawling diverse Halifax County. The author, Bill Hayward, declared that "Unitary government is the most appropriate form of municipal government in Halifax County. Community participation through community councils and other mechanisms will allow for greater representation but at the same time utilize the administrative

Figure 4.1: Municipal boundaries before the 1996 amalgamation

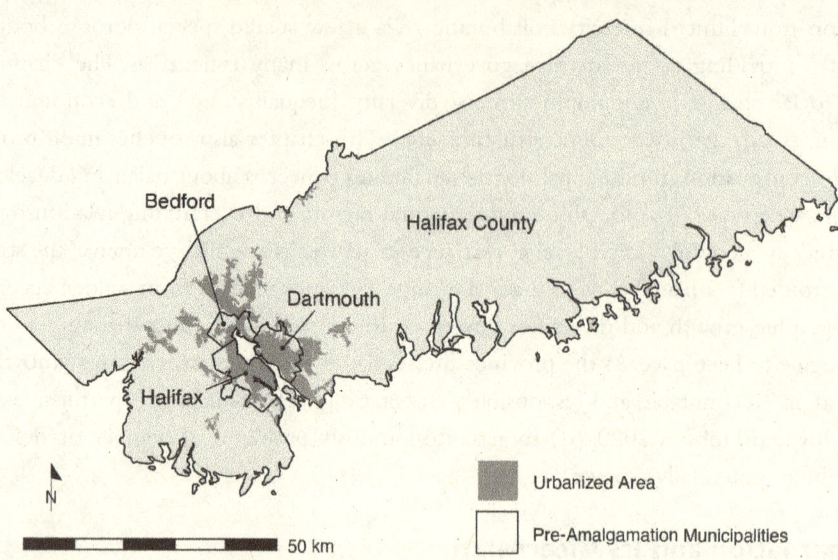

Notes: The 1996 amalgamation combined Halifax County with the cities of Bedford, Dartmouth, and Halifax. The former county and Halifax Regional Municipality today contain not only an urban core but also suburban and rural areas beyond it. Source: Municipal boundaries – 1991 Census Subdivisions, Statistics Canada; Contiguous Settlement Area – Statistics Canada.

efficiencies of the larger unit" (Hayward 1993, i). It was in keeping with the tenor of the times, as austerity policies at federal and provincial levels were accompanied by downloading of key responsibilities to municipalities (Vojnovic 2000, 412). A service exchange close to the time of amalgamation saw municipalities assuming responsibility for policing and roads and the province taking over social services, health, and justice (Vojnovic 1999, 518). Efficiencies and cost savings were promised (Sancton 2001, 548), and elimination of "overlap and duplication" from "ad hoc" division of provincial and municipal responsibilities (Vojnovic 1999, 514). The separate units had a total of 60 elected representatives whereas 24 councillors were proposed for the amalgamated city. There were also numerous agencies and special purpose bodies in the separate municipalities that could be more efficiently administered in a unicity (Hayward 1993, 13–16).

A notable goal was elimination of "dysfunctional business park competition" (Poel 2000, 33) given competing efforts by Halifax, Dartmouth, and Bedford. A merger allowed for more cohesive pan-regional planning though the relatively "vague" criteria for adjusting service provisions across levels of government and inattention to equity across municipalities makes it difficult to assess (Mavroyannis 2002). HRM appears to have fared better in the trade-offs than other cities and towns in the province (Vojnovic 1999, 530). Savings did not materialize, with amalgamation costs up to four times the projection, leading to higher user fees and property taxes (Bish

2001, 35). As Sancton suggested, pre-amalgamation predictions of savings were not informed by academic debates nor by comparisons with municipalities elsewhere. In addition, Hayward, not the council, selected the first chief administrative officer (CAO), Ken Meech, who reported that the "rushed" process created problems while some component units ran deficits (Sancton 2000, 94). While some service integration proceeded better in the new framework, such as solid waste disposal, cost overruns were substantial, notably from pressure to equalize wages and benefits among rural and urban workers and the limited ability to trim numbers of public employees. Meech also reported conflicting cultures, communications and computer systems, and job cuts (e.g., in finance) that required expensive consultants and overtime hours. With transition costs doubling without provincial compensation, higher debt servicing costs added to the new city's burdens (Sancton 2000, 95).

The budget situation has improved in subsequent years, though tensions around taxes and service levels remain (Amalgamation Yes n.d., 3–3). Moreover, resentment of the shotgun marriage lingers; residents in communities like Dartmouth insist on distinct nomenclature and signage wherever possible, while each rural hamlet remains distinctly marked within the larger unit. Halifax contains some "200 named 'communities' that range from historic fishing villages to exurban developments, to urban neighbourhoods" (Grant and Ramos 2020, 172). While the urban–rural merger bred continuing resentments, these established communities lack self-governance and are covered by common city policies and services, though elements of taxation and land use planning remain distinct. Major mergers of services included Halifax Water, solid waste management, and Halifax Police, though the latter shares jurisdiction with provincially contracted RCMP outside the urban core. Studies suggest that some of these mergers did not cut costs or improve services (McDavid 2002) and have not appreciably reduced costs in council operations. Notably, the city designed moderately different tax rates based on variations in service provision in urban, suburban, and rural districts. Tax per $100 of assessments ranged from $0.6260 in the urban core, where neighbourhoods were provided with sidewalks (including snow ploughing), to $0.5930 where sidewalks were not covered. Otherwise, the full range of municipal services is provided to all these zones. Commercial rates were $3.0160 per $100 of assessment, which were reduced to $2.6620 in rural areas. Community area rates are also used to pay for specific community services and facilities, usually through a nongovernmental agency, which can fine-tune some modest spending according to community priorities and needs.

MUNICIPAL POLITICAL INSTITUTIONS

Halifax has a mayor-council system wherein the mayor and councillors are elected every four years. At amalgamation, there were 23 councillors to encompass the vast geographic area. There are now only 16 councillors elected across districts that were enlarged before the 2012 elections (see Figure 4.2). This change, resisted by the council, was at the initiative of

Figure 4.2: Municipal wards, Halifax Regional Municipality

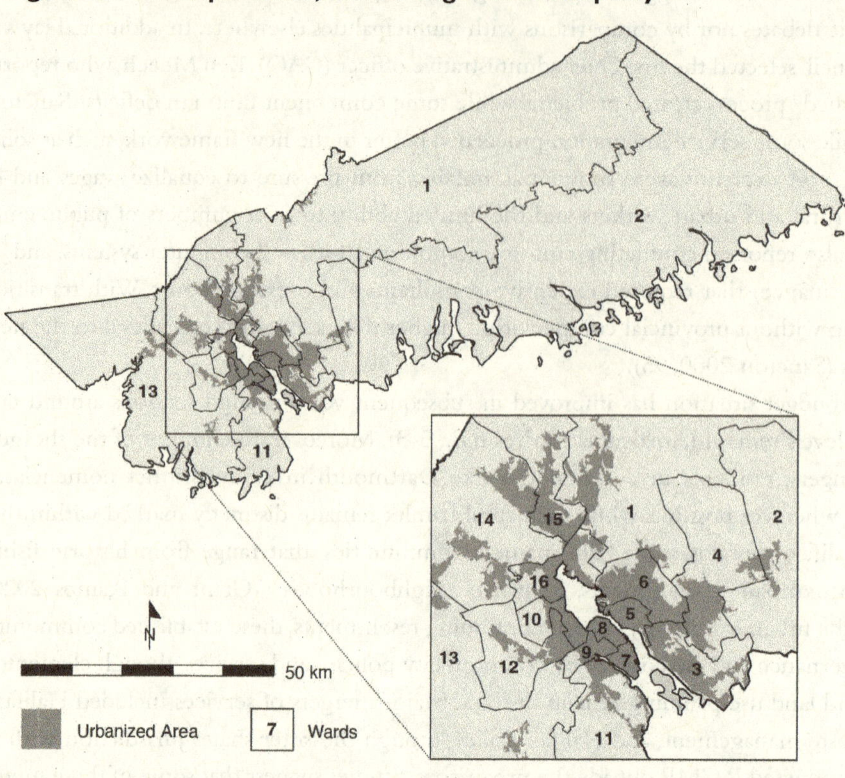

Notes: Most wards are urban or suburban, but several peripheral wards are almost entirely rural.
Source: Wards – Halifax Regional Municipality; Contiguous Settlement Area – Statistics Canada.

the Chamber of Commerce, reflecting the business sector's impatience with the slow pace of decisions on economic and development matters. Some councillors worried that this would mean a reduction in democratic debate (Bousquet 2010a,b). Others supported the streamlined decision making in a municipality that had many protracted debates over development, zoning, and other matters. Nonetheless, this reduction was imposed on order from the unelected Utilities and Review Board. For critics, this reduced the community character of districts, some of which cut across interest communities, urban and rural or on and off peninsula neighbourhoods. This might (as one advocate put it) reduce "parochial thinking" (Outhit 2010) but could run counter to principles of community representation in municipal governance. The rule that districts should not exceed 10 per cent variation from the average is generally well applied, with district electorates ranging from 21,000 to 29,000, though there are disparate population sizes as the pace of growth varies in parts of the city. However, the sprawling metropolis now has fewer districts and councillors municipally than it does in the provincial legislature.

The reforms also included creation of three community councils, covering five to six districts each, composed of the councillors elected from each district. Initially resisted by some councillors as an unnecessary complication, others found them to be useful instruments for fine-tuning on land use policy in particular districts, though there were some concerns about transparency (Sancton 2000, 96). These are advisory bodies, which convene to discuss local issues (though across multiple districts) and make non-binding recommendations to Council. The Halifax and West, Harbour East–Marine Drive, and Northwest community councils do create an additional opportunity for public input with some revenue raising capacity. Yet, as originally designed, their cross-district composition diluted community responsiveness. They are quite large and do not conform to historical community boundaries (Bish 2001, 17). Some of the councils cut across the urban–rural and peninsula–mainland divisions. As a partial response, the Regional Centre Community Council was created (overlapping existing councils), covering those areas of districts 5, 6, 7, 8, and 9 proximate to the cross-harbour urban core of Halifax and Dartmouth. This permitted a more focused consultation on core centre issues; on the other hand, the need for community councils covering urbanized areas of sprawling rural districts may illustrate the lack of community character of those broader districts in the first place.

The administration is headed by a CAO, who acts as single point of coordination of fiscal management and public policymaking. The CAO works with a core staff and coordinates provision of supports to the mayor and councillors. The CAO also heads the various units and agencies designing and implementing public policy for Halifax. Core areas that the CAO is responsible for include finance, asset management, human resources, diversity and inclusion, legal, municipal clerk, external affairs, public services, corporate and customer services, transit, fire and emergency, police, parks and recreation, planning and development, and transportation and public works. Compared to other cities, Halifax has fewer accountability officers, with only an auditor general appointed by council and no public ombudsperson, though Council has considered appointing an integrity commissioner to address potential conflicts of interest.

Council employs standing committees to focus on core topics such as audit and finance, community planning and economic development, budget, environment and sustainability, and transportation. Committees consist of six councillors with one member from each of the three original community councils and three appointed at large. There is also an executive standing committee, which included the mayor and deputy mayor with five other members appointed by each standing committee. This committee has a general grant of powers to fulfil the self-governance functions of the Council, including emergency management and planning, and nominating committee for boards and committees (Halifax n.d.). Adjustment has not always been smooth, and frequent changes at the top have caused problems. One CAO departed to work for a developer. Another CAO left after questionable payments to a concert promoter (Canadian Press 2011). Such frequent changes at the top were of concern to staff members, who suffered from disruptions from changing leadership and approaches. Some also complained of

Table 4.1: Halifax elections, 1995–2024

Election year	# of councillors	Acclaimed	Incumbents running	Incumbents winning	Incumbent success %	Mayor elected	Voter turnout*
1995	23	0	**	**	**	Walter Fitzgerald	41.9%
2000	23	0	14	8	57%	Peter Kelly	36.0%
2004	23	0	18	16	89%	Peter Kelly	48.4%
2008	23	4	19	17	89%	Peter Kelly	36.2%
2012	16	0	**	**	**	Mike Savage	36.9%
2016	16	4	12	12	100%	Mike Savage	33.6%
2020	16	1	11	10	91%	Mike Savage	39.8%
2024	16	0	11	9	82%	Andy Fillmore	36.8%

* Based on percentage participation in citywide mayoral elections – district totals vary. ** no incumbents due to first election or reduction in the size of council. While districts were redrawn in 2024, most substantially overlapped with their predecessors. Due to limitations in available data, incumbency does not reflect seats won in general elections by candidates previously elected in byelections. Source: https://www.halifax.ca/city-hall/elections/past-election-results.

lack of communication or encouragement of innovation by administrative leaders (Grant et al. 2017, 679–80).

Halifax was a pioneer among cities its size introducing online internet voting from 2008 for civic and school board elections, expanding subsequently to council elections. The experiment was seen as a means to engage new voters and reduce logistical costs of traditional elections (Goodman 2010). The Halifax model was credited with reducing barriers to participation, addressing the digital divide by allowing a choice of telephone or internet voting, while maintaining electoral security by multifactor identification and allowing candidate monitoring of electoral participation (Goodman et al. 2010, 19). Nonetheless, voter turnout remained on a downward trend, with lower in-person voting offset by higher online voting (Pammett and Goodman 2013, 28) (see Table 4.1). Turnout increased in 2020 during the COVID-19 pandemic, when more residents used online voting (Campbell 2020). However, turnout was very low in some districts, where councillors were elected with limited popular support. While incumbent victories are most common, the election of 2000 did bring changes on Council, including election of a Bedford-based critic of amalgamation, Peter Kelly, as mayor. Unlike cities like Vancouver, municipal political parties in Halifax have not emerged, even though councillors tend to be grouped on priorities related to development, heritage, and so on. The composition of council was slow to diversify; no women were elected to the first amalgamated council (Canadian Press 1995). Election of racially diverse councillors also took time; Graham Downey served from 1996 to 2000, but no other African Canadian was elected until 2016, when Lindell Smith was elected (Domise 2016).

POLITICAL TRANSITIONS

Since 1996, there have been three mayors with distinct styles but with an increasing focus on planning for future economic and demographic challenges. During their tenure, the city's approach to growth and development transitioned from conservative and cautious to progressive and dynamic. The amalgamation process was led by a veteran politician, Walter Fitzgerald, who had served twice as mayor of Halifax, interspersed with stints as a provincial member of legislative assembly and cabinet minister. He was re-elected as mayor of Halifax after a 30-year gap and held the position from 1994 to 1996 as amalgamation was implemented. He was initially a vigorous opponent of the unilateral amalgamation plan, condemning it as imposed "by a premier who could not run a small city," (Canadian Press 1994). Nonetheless, in the 1995 election for the first mayor of the new HRM, he defeated the mayor of rural Halifax County, Randy Ball (another opponent of amalgamation), with 46,333 votes to 40,847. He used his communication skills to overcome the charge of being a "city boy" and pledged a collaborative effort focused on communications among the new city's diverse components (Canadian Press 1995). However, despite his stature and experience, he failed to implement a unified vision among the diverse communities and finished in a distant and unexpected third place in the 2000 election.

Peter Kelly, from suburban Bedford, had also opposed the amalgamation of the cities but took charge in a sweeping victory, more than doubling the vote of the second-place candidate. Kelly ran a well-organized campaign and pledged to focus on the needs of outlying areas. He championed important initiatives as mayor, notably the harbour cleanup project and hosting the Canada Winter Games in 2011. He focused on making Halifax a major metropolis, through initiatives like the bid for the 2014 Commonwealth Games, which he argued would be a "once-in-a-lifetime opportunity to invest in a lasting legacy of infrastructure and community and economic development for the region," including "world-class sport facilities and programs, affordable housing units, improved transit systems, and economic growth" (Canada, 2014). Despite winning endorsement as Canada's official bid, the initiative was eventually abandoned given the high price tag and lack of agreement on funding, dealing a blow to Kelly's civic boosterism. His time as mayor ended in controversy when he was criticized after $400,000 was paid to a concert promoter to attract world famous performers for concerts on the Halifax Common without knowledge of council. He was also accused of taking funds from an estate as executor and failing to fulfil his duties to the beneficiaries in a timely matter. Despite raising over $80,000 for a 2012 campaign, he withdrew from the race, acknowledging his personal missteps (Bousquet 2014).

Former Dartmouth Liberal MP Mike Savage, son of premier John Savage, who amalgamated the cities, became the next mayor in 2012, a post he held until 2024, when he announced that he would not reoffer in the next election. This meant that the three largest pre-amalgamation urbanized communities had taken turns providing a mayor, perhaps cementing integration. Like his father, who had brought the healthy community movement to the province and initiated programs to favour Black businesses in bidding for construction projects, Mike Savage pivoted city governance in largely progressive directions. He placed substantial emphasis on economic growth and development. He presided over a reinvigoration of formalized planning processes. He initiated "state of the city" addresses – delivered not to council, the public, or the press, but to the chamber of commerce (e.g., Savage 2015). Summarizing his accomplishments in 2019, he emphasized attracting new professional sports franchises, along with "improvements in transit and active transportation; and strong economic, population and construction growth" (Woodford 2019) consistent with an inclusive and sustainable city. In launching a successful third-term bid, he noted his strengths as a "consensus builder" focused on affordable housing, which grew scarcer during his third term. His re-election campaign, which also emphasized race relations, was announced on the site of Africville, where he emphasized policies such as strengthening the diversity and inclusion office (Woodford 2020). Hence, while there was continuity of overarching goals for urban growth and development, the three mayors differed in style, with the most recent mayor adopting the inclusive rhetoric of the changing times. How former MP Andy Fillmore – elected in October 2024 with no prior elected municipal experience – will operate in style and substance remains to be seen.

SOCIAL TRANSFORMATION AND DIVERSIFICATION

Amalgamation created a diverse and changing urban community. By the 2021 census, metropolitan Halifax comprised 439,819 residents, 45.4 per cent of the provincial total. Growth has accelerated since, making the city an agglomerative magnet economically. Since the 2021 census, Halifax has become one of the faster growing cities in the country, at 4.4 per cent in 2022 (Patil 2023), bringing the population to 480,582. This reversed a trend of slow increases and overall population aging; primary school enrolments have grown for the first time in half a century. It has long been recognized that diversification of the population with international immigration was beneficial and essential to provincial development (Tetreault 2004). For many years, the ability of the city to attract and retain international residents was more limited than larger centres in Canada. Until the mid-1990s, the source for most migrants remained the United Kingdom and the United States, but the diversity of sources increased after that (Murray and Caverhill 2008, 185). There are notable increases in new residents from China, India, South Korea, South Africa, the former Yugoslavia, Taiwan, Vietnam, and the Philippines.

The Ivany Report on provincial economic prospects placed high emphasis on the attraction of new Canadians both as skilled labour and as investors and job creators (2014, 4). Retention rates have steadily improved since the turn of the twenty-first century. Recent data indicates that population increase is driven not only by interprovincial migration but increasingly (with the exception of the pandemic-disrupted years) by international arrivals. Referencing the rise in house prices and decline in apartment vacancies, one business contact commented that the

Ivany strategy had been "too successful." But the relatively high rates of investment in real estate by non-resident, out-of-province actors (around 5 per cent) makes the situation worse (Canada 2023). The demographic composition of the city shows significant change with a higher percentage of new Canadians from outside the traditional linguistic groups. The francophone population is decreasing. The 2021 census indicates the presence of small numbers of speakers of Indigenous languages. Among newcomers, several linguistic groups are evident, notably from Middle Eastern and South and East Asian backgrounds. The top source countries since 2011 have been India and China (Halifax Partnership 2022), though refugees from Syria and Ukraine have recently added to diversity. Gosse et al. (2016) indicate that contrary to stereotypes of conservatism in the region, those Haligonians who noticed neighbourhood change are either unbothered or supportive.

While adjustment to diversity has been welcomed by many, historical atonements remain necessary (as we discuss below on racial conflict and efforts at reconciliation). With an aging population, the ratio of dependent to working people has increased. While youth retention is essential to long-term growth (Jacob 2015), the recent period of augmented attraction of the young (Nova Scotia 2019) has been anomalous. This trend was also driven by the COVID-19 pandemic, when a reputation for safe and more affordable communities contributed to more rapid population growth. But the housing crunch, lack of access to medical services, and poor-quality transit threaten to reverse this. Whether growth can continue depends in part on the problematic issue of housing, which has made Halifax among the most expensive Canadian cities, despite historically lower wage rates (RBC 2022). While inequalities in the city are considered less "stark" than in other Canadian centres, gentrification of the urban core and declines in income and status in some major suburban areas have created a polarized social climate, which is accentuated by more rapid growth and housing price escalation (Grant and Ramos 2020, 190).

ECONOMIC GROWTH AND STRUCTURAL CHANGE

Halifax has emerged as the key economic engine of the region, with GDP growth and sectoral diversification outstripping the rest of the province. It is the administrative heart of the Atlantic region, with multiple universities and healthcare facilities. Alongside provincial government offices, military establishments, and regional headquarters for federal departments, public sector employment remains significant as a percentage of economic activity. It also serves as the primary retail and hospitality centre for the region, with significant income from tourism driven recently by an increase in cruise ship arrivals (Grant 2016, 267). Much like most other urban centres in the developed world, the move to a post-industrial economic form is in full swing, with decline in primary sectors and stagnation in secondary manufacturing. Employment growth is concentrated in tertiary services sectors. These are diversified, ranging from lower wage retail and food service to high-end technical, medical, and research activities. Halifax has been classified as a "service replacement economy" where strengths in "knowledge intensive business services" and other service sectors allow moderate diversification and offset job declines in manufacturing and primary resource sectors (Burnett and Brunelle 2019, 118).

Traditional sectors in manufacturing and primary extraction declined, especially after the collapse of important groundfish stocks and scaling back of that sector in the 1990s (Kaida et al. 2020, 195). At the same time, the economy benefited from infusions of investment in the oil and gas sectors, which boosted spinoff activities from this sector. The producing life of natural gas fields on the offshore proved relatively short, however, and production declined steadily from brief peaks. A significant percentage of the resource was exported to the US northeast, though local homes and businesses did connect to the pipelines from the Sable and Deep Panuke projects (Canada 2021). Naval shipbuilding orders, though trimmed, provided a major infusion of investment, which created some employment and helped sustain Irving Shipyards in an era when most such work had moved to developing nations (Erskine 2014). Government reports noted significant contributions to investment and employment as well as collaboration with local educational institutions to increase the amount of skilled labour available in the shipbuilding sector (Canada 2022).

Major activities did limit the dynamism of Halifax as an economic cluster, as its focus on "maritime, logistics, business services, ICT services, and education clusters" did not produce "patent-intensive industry" (Spencer et al. 2010, 711). However, in subsequent years, new clusters in cultural and biomedical industries created new sources of dynamism (Grant 2016, 267). Health and life science firms have emerged with innovative capacity, drawing on the multiple educational and health care centres with access to skilled labour, high-end infrastructure and global connections (Halifax Partnership 2023). Government research support via Genome Canada, National Research Council, the Healthcare Venture Fund, and the province's Innovacorp supported firms' conversion of laboratory results into viable commercial projects, helping to fuel the sector (Atlantic Canada Opportunities Agency 2023). Governments, universities, and industries have also concentrated on ocean-based technology to make Halifax and the wider region a "supercluster." Government, alongside its partners, committed to "gather a critical mass of growth-oriented firms and strengthen collaborations between private, academic and public sector organisations" (Doloreux and Shearmur 2018, 34). While the consistency of government commitment and investment has been questioned (Crane 2022), investments for startups and research and development initiatives in Halifax —and elsewhere – fuelled activity in ocean technology, communication and navigation systems, robotics, sensors and surveillance, defence and security, biofuels and wind energy, seafood and fish harvesting and processing, aquaculture, biomedical information and artificial intelligence, and other cutting edge economic activities (Canada's Ocean Supercluster n.d.; Halifax Partnership – Oceans Economy n.d.).

Grant (2016, 265) reported on the efforts in Halifax to focus on smart city and creative industries as championed by Richard Florida. Some studies indicated the city performed well for its size on indicators regarding skilled, creative, and technologically minded individuals (Spencer 2009). It placed above where it would be expected, given the lower percentage of foreign-born individuals in the population (Gertler et al. 2002). While at the time (2000s) the city was not drawing or retaining as many international immigrants as might be required for synergies in the smart city direction, it appeared attractive to creative-class migrants from other parts of the country. To a degree, "the appealing lifestyle and affordability of the city-region may compensate for other missing social dynamics valued by contemporary urban and creative class theorists" (Grant 2016, 279), although the rapid influx of migrants from other provinces during the pandemic may indicate change in these regards. Likewise, the Halifax Partnership (2023) noted that the lower costs of operating a business helped attract investors in innovative sectors like health and life sciences. These attributes have recently come under stress from densification. As the city retains more immigrants, there are increased problems of access to affordable housing and core services, including transit and medical services. Yet the economy is now significantly diversified, and Halifax can no longer be considered mainly a public administration hub. Moreover, barring continuing inflation and low vacancies in housing, the city appears to be building momentum in attracting migrants, both domestic and international, increasing the smart city potential.

DEVELOPMENT PRIORITIES – ECONOMICS, HERITAGE, AND SUSTAINABILITY

After 1945, the city focused efforts on development policies to foster economic growth, improved quality of life, slum clearance, beautification, transportation infrastructure, and industrial development (Dube 2014). Since then, regional development planning has been stop-and-start, with reflection on urban design during the 1960s and 1970s, followed by laissez-faire in the 1980s and 1990s, which favoured less regulated sprawl development (Millward 2002, 33). Gregory (2014) summarized post-war development planning with a consistent series of initiatives for the first thirty years focused mostly on the downtown as an economic unit. A complex combination of issues, including restricted access to the urban core, a problematic rocky geography, and political self-interest distorted planning. Without explicit plans in place, suburban sprawl occurred, and continued post-amalgamation, with spread to outer areas of Dartmouth, Bedford, and Sackville.

Tensions between growth and modernization versus heritage and community were evident throughout. A 1960s plan for ring roads, bridges, and harbourside expressways was thwarted to preserve much of the historic downtown and waterfront district; contemporary densification development is eroding some that architectural and community capital. The Waterfront Development Corporation guided a renewal of the harbourside district, with some historical preservation and improved waterside access but some development replacement (Hoyle 2000,

236). This success was credited to community groups and individuals resisting the expressway, yet other citizen victories (e.g., on view planes from Citadel Hill to the harbour) were gradually abandoned as views were increasingly extinguished by downtown high-rise development. Developers played a big role in reshaping the downtown core, displacing some street patterns in the creation of core projects like Scotia Square (Grant 1994). The wealthy South End was also privileged in decision making, protected from encroachment by arterial roads (Keuper 1987) despite the problems of moving people to the educational and medical centres there via choke points like the Armdale roundabout. Instead, road widening projects on Chebucto and Bayers Roads proceeded at the expense of more modest homeowners and communities (Finbow 2012).

Business–government collaboration was cemented by the creation of the Halifax Partnership, which speaks for businesses in the core. The partnership is funded jointly by businesses and government, with participation from universities. A companion organization based in social constituencies, the Halifax Regional Development Agency, was more short-lived (Wolfe 2009, 171). Such organizations created an interactive planning process with ongoing input, especially from businesses. More recently, planning centred around efforts to make Halifax a "smart city," drawing on the approaches coined by Richard Florida, who was connected to city administrators via the Greater Halifax Partnership (Grant et al. 2008). The focus shifted from industrial and resource extraction, offshore energy spinoffs, and industrial and commercial parks as the key drivers of the economy. Instead, the city prepared to attract more creative and knowledge sector activities as the post-industrial shift to service sectors gathered steam in the region (Halifax Partnership 2021). Over time, the Smart City ideal became the core focus of planning in various incarnations, which emphasized a city that is attractive to creative and information-based industries. The HRM by Design plan of 2008 appeared encouraging as it permitted public consultation on heritage protection and development via design review committees albeit with appeals to council from applicants seeking zoning variances. The Plan was often bypassed for specific projects (Bousquet 2019). Its effectiveness was limited by developers who rushed non-conforming projects just before its implementation, with exceptions granted for major projects like a new convention centre (Finbow 2012, 90).

A comprehensive planning process led to the development of the Centre Plan (Halifax 2017) with ambitious intentions of reshaping the city based on maintaining "vibrant places" and enhancing "quality of life." As envisioned, the "Regional Centre will accommodate strategic growth, foster complete communities with access to multiple services and attractions, and place pedestrians first in a human scaled environment" (Halifax 2017, 3). It focused on core principles to cover development in the centre (spanning the downtowns of Dartmouth and Halifax): "complete communities, human-scale, pedestrians first and strategic growth" (Halifax 2017). Critics suggest that these plans are accompanied by "innumerable 'exemptions,' 'grandfathers,' and 'public purpose considerations'" to permit developments to proceed outside agreed upon planning parameters; "should any regulations with merit be mistakenly left in the actual document, the whole thing can be ignored at any time because councillors can instead enter into a development agreement with favoured developers – that is to say, all developers – and sidestep the plan entirely" (Bousquet 2019).

Tensions over development priorities across the urban–rural divide hampered optimal adoption of a core-focused development model, with rural actors making demands for spending on their communities (Grant 2016, 278). Overall, the efforts to reconfigure development priorities ran into "path dependencies" based on multiple factors: "development patterns are constrained by geography and spatial configuration, the legacy of institutions and decisions, transportation technologies, political culture, the development industry, population growth and characteristics, economic conditions, and cultural and lifestyle preferences" (Grant et al. 2019, 240). Hence, much construction remains in sprawling suburban areas, though densification at the core has also proceeded. The late spring fires of 2023 in suburban Tantallon and Hammonds Plains revealed the perils of such intrusions into forested areas, as districts became more vulnerable to catastrophic destruction from climate change, coupled with a lack of suitable escape routes and water and hydrant infrastructure in wooded districts.

Halifax has addressed sustainability and environmental ambience in the Halifax Green Network Plan (HGNP), a municipal strategic plan to preserve and promote connected open or green spaces throughout the sprawling city. The goals include preservation of important ecological terrain and sites, including aquatic spaces, parks and connected corridors, and promotion of sustainability in resource-based activities (Halifax 2018). Multilevel governance of ecological space, shared across the three levels of government, complicates municipal efforts to balance the search for vigorous economic growth with preservation of sensitive natural, cultural, and historic spaces. The city requested authority to acquire environmentally sensitive sites and create parkland, but the province delayed (Sivak 2021, 170). Recent provincial impositions make implementation of this vision more difficult, as housing development has been prioritized over sensitive ecological spaces (Woodford 2022).

There has been a commitment to active transportation, notably biking, with Mayor Savage asserting that the city would be a cycling city by 2022 (Bauman 2022). Additional biking infrastructure has been developed with more committed for the near future, despite pushback based on concerns respecting street capacity, parking, and tree cover (Arif 2023). Some investments and improvements have occurred, but the progress on biking infrastructure has lagged with safety issues from incomplete, stop-and-go routes. Transit routes are insufficient in frequency, with inconvenient and often dysfunctional transfers through a limited number of nodes. Service in established areas has been reduced in places to provide more coverage to new developments. Additionally, inequality and polarization have increased, with many peninsular and downtown districts becoming pricier (Grant and Gregory 2016). This limits affordability for many middle-class and working-class families, contributing to a lack of workers in several sectors. Those social strata will be forced to live in ever more distant suburbs with underbuilt transit, generating more demand for bigger, more ecologically disruptive highways.

INDIGENOUS RECONCILIATION: ADDRESSING COLONIAL LEGACIES

Recently, the arrival of more diverse immigrants appears to be accelerating, despite the housing and services crunch. However, while the laid-back pace of life and personalistic culture attracted newcomers from English-speaking countries and other provinces, the city has not always been as accommodating to new Canadians of other backgrounds (Grant and Kronstal 2013). Moreover, issues arising from colonialism, structural racism, and ethnocultural discrimination have been evident throughout its history. These legacies remain unaddressed and are now taking more prominence in urban politics.

The Indigenous community in Halifax includes both substantial numbers of off-reserve individuals as well as the communities with reserve lands within the sprawling city: Acadia, Sipekne'katik, and Millbrook. The Indigenous identity population rose to 16,615 in the 2021 census, comprising First Nation, Métis, and Inuit peoples (Statistics Canada 2021). Compared to other Canadian cities, Halifax has a mixed record in taking actions to redress colonial-era legacies, having yet to create a specific Indigenous affairs office or act on the UN Declaration on the Rights of Indigenous Persons (though provincial legislation in 2023 did mandate this). While Indigenous relations are managed by the Office of Diversity and Inclusion, the city did create a committee for partnership with First Nations communities and commit to staff training on Indigenous matters (Anderson and Flynn 2021, 17–18).

Following the report of the Truth and Reconciliation Commission (TRC), the city adopted a Statement of Reconciliation (Halifax 2015) and committed to address concerns about access to municipal services by the urban Indigenous community. The city participated in the Federation of Canadian Municipalities (FCM) initiative to build partnerships with Indigenous communities to promote economic opportunity and implement the TRC recommendations (FCM n.d.). Halifax employed an advisor, Indigenous Community Engagement from the community to develop integrated policy in conjunction with the community and assist with training and education for municipal employees. The city has also built relationships with organizations like the Kwilmu'kw Maw-klusuaqn (the Mi'kmaq Rights Initiative), Mi'kmaw Kina'matnewey (the Mi'kmaw education authority), and the Mi'kmaw Native Friendship Centre. The latter long-standing institution has been central to provision of social services to Indigenous residents and is renewing its facilities to better serve the growing Indigenous population, with support from the federal Indigenous Community Infrastructure Fund (Lalonde 2022). Improvements in opportunity and access to services remain essential as Indigenous residents have long trailed city averages on human development indicators in education, employment, earnings, housing, and health (Milligan 2009).

Yet contentious issues remain to be addressed, some coming to a head around colonial-era symbols considered insensitive to the Indigenous community. Halifax's first governor, Edward Cornwallis, is now discredited for his infamous bounty on the First peoples of the region, despite negotiation of Peace and Friendship treaties by 1752. At the time of the 250th anniversary of the "founding" of the European city on Mi'kmaqi territory, HRM mayor Fitzgerald apologized to First Peoples for atrocities committed by the first British governor (Galloway 1999). Under pressure from the Indigenous community, the city removed a statue of Cornwallis in 2018. When the statue was first covered pending removal, there was grumbling by some councillors (Boon 2017). Eventually there was agreement on the desirability of removing this tribute to a controversial racist figure, though one councillor did retweet a group critical of removing the statue and was sanctioned in camera by council (Boon 2018a). The city created a task force jointly with the Assembly of Nova Scotia Mi'kmaq Chiefs to examine commemoration of Cornwallis at various city sites and larger issues of representation, including "outdoor spaces for the recognition and commemoration of Indigenous history" (Halifax 2020, iv). The city renamed Cornwallis Park as Peace and Friendship Park in honour of the treaties with Mi'kmaq people. A street named for Cornwallis was renamed for Mi'kmaq activist Nora Bernard, a residential school survivor, whose activism led to compensation for those victimized in residential schools (Seguin 2022). A local junior high school also removed Cornwallis's name, while a high school ended association with Sir John A. Macdonald for similar reasons.

AFRICAN NOVA SCOTIANS: COMBATTING STRUCTURAL RACISM

The long-standing African Canadian community in Halifax (Sehatzadeh 2008, 408) deals with continuing structural racism. Dating back to the colonial period, Black settlers received marginal land grants on the periphery of Metro, with uncertain title and a legacy of continuous poverty and marginalization. Most notoriously, rather than extend modern urban services to the community of Africville, at the far north end of peninsular Halifax, the long-established community was razed in the late 1960s for roadways connecting to the MacKay bridge (Nelson 2008). The area was marginalized by city decision makers despite its strong social capital and community institutions. It was divided by railways in mid-1800s and was next to the sites chosen for a prison, waste disposal, infectious disease hospital, power transmission lines, oil and coal storage, and other industrial activities. When redevelopment was considered, outside consultants suggested the population be rehoused elsewhere as "land which they now occupy will be required for the future development of the city" (Murray 2007, 137). Requests for the extension of city services to the area were routinely denied, despite tax levies on Africville residents. Employment prospects in nearby industries were limited because of racial biases in hiring by firms (Murray 2007, 138). A similar community, dubbed "the Avenue" in Dartmouth near Crichton Avenue, was more integrated into the wider community via schooling and interactions, but it also was deprived of services with unpaved roads and no sewage line until the mid-1980s (Sehatzadeh 2008, 409). Though not formally relocated, it was gradually dispersed by development projects, with many families induced to sell land at low rates (Foster 2022). A relocation contemplated for the suburban community of Beechville was managed more effectively by creating a cooperative supported by Canada Mortgage and Housing Corporation (CMHC), reflecting changing views on community vibrancy (Clairmont and Magill 1999, 10).

The relocation and decimation of Africville is now regarded as a human rights catastrophe. Mayor Kelly issued a formal apology to former residents and their descendants in 2010. Some efforts at making amends and at reconstruction of social capital have occurred, such as reconstruction of Seaview Baptist Church and a dedicated space in Seaview Park (Barber 2010). But city governance has not been fully reconciled with this and the nearby Black communities. The core elements of the planning process were motivated by a modernization and neoliberal emphasis, which may seem a well-intentioned effort to eliminate "segregation" (Loo 2019, 9). But relocation efforts ultimately disregarded the unique character of the African Canadian community, in pushing integration into majoritarian communities rather than providing supports to existing, differentially functioning ones (Rutland 2018, summarized in Burrill 2020). Under-representation of this community – as with women – has affected public policy in general (Murray and Caverhill 2008). Even where efforts are made to address sites with a legacy of race-based tragedy, like the redevelopment of the lands of the Nova Scotia Home for Coloured Children, the lack of sufficient consultation with the affected populations, coupled with an approach that still does not fully address issues of racial trauma, renders problematic outcomes (Kitson and Berglund 2021). Meanwhile issues becoming endemic across the city, such as homelessness and a lack of affordable quality housing, continue to disproportionately affect minority communities, given the persistent overlap of race and class (Luck 2023). A proposed 2022 amendment to district boundaries would unite some historically Black communities near Dartmouth, which would improve chances of representation from that demographic on council (Ryan 2022).

Periodic escalations such as confrontations downtown over mistreatment of Black people in bars or tensions in high schools along racial lines have increased the pressure for restorative measures. According to Ajadi (2023), downtown protests in 1991 signified a generational shift within the Black community leadership, which changed strategies and articulated new demands. Under federal pressure, the province created an advisory group on race relations tasked with making recommendations to address racism, including reforms to public administration, antiracism education, employment equity, and police and justice systems (Nova Scotia 1991). However, despite acceptance of the recommendations by all levels of government, decades later, the problems persist. Ongoing concerns with racial biases in policing and criminal justice were identified in the Wortley report (2019) on street checks. The report indicated that Halifax was above average in the number of street checks conducted and that Black civilians were five times more likely to face street checks than other residents and six times more likely than white residents (Wortley 2019, 104). Critics suggested that the recommendations were too slow to influence city decisions.

In the wake of the transnational movement around Black Lives Matter in 2020, the province adopted some recommendations from the report – notably a moratorium and subsequently a ban on street checks – but Wortley worried that slow implementation would potentially derail the process (Smith and Ryan 2020). Recommendations on monitoring compliance, disseminating information about rights, strengthening the complaints process, and addressing racial bias elsewhere in the criminal justice system remained works in progress (Nova Scotia 2021). Moreover, in response to Black Lives Matter, the Halifax Board of Police Commissioners issued a report on reassigning some police functions to non-law enforcement agencies. The report on *Defunding the Police* noted long-standing problems with racial profiling and inequitable treatment by law enforcement (Ajadi et al. 2021, 34–37). It recommended reassigning responsibility for tasks such as mental health checks and traffic enforcement, among others, to non-police agencies (Ajadi et al. 2023, 106). The recommendations faced opposition from police associations and some politicians. By late 2022, council rejected the commissioners' proposal for a committee to consider implementation, and instead called for an in-house staff report (Woodford 2022b).

INTERGOVERNMENTAL RELATIONS: CONTENTIOUS COLLABORATIONS

Halifax is a charter city, with powers specified by provincial legislation. The Halifax Regional Municipality Charter governs the operation of the city and determines where it can legislate and raise and spend revenues (Nova Scotia Legislature 2023). Halifax is governed under different legislation than other municipalities since the passage of the Halifax Regional Municipality Charter Act in 2008. Most of its powers resemble those exercised in other communities, and the "constitutional relationship of municipalities to the province remains unchanged" (Taylor and Dobson 2020, 62). The charter limits how the municipality can spend money, covering core city activities, including snow and ice removal, garbage collection, road and sidewalks, recreation, and playgrounds. As in much of the country, "despite the inherent power imbalance, provinces have supported cities as close partners and allies. But since the late twentieth century, provinces have sought instead to impose their will on cities" (Sewell 2021, 135). This includes forced amalgamations, interference with specific policies, and downloaded costs with insufficient revenue collection potential.

Under Mayor Savage, the city has pressed for greater autonomy from the province and conducted a review of the charter seeking changes in recognition of the city's "maturity," including a grant of "natural person powers" and "peace, order, and good government provisions":

> The intent of a new Charter is to move HRM towards a more permissive statutory regime that would broadly empower HRM to act on behalf of the interests of residents, subject only to such limits as are necessary. The scope of work approved by Council in 2014 includes

> natural person powers, broader and more permissive bylaw powers, ability to determine some
> governance structures, greater flexibility in taxation powers, research on partnering with the
> non-profit and private sectors, and ability to establish tax agreement. (Halifax, 2017, 2)

These powers would allow the city to take initiative on matters like campaign finance regulation and "do the normal things expected of a municipality on a day-to-day basis" (Berman 2016b). The province did not accede to this request (Taylor and Dobson 2020, 54). While the province does grant some amendments (as on campaign finance regulations), these take time and slow down decision making. Therefore, development priorities and social accommodations have depended on the whims of changing provincial governments. The province has not adjusted the Charter to address changing governance requirements. There are no arrangements to change Halifax's boundaries or promote more effective collaboration with its neighbours, as development diffuses out of the city to the west and north while much of the eastern hinterland within city boundaries remains mainly rural and disconnected from the urban economy (Taylor 2020).

Instead of addressing these needs for flexible, evolving governance, the relations with the city became fraught as a result of changing approaches by provincial governments. The Liberal government of Stephen MacNeil folded the powerful Waterfront Development Corporation into a province-wide agency, Develop NS, which shifted focus from Halifax and Bedford to rural broadband networks (Nova Scotia 2018). Consolidation continued after the election of Tim Houston's Progressive Conservative government in 2021. A new agency, Build NS, was created, led by a personal friend of the premier (Thomas 2022). This was but one of the personal connections that have influenced the policymaking context (the premier's chief of staff was also involved in a major waterfront development project). An interventionist approach was adopted by the Houston government, with the aim of doubling the province's population from 1 to 2 million by 2050, a drastic change from the more gradual growth typical of the region. Halifax has been growing significantly faster since 2016 – especially since the COVID-19 pandemic – but there has been too little investment, and population growth has brought increased transit and traffic woes (Bousquet 2022).

In response to a critical housing shortage and rental cost inflation, the province asserted direct control over development approvals, setting aside municipal planning and zoning procedures. The province used an accelerated process that privileged "increased density (i.e. multi-unit development) 'as-of-right' along the existing and potential future transit corridors throughout HRM (beyond Regional Centre)," where water, wastewater, and other infrastructure were suitable (Nova Scotia 2022a). City decisionmakers feared that roadways and transit infrastructure are inadequate. Capacity and effectiveness issues plague the transit system as the population spiked with the rapid growth promoted by the province (MacInnis 2022). Yet multiple high-rises were approved for major streets (often adjacent to existing neighbourhoods of single-family dwellings) that currently lack the capacity for moving people effectively or sustainably. The mayor and CAO demonstrate an awareness of these challenges, though whether or not they will find the resources to implement the required improvements remains to be seen (Taplin 2023).

A Halifax-only charter amendment gave the provincial minister the power to override municipal zoning and development policies in the interest of expediting approvals for new homes in the province's major metropolis (Nova Scotia Legislature 2022). Bill 225 amended HRM's charter to ensure municipal bylaws do not prevent "fast-tracking of housing development and increasing supply" (Taplin 2022). Minister of municipal affairs John Lohr created the "Executive Panel on Housing in the Halifax Regional Municipality" (chaired by an MLA from Glace Bay, not Metro), which was empowered to take in-camera decisions on development priorities, focused on so-called special planning areas (Nova Scotia 2022). The initial three areas were all undertaken for proposals by Clayton Developments for Penhorn and Southdale–Mount Hope in Dartmouth and Bedford West. Clayton Developments also received a $22 million provincial loan to incorporate affordable units in the developments (Doucette 2022). Councillors decried the province's dictatorial tone, distant decision making, and damaging of relationships with stakeholders in the housing sector (Taplin 2022). Nova Scotia's chronic urban–rural division was on display as the premier, minister of municipal affairs and housing, and chair of the executive panel all hailed from small communities distant from the metropolis. The initial case involved the Southdale Future Growth Node, which was in the process of being considered using the municipality's Regional Centre Secondary Planning Strategy and Regional Centre Land Use By-law – "a comprehensive planning process needed to effectively guide the development of mixed-use communities with supporting public infrastructure" (Halifax 2022). The municipal process was simply displaced by the province's unilateral declaration of Special Planning Areas with permissions granted in secret to developers who were provided a subsidy to make some of the units "affordable" at 60–80 per cent of market rates for 20 years. The new process eliminated several steps of public and community involvement. There would have been "public hearings on both the municipal planning strategy amendments and the development agreement. Instead, public input comprised information meetings and an online survey" (Woodford 2022). This was despite the potential impact on the ecologically sensitive Eisner wetlands. Affordability of the proposed units was also disputed. Councillors similarly complained of unilateralism in the site choice for new facilities, like schools, in established city parklands (Ryan 2023a). Hence, city policies on development and sustainability were hobbled, not enabled, by the province.

Relations with the federal government – a major owner of property, including military bases and public buildings – are similarly complex. There has been much discussion over the years of converting surplus federal lands, especially surplus defence facilities, into intermodal transport hubs (at Shearwater air base), or recreational facilities (a sports stadium at Shannon Park), but these projects never materialized (Finbow 2012, 74). Many federal properties have languished while the intergovernmental bargaining lagged. For instance, an intermodal shipping facility was thwarted when the Department of National Defense (DND) decided not to relinquish the property for civilian use; the abortive bid for the Commonwealth games and a lack of a credible ownership group for a CFL franchise – combined with the high public subsidies required – undermined the stadium's viability. (Finbow 2012).

The federal presence continues to be economically significant, especially through military contracts. In 2011, Irving Shipbuilding was awarded a $25 billion shipbuilding contract for six arctic and offshore patrol ships (AOPS) for the Canadian Navy, two AOPS for the Canadian Coast Guard, and 15 Canadian surface combatants for the Navy. Proponents predicted some 11,500 direct and indirect jobs for the regional municipality, a sizeable stimulus. Three AOPS ships were completed, and another was under construction at end of 2021, although slowed by the COVID-19 pandemic (Canada Public Works 2022). DND acknowledged that the effort to provide affordable ships in relatively short order was not achieved, with both cost overruns and numerous delays plaguing the project. While the arctic patrol ships have arrived, the supply and combatant ships are years away from delivery. The government still touted the employment and economic benefits to the region and country (Canada DND, 2021). Private sector groups also lauded the revival of Canada's shipbuilding industry and its economic benefits, including some 4,200 jobs locally and training of many skilled shipbuilders to boost the industry in future (Conference Board 2023), even though such growth contributed to housing scarcity.

Port facilities have been important to the local economy, but also problematic because the locations on areas of the peninsula create challenges of downtown traffic by the presence of large trucks and containers. The Halifax Port Authority (HPA), a self-funded federal Crown corporation, manages the marine facilities at the port, including container facilities near Point Pleasant downtown and Fairview Cove in the northwest. The HPA includes representatives from the federal, provincial, and municipal governments as well as federally appointed members nominated by port users (Government of Canada 2023). While management has worked well in some instances, funding issues can be contentious. For instance, the federal government at first rejected a request for funding to expand the Point Pleasant facility to prevent further increase in truck traffic in the already-congested downtown core; super container facilities elsewhere in the province and country seemed to become federal priorities instead (Withers 2018). Eventually the federal government did help fund pier extensions (Port of Halifax 2022) and road and railway projects, as well as Fairview Cove Container terminal upgrades, which would benefit the port and reduce truck traffic (Public Services and Procurement Canada 2022). Collaboration at the port reflects the challenges of multilevel governance in the contemporary urban space but has permitted at least incremental advances. Agreement on areas of trilevel collaboration, required for major projects like transit upgrades, is sometimes more elusive (Patil 2020). At the time of writing, Ottawa pledged funding from its Housing Accelerator Fund tied to approvals for increased residential density and affordability strategies; while generally onside, council resisted calls for blanket approval of four-storey units across the core (Ryan 2023b). Meanwhile CMHC prioritized neo-liberal private sector solutions, such as "filtering" (Grant 2023), expecting upscale construction to eventually free up more modest units, despite evidence to the contrary. Significant incentives for more affordable housing, including public housing, are required to rectify the escalating crisis in affordability (Grant 2023).

CONCLUSION: DEMOCRATIC DEFICITS?

As Cameron and Aucoin (1983) predicted, accelerating geographic, demographic, and economic growth have disrupted elements of accommodation and incrementalism in urban policymaking in Halifax. The city is still grappling with legacies of structural racism and colonialism, although such issues are increasingly a policy priority. Post-amalgamation, the city has a complex and often contentious governance system, presiding over a diverse, dynamic, and evolving social and economic community. While elements of the merger may have improved efficiency and region-wide provision of services, issues of accountability and responsiveness to constituents do arise. Like many urban communities, modest voter turnout and frequent incumbent victories limit pressures for councillors to address community concerns. Council reduction and district boundaries – pushed to "reduce parochialism" and help move the community forward – might have increased this problem. A smaller council might be justified if Halifax was entirely urbanized, but with districts incorporating hamlets – and, therefore, many residents not facing urban concerns or sharing urban values – it does seem to dilute the mix. A councillor whose district spans urban, suburban, and rural spaces, or on and off peninsula residents is not always able to reconcile differences in needs related to transit, commuter roadways, parking, urban amenities, etc. Community councils likewise span multiple districts whose residents have different or even competing interests. Rural distrust of the urban and resulting "misunderstandings" may obscure possible shared interests or compromises (Nova Scotia 2014, 9). But under provincial governments, which are predominantly based in rural areas, the tendency to align against core urban interests might be enhanced. With the majority Conservative provincial government shut out of most of the urban core in elections in 2021 and 2024, this problem will continue to fester.

A general sense of disconnect between decision makers and social constituencies, evident in several policy areas (Finbow 2012), persists, while activists on issues like homelessness or racially biased policing suggest that their entreaties are often ignored. Grant notes, regarding the creative sectors, that "social and cultural associations and interests felt excluded from the development agenda and from civic governance mechanisms" if not deemed "politically acceptable" (2016, 277). The council boasts of numerous consultation exercises on development and planning issues, though these sometimes resemble dissemination of information, as set pieces and entrenched positions seem to prevail. Occasionally councillors have let slip on social media their contempt for certain political positions, such as opposition to clearing homeless encampments, dismissed infamously by one councillor as "far left" (Macdonald Notebook 2024). In-camera meetings have become more commonplace, with the city charter permitting discussions outside the public eye for a range of issues including "personnel matters, labour relations, contract negotiations, litigation or potential litigation, legal advice eligible for solicitor–client privilege, public security," and so on (Woodford 2022a). Some contentious issues have been discussed behind closed doors, notably increases in police funding, which had been rejected by the public because of the controversies over homeless encampments and treatment of Black and other visible minorities. Such in-camera discussions were "not typical for the budget committee" (Woodford 2022a). Sometimes, proposals with potentially significant public expenditures remained closed (Woodford 2018), including naming arrangements for public–private projects. Such occurrences do little to improve public trust in council's accountability and responsiveness.

The capture of the political process by the development sector is concerning to many. Developers are prominent political campaign contributors, both provincially (Henderson 2021) and municipally (Ward 2015), and have long influenced urban development priorities (Grant 1994). This situation was facilitated by the weakness of Halifax's campaign financing laws, which until recently provided for no limits or disclosure rules. Concerns included Mayor Savage paying himself a salary from contributions (DuBreuil 2012). In a public consultation held to design better rules, concern with corporate and developer influence and perceived lack of transparency in decision making were common themes. The idea that "developers should not contribute" was by far the most common response to the issue of donor eligibility (Densmore 2017, 9). Overall, the public was concerned that "The greatest potential conflict of interest involves developers who propose developments that Regional Council is expected to evaluate. Such developers should not be permitted to contribute to campaigns: either as individuals or as companies" (Densmore 2017, 8). Some asked that developer contributions be spelled out in disclosures, and that online tools should permit "real time reporting on donations" (Densmore 2017, 49), though these provisions were not included in the new by laws. By-law C-1100 included a ban on corporate or union donations, with individual contributions capped at $1,000 per candidate (and $2,500 for mayor) and $5,000 in total. Spending limits were set at $300,000 for mayoral candidates and $30,000 for councillors (Halifax 2018). It would be challenging to control contributions from individuals in the property development sector as the personal nature of donations makes tracing connections to developers or other funders more difficult. Anonymous or cash donations, as reportedly offered to councillors by developers, are hard to regulate (Boon 2018b), and enforcement was problematic because city reports were covering up violations in amounts and donors (Woodford 2021).

Developers have been central to the city planning process throughout the city's existence (Halifax 2005). Some councillors bristled at the need for transparency and rejected the idea that their votes could be bought by developers, whom they described as "pillars" of the community. As Boon noted, "Councillors don't build cities. It's the development community" (Boon 2018b). Planning criteria, such as Halifax by Design and the Centre Plan, and broader planning strategies are often sidelined to accommodate developer requests (Bousquet 2019). Numerous disruptions from major development projects – like noisy construction and closed streets, sidewalks, or lanes – persist for months or years, at the expense of mobility and a community atmosphere (Williams 2021; MacLean 2020). Small businesses found these disruptions costly (O'Kane

2017) and, in some cases, threatened legal action (Boon 2016). Career movement between the development sector and government positions is routine, notably recently with the selection of a prominent developer to Premier Houston's inner circle (Henderson 2021). One CAO, immediately after resigning, moved to a major developer with substantial investment in the city. While Mayor Savage denied this was a direct conflict of interest, he acknowledged that council should enact "a non-compete clause for future hires that stipulates 'what kind of employment people can go to after the fact'" (Zaccagna 2015).

While many developments are moving the city forward and rapid growth is prioritized, the concentration of construction in high-end retail and condominium development has increased polarization through gentrification, making the urban core unaffordable for existing and new residents. The neo-liberal approach of the province, combined with support by some councillors, creates a misguided reliance on "filtering" to fix the problem (Grant 2023). The housing shortage reflects many pressures, but private sector supply alone is an insufficient response because most construction is for high-end expensive units, and investment and speculation add to inflationary pressures (Canada 2023). Nova Scotia has climbed to second place (after British Columbia) in rental costs despite the region's lower wages, and Halifax is ranked as one of the least affordable cities in Canada for young people (Armstrong 2022). Hence, the agglomerative economic potential could be lost as young people are forced to "go down the road" again to seek more affordable circumstances. For Halifax and some other major cities, "it is possible that young Canadians may move out of city centres or Canada altogether causing a potential drain on talent, decreasing the vibrancy of the city, limiting young people's options and therefore decreasing sense of belonging" (Youthful Cities 2022). The affordability crisis has resulted in increased homelessness and tensions. Halifax Police have been used to dismantle homeless encampments, once during the Remembrance Day observances in 2011 and again the day after the election of the Progressive Conservative provincial government in 2021 (Al-Hakim and MacLean 2021). Ugly confrontations between the homeless, their supporters, and law enforcement have recurred while council and the province debate temporary housing solutions. This damages Halifax's image of affordability, liveability, and hospitability, essentials for an aspiring "smart city." In addition to infrastructure inadequacies, rapid densification disrupts communities, robbing Halifax of much of its ambience, which attracted "smart city" migrants in the first place. "Tall towers are out of scale with the type of building that used to make Halifax/Dartmouth attractive" (Epstein 2022). European-style densification would require decentralization of major employers outside the core, provision of amenities, and improved, more frequent public transit. Efforts to create more options for active commuting, notably biking, are proceeding with delays, but mass transit is underfunded in favour of roadways, thus limiting the sustainability promised by densification.

The ability of the province to intervene decisively might lead to effective action, but it also threatens long-term consequences for urban forms and democratic governance. While some councillors push for more open and consultative processes, the province at times imposes arbitrary decisions in pursuit of perceived public need or private interest. As Taylor notes, "Closed decision-making processes potentially restrict the scope of input from external voices and the airing of alternative ideas and proposals, sometimes resulting in mistakes, unintended consequences, and harmful outcomes with long-term effects" (Taylor 2019, 309). Halifax may be an example of this, with historic decisions like modernization and automobile-focused development, slum clearance and destruction of neighbourhoods like Africville, imposed unification of municipalities without consultation or plan, deviations from planning criteria, and recently, the province overriding city charter and autonomy in development and housing, perhaps tinged with Ontario Greenbelt-style developer-focused land policies. Such unresponsiveness may reflect a decision-making approach – both from the province and the municipality – that pays "far too little attention to the perceptions, needs, and aspirations of those who would ostensibly benefit from their plans – ordinary people in their neighbourhoods" (Taylor 2019, 309). While amalgamation and the creation of a metropolitan unicity should produce more accountable governance, interest group capture and provincial policy overrides have created a confused system where both staff and council appear out of touch with their constituents.

Three-term mayor Mike Savage did not re-offer in 2024. The election of former Halifax Liberal MP Andy Fillmore as mayor heralds not so much transformation as continuity. The new mayor follows his predecessor's footsteps in moving from federal to municipal politics. And his long-standing connections to the development sector in Halifax suggest, many major elements of urban development will continue along similar trajectories to the past couple of decades. He played a significant role in planning policy as manager of urban design for Halifax and vice president of planning & development of the Waterfront Development Corporation. Fillmore was elected – with bipartisan support, including from prominent Conservative party insiders (Delorey 2024) – despite a problematic campaign in which he alienated civil society actors by taking a tough line on encampment sites for the homeless. That position became among his initial priorities after the election, despite continued shortages and inflation in housing sector. He was rebuffed by council on some points, which portends a confrontational relationship. The new mayor has supported provincial priorities over long-standing municipal approaches; some insiders suggest he might like to have "strong mayor" powers similar to those introduced in Ontario that would enable him to override council (Macdonald Notebook 2024).

With the loss of local ownership and extensive cutbacks at the local broadsheet newspaper, coverage of the election and local governance issues is diminished despite strong online journalistic outlets. Participation in the election continued a downward trend, despite the city's flexible online and advance voting provisions. Overall, it seemed an apt transition for the relatively diminished quality of liberal democracy in the province, country, and beyond. With several councillors not reoffering, there are some new voices in city council, but whether they can alter the balance between permanent staff, lobbying interests, and popular, civil society sectors remains unclear. The city retains its dynamism and has many educated and engaged residents, but whether it can continue to attract and retain younger professionals given the shortages in housing, health care, and transit remains uncertain.

REFERENCES

Ajadi, Ifeoluwatari. 2023. *Power in Presence: Understanding Black-Led Coalitions and Policy Change in Halifax, Nova Scotia.* PhD dissertation, Dalhousie University. https://dalspace.library.dal.ca/handle/10222/82529.

Ajadi, Tari, Harry Critchley, El Jones, Julia Rodgers, Francesco Bruno, Mariah Crudo, Skyler Curtis, Noel Guscott, Nicola Hibbard, Sydney Keefe, Madeleine McKay, Jennifer Taylor, and Sophia Trinacty. 2021. *Defunding the Police: Defining the Way Forward for HRM.* Report prepared for the Halifax Board of Police Commissioners. https://www.halifax.ca/sites/default/files/documents/city-hall/boards-committees-commissions/220117bopc1021.pdf.

Al-Hakim, Aya, and Alexa MacLean. 2021. "Halifax Police Pepper Spray Crowd During Protests Over Torn-Down Shelters." *Global News,* August 18. https://globalnews.ca/news/8121705/halifax-homeless-camps-spring-garden-road/.

Anderson, Doug, and Alexandra Flynn. 2021. "Indigenous-Municipal Legal and Governance Relationships." *IMFG Papers on Municipal Finance and Governance* 55. http://hdl.handle.net/1807/107492.

Arif, Hafsa. 2023. "Halifax to Expand Bike Lanes in Downtown Core." *CTV Atlantic,* January 11. https://atlantic.ctvnews.ca/halifax-to-expand-bike-lanes-in-downtown-core-1.6227466.

Armstrong, Lindsay. 2022. "Young People Living in Halifax Say They Are Being Priced Out of the City They Love." *The Globe and Mail,* May 25. https://www.theglobeandmail.com/canada/article-young-people-living-in-halifax-say-they-are-being-priced-out-of-the/.

Atlantic Canada Opportunities Agency. 2023. "Halifax Health Science Firms Pursue Innovative Solutions to Medical Challenges." News release, January 24. https://www.canada.ca/en/atlantic-canada-opportunities/news/2023/01/halifax-health-science-firms-pursue-innovative-solutions-to-medical-challenges.html.

Barber, Mike. 2010. "Africville Apology Is a Start, Not an End." *Huffington Post,* May 10. https://www.huffpost.com/entry/africville-apology-is-a-s_b_480361.

Bauman, Martin. 2022. "Halifax Planned to Become "A Cycling City" in 2022. How's That Coming?" *The Coast,* November 14. https://www.thecoast.ca/halifax/halifax-cycling-network-lags-behind-region-promises/Content?oid=29704322.

Berman, Pam. 2016a. "Council Votes to Ease Height Restrictions on Willow Tree Tower." *CBC News,* September 7. https://www.cbc.ca/news/canada/nova-scotia/halifax-robie-quinpool-willow-tree-tower-1.3751743.

Berman, Pam. 2016b. "Halifax Regional Council Looking to Ask Province for More Power." *CBC News,* November 10. https://www.cbc.ca/news/canada/nova-scotia/halifax-municipal-council-natural-person-powers-1.3843984.

Bish, Robert L. 2001. "Local Government Amalgamations: Discredited Nineteenth Century Ideals Alive in the Twenty First." *Commentary* 150, 1–29. Toronto: C.D. Howe Institute. https://cdhowe.org/publication/local-government-amalgamations-discredited-nineteenth-century-ideals-alive-twenty-first/.

Boon, Jacob. 2016. "Downtown Businesses Launch Legal Action Against Nova Centre." *The Coast,* June 27. https://www.thecoast.ca/news-opinion/downtown-businesses-launch-legal-action-against-nova-centre-5470403.

Boon, Jacob. 2017. "David Hendsbee Surprised Halifax Didn't Use a Red Tarp to Cover Cornwallis Statue." *The Coast,* October 25. https://www.thecoast.ca/halifax/david-hendsbee-surprised-halifax-didnt-use-a-red-tarp-to-cover-cornwallis-statue/Content?oid=10335984.

Boon, Jacob. 2018a. "Mayor 'Prepared to Revisit' Integrity Commissioner after Whitman Debate." *The Coast*, February 28. https://www.thecoast.ca/halifax/mayor-prepared-to-revisit-integrity-commissioner-after-whitman-debate/Content?oid=12966660.

Boon, Jacob. 2018b. "Council Approves New Campaign Finance Rules." *The Coast*, October 3. https://www.thecoast.ca/news-opinion/council-approves-new-campaign-finance-rules-17959761.

Bousquet, Tim. 2010a. "'Appalling' Democracy: Decreasing the Number of Halifax Councillors Would Decrease Representation and Hurt Voter Turnout." *The Coast*, August 12. https://www.thecoast.ca/halifax/appalling-democracy/Content?oid=1768565.

Bousquet, Tim. 2010b. "Halifax Council Rejects Reducing the Number of Councillors." *The Coast*, August 5. https://www.thecoast.ca/halifax/halifax-council-rejects-reducing-the-number-of-councillors/Content?oid=1756175.

Bousquet, Tim. 2012. "Halifax Council Looks to Make Highly Paid Employees' Salaries Public." *The Coast*, August 15. https://www.thecoast.ca/news-opinion/halifax-council-looks-to-make-highly-paid-employees-salaries-public-3255731.

Bousquet, Tim. 2014. "A Trust Betrayed: Peter Kelly and the Estate of Mary Thibeault." *Halifax Examiner*, February 16. http://www.halifaxexaminer.ca/investigation/a-trust-betrayed-peter-kelly-and-the-estate-of-mary-thibeault/.

Bousquet, Tim. 2019. "The Centre Plan Is a Colossal Waste of Time, Money, Public Attention, Newsprint, and Reporter Energy." *Halifax Examiner*, September 18. http://www.halifaxexaminer.ca/uncategorized/the-centre-plan-is-a-colossal-waste-of-time-money-public-attention-newsprint-and-reporter-energy/.

Bousquet, Tim, and Jennifer Henderson. 2021. Scott McCrea and His Family Gave a Lot of Money to the PC Party In 2020; Now McCrea Oversees PC Premier-Designate Tim Houston's Transition Team. *Halifax Examiner*, August 27. http://www.halifaxexaminer.ca/morning-file/scott-mccrea-and-his-family-gave-a-lot-of-money-to-the-pc-party-in-2020-now-mccrea-oversees-pc-premier-designate-tim-houstons-transition-team/.

Bousquet, Tim. 2022. "Nova Scotia Aims to Double the Population but Has No Realistic Plan for Moving Around a Million More People." *Halifax Examiner*, December 12. http://www.halifaxexaminer.ca/morning-file/nova-scotia-aims-to-double-the-population-but-has-no-realistic-plan-for-moving-around-a-million-more-people/.

Burnett, James, and Cédric Brunelle. 2019. "The Uneven Economic Diversification of Small and Mid-Sized Canadian Cities, 1971–2016." *Canadian Journal of Regional Science* 42 (2): 113–22. https://doi.org/10.7202/1083620ar.

Burrill, Fred. 2020. "On Good Intentions: A Critical Note on Recent Studies of State Planning in Canada." *Acadiensis* 49 (1): 171–80. https://doi.org/10.1353/aca.2020.0006.

Cameron, David M., and Peter Aucoin. 1983. In *City Politics of Canada*, edited by Magnusson, Warren, and Andrew Sancton, 166–89. Toronto: University of Toronto Press.

Campbell, Francis. 2020. "Voter Turnout Surpasses That of Recent Halifax Regional Municipality Elections." *Saltwire*, October 18. https://www.saltwire.com/nova-scotia/news/voter-turnout-surpasses-that-of-recent-halifax-regional-municipality-elections-510570/.

Canada Energy Regulator. 2017. "CER – Market Snapshot: 25 Years of Atlantic Canada Offshore Oil & Natural Gas Production." October 12. https://www.cer-rec.gc.ca/en/data-analysis/energy-markets/market-snapshots/2017/market-snapshot-25-years-atlantic-canada-offshore-oil-natural-gas-production.html.

Canada Department of National Defence. 2021. "Key Issues Notes." https://www.canada.ca/en
/department-national-defence/corporate/reports-publications/proactive-disclosure/pacp-national
-shipbuilding-strategy-25-may-2021/key-issues-notes.html.

Canada's Ocean Supercluster. n.d. Project List. https://oceansupercluster.ca/what-we-do/project
-portfolio/.

Canadian Press. 2011. "Top Bureaucrat for Halifax Retiring Amid Controversy over Concert Loans:
Halifax CAO Retiring Amid Controversy." March 16 Retrieved from https://www-proquest-com
.ezproxy.library.dal.ca/printviewfile?accountid=10406.

Canadian Press. 1995. "Fitzgerald and All-Male Council Will Run New Regional City." https://dal
.novanet.ca/permalink/01NOVA_DAL/ev10a8/cdi_proquest_wirefeeds_356732486.

Canadian Press. 1994. "Savage Angers More Liberals with Amalgamation Plan (Halifax, Dartmouth,
Bedford, Super-City Plan)." www.proquest.com/wire-feeds/savage-angers-more-liberals-with
-amalgamation/docview/346102415/se-2.

Clairmont, Donald, and Dennis W. Magill. 1999. *Africville: The Life and Death of a Canadian Black
Community*. Toronto: Canadian Scholars' Press and Women's Press.

Conference Board of Canada. 2023. "Value for Money 2023: The Conference Board of Canada
Assesses the National Shipbuilding Strategy's Impact." News release, May 30. https://www
.newswire.ca/news-releases/value-for-money-2023-the-conference-board-of-canada-assesses-the
-national-shipbuilding-strategy-s-impact-834969479.html.

Crane, David. 2022. "Will the 'Superclusters' Ever Live Up to Their Superstar Status?" *Hill Times*, July 11.
https://www.hilltimes.com/story/2022/07/11/will-the-superclusters-ever-live-up-to-their-superstar
-status/270892/.

Delorey, Fred. 2024. "I'm Proud to Have Served as Strategic Advisor to Andy Fillmore in His
Successful Campaign to Become Mayor of Halifax." LinkedIn. https://www.linkedin.com/posts
/fred-delorey-667188109_im-proud-to-have-served-as-strategic-advisor-activity
-7253839471341715456-p6fe/.

Densmore, Karen. 2017. *Public Engagement: Municipal Campaign Finance Accountability*. Halifax: Halifax
Regional Municipality and Karen Densmore Communications. http://www.shapeyourcityhalifax
.ca/3492/documents/7824.

Doloreux, David, and Richard Shearmur. 2018. "Moving Maritime Clusters to the Next Level:
Canada's Ocean Supercluster Initiative." *Marine Policy* 98: 33–36. https://doi.org/10.1016/j
.marpol.2018.09.008.

Domise, Andray. 2016. "Why Lindell Smith's Election in Halifax Was So Remarkable." *Macleans*, October
19. https://macleans.ca/news/canada/why-lindell-smiths-election-in-halifax-was-so-remarkable/.

Doucette, Keith. 2022. "Nova Scotia Announces $22 Million to Help Build Affordable Units in
Halifax Area. *Global News*, March 28. https://globalnews.ca/news/8715667/nova-scotia
-investment-affordable-housing-units-dartmouth/.

Dube, Sarah. 2014. *The Master Plan for The City of Halifax: Its Origin and Impact*. Research project, School
of Planning, Dalhousie University. http://theoryandpractice.planning.dal.ca/_pdf/history/sdube
_masterplan_2015.pdf.

DuBreuil, Brian. 2012. "Savage's Campaign Salary." *CBC News*, November 5. https://www.cbc.ca/ns
/beyondtheheadlines/2012/11/savages-campaign-salary.html.

Epstein, Howard. 2022. "What Happened to Owner-Occupied Houses in Halifax?" *SaltWire*, November
5. https://www.saltwire.com/atlantic-canada/opinion/howard-epstein-what-happened-to-owner
-occupied-houses-in-halifax-100790414/.

Erskine, Bruce. 2014. "Navy Program: 'It's All Systems Go'; Irving Shipbuilding President Insists Vaval Contract Firm. *The Chronicle Herald*, October 23. https://www.proquest.com/newspapers/navy -program-all-systems-go-irving-shipbuilding/docview/1774045456/se-2.

Federation of Canadian Municipalities. n.d. "Indigenous Partnerships." https://fcm.ca/en/focus-areas /Indigenous-partnerships.

Finbow, Robert. 2012. "Submerging the Urban: Halifax in a Multilevel Governance System." In *Sites of Governance: Multilevel Governance and Policy Making in Canada's Big Cities*, edited by Martin Horak and Robert Young, 73–103. Montreal and Kingston: McGill-Queen's University Press. https://doi .org/10.1515/9780773586918-005.

Foster, Kate. 2022. "'Where I Belong': Tracing the History of Dartmouth's 'The Avenue.'" Halifax Public Libraries (blog). https://www.halifaxpubliclibraries.ca/blogs/post/where-i-belong-tracing-the-history -of-dartmouths-the-avenue.

Galloway, Gloria. 1999. "Mayor of Halifax Apologizes for Atrocities Against Mi'kmaqs: Natives Threatened to Disrupt Celebration of 250th Anniversary." *National Post*, February 4, A1. https: //www-proquest-com.ezproxy.library.dal.ca/newspapers/mayor-halifax-apologizes-atrocities -against/docview/329359186/se-2.

Gertler, Meric, Richard Florida, Gary Gates, and Tara Vinodrai. 2002. *Competing on Creativity. Placing Ontario's Cities in North American Context*. Report prepared for the Ontario Ministry of Enterprise, Opportunity and Innovation and the Institute for Competitiveness and Prosperity. https: //policycommons.net/artifacts/636666/competing-on-creativity/1617983/.

Gosse, Meghan, Howard Ramos, Martha Radice, Jill L. Grant, and Paul Pritchard. 2016. "What Affects Perceptions of Neighbourhood Change?" *The Canadian Geographer* 60 (4): 530–40. https://doi.org /10.1111/cag.12324.

Government of Canada, Federal Organizations. 2023. "Organization Profile - Halifax Port Authority." https://federal-organizations.canada.ca/profil.php?OrgID=PHA&lang=en.

Graham, John F. 1975. "An Introduction to the Nova Scotia Royal Commission." *Canadian Public Policy* 1 (3): 350–4. https://doi.org/10.2307/3549382.

Grant, Jill L. 1994. *The Drama of Democracy: Contention and Dispute in Community Planning*. Toronto: University of Toronto Press. https://doi.org/10.3138/j.ctt2ttz9j.

Grant, Jill L., and Karin Kronstal. 2013. "Old Boys Down Home: Immigration and Social Integration in Halifax." *International Planning Studies* 18 (2): 204–20. https://doi.org/10.1080/13563475.2013 .774148.

Grant, Jill L. 2016. "The Social Dynamics of Economic Performance in Halifax." In *Growing Urban Economies*, edited by David A. Wolfe, Meric S. Gertler, 265–84. Toronto: University of Toronto Press. https://doi.org/10.3138/9781442629455-014.

Grant, Jill L., Pierre Filion, and Scott Low. 2019. "Path Dependencies Affecting Suburban Density, Mix, and Diversity in Halifax." *The Canadian Geographer/Le Géographe Canadien* 63 (2): 240–53. https://doi.org/10.1111/cag.12496.

Grant, Jill L., and Will Gregory. 2016. "Who Lives Downtown? Neighbourhood Change in Central Halifax, 1951–2011." *International Planning Studies* 21 (2): 176–90. https://doi.org/10.1080/1356347 5.2015.1115340.

Grant, Jill L., Robin Holme, and Aaron Pettman. 2008. "Global Theory and Local Practice in Planning in Halifax: The Seaport Redevelopment." *Planning, Practice & Research* 23 (4): 517–32. https://doi .org/10.1080/02697450802522848.

Grant, Jill L., and Howard Ramos. 2020. "Halifax: Scaling Inequality." In *Changing Neighbourhoods: Social and Spatial Polarization in Canadian Cities*, edited by Jill L Grant, Alan Walks, and Howard Ramos. Vancouver: UBC Press. https://doi.org/10.59962/9780774862042.

Grant, Jill L., Amanda Taylor, and Wheeler, C. 2017. "Planners' Perceptions of the Influence of Leadership on Coordinating Plans." *Environment and Planning C: Politics and Space* 36 (4): 669–88. https://doi.org/10.1177/2399654417720798.

Grant, Jill L. 2023 "10 Reasons Affordable Housing Is Hard to Deliver." *Halifax Examiner*, August 31. https://www.halifaxexaminer.ca/commentary/10-reasons-affordable-housing-is-hard-to-deliver/.

Goodman, Nicole, Jon H. Pammett, and Joan DeBardeleben. 2010. "Internet Voting: The Canadian Municipal Experience." *Canadian Parliamentary Review* 33 (3): 13–21.

Goodman, Nicole. 2010. "The Experiences of Canadian Municipalities with Internet Voting." *CEU Political Science Journal* 5 (4): 492–520.

Gregory, William. 2014. *Who Lives Downtown? Population and Demographic Change in Downtown Halifax, 1951–2011*. Planning project, School of Planning, Dalhousie University. https://canadacommons.ca/artifacts/1199901/who-lives-downtown/1753021/.

Halifax. n.d. *Executive Standing Committee Terms of Reference, Schedule 6*. https://www.halifax.ca/sites/default/files/documents/city-hall/legislation-by-laws/AO1.pdf#page=72.

Halifax. 2005. "HRM To Set a Vision for the Region's Economic Future." News release. *Economic Strategy Vision*. https://legacycontent.halifax.ca/mediaroom/pressrelease/pr2005/050622EconomicStrategyVision.php.

Halifax. 2017. *Centre Plan*. https://www.shapeyourcityhalifax.ca/1041/widgets/67845/documents/41248.

Halifax. 2018. *By-Law Number C-1100 Respecting Campaign Financing*. https://cdn.halifax.ca/sites/default/files/documents/city-hall/legislation-by-laws/By-lawC-1100.pdf.

Halifax. 2022. *Case 23820: Southdale Future Growth Node Planning Process. Shape Your City Halifax*. https://www.shapeyourcityhalifax.ca/southdale-planning.

Halifax Partnership. 2021. "Celebrating a Quarter Century of Partnership." https://halifaxpartnership.com/news/article/celebrating-a-quarter-century-of-partnership/.

Halifax Partnership. 2023. "Ocean Economy." https://halifaxpartnership.com/key-sectors/oceans/.

Halifax Partnership. 2022. "Halifax Index 2022: People." https://halifaxpartnership.com/research-strategy/halifax-index-2022/people/.

Halifax. 2015. "Statement of Reconciliation." http://legacycontent.halifax.ca/council/agendasc/documents/151208ca1442.pdf.

Halifax. 2017. "HRM Charter Review: Natural Person Powers. Item No. 14.3.1." Report submitted to Halifax Regional Council, August 1, 2017. https://www.halifax.ca/sites/default/files/documents/city-hall/regional-council/170801rc1431.pdf.

Halifax. 2018. "Halifax Green Network Plan." https://www.halifax.ca/about-halifax/regional-community-planning/regional-plan/halifax-green-network-plan.

Halifax. 2020. "The Task Force on the Commemoration of Edward Cornwallis and the Recognition and Commemoration of Indigenous History." Report submitted to Halifax Regional Council. https://www.halifax.ca/sites/default/files/documents/city-hall/regional-council/200721rc11110.pdf.

Halifax Partnership. 2022. *People, Planet, Prosperity*. https://halifaxpartnership.com/sites/default/uploads/Documents/People.Planet.Prosperity.Halifaxs-Inclusive-Economic-Strategy-2022-27-Full-Document.pdf.

Halifax Partnership. 2023. "Life Sciences & Health." https://halifaxpartnership.com/why-halifax/key
-sectors/life-sciences/.

Hayward, C. William. 1993. *Interim Report of the Municipal Reform Commissioner Halifax County (Halifax
Metropolitan Area)*. Halifax, NS: Department of Municipal Affairs.

Hoyle, Brian. 2000. "Confrontation, Consultation, Cooperation? Community Groups and Urban
Change in Canadian Port-City Waterfronts." *The Canadian Geographer* 44: 228–43. https://doi.org
/10.1111/j.1541-0064.2000.tb00706.x.

Jacob, Paul A. 2015. *A Generation of Change. Youth as Nova Scotia's Defining Moment.* Report prepared for
Halifax Partnership. https://halifaxpartnership.com/sites/default/uploads/Research-Strategy-Section
/A-Generation-of-Change-Youth-as-Nova-Scotias-Defining-Moment-Halifax-Partnership
-Report-October-2015.pdf.

Kaida, Lisa, Howard Ramos, Diana Singh, Paul Pritchard, and Rochelle Wijesingha. 2020. "Can
Rust Belt or Three Cities Explain the Sociospatial Changes in Atlantic Canadian Cities?" *City &
Community* 19 (1): 191–216. https://doi.org/10.1111/cico.12424.

Keuper, Eric. 1987. *The Northwest Arm Bridge: A Case Study of Locational Decision-Making*. MA Thesis,
Department of Geography, Saint Mary's University, Halifax.

Kitson, Alexandra, and Lisa Berglund. 2021. "Heritage and Trauma: Reimagining the Preservation
Planning Process for the Nova Scotia Home for Colored Children." *Canadian Journal of Urban
Research* 30 (2): 94–108.

Lalonde, Marc. 2022. "New Halifax Mi'kmaw Native Friendship Centre Building Gets Funding." The
Canadian Press, November 3.

Loo, Tina. 2019. "The View from Jacob Street: Reframing Urban Renewal in Post-War Halifax."
Acadiensis 48 (2): 5–42. https://doi.org/10.1353/aca.2019.0009.

Luck, Shaina. 2023. "Halifax ranks worst in country for adequate housing for racialized people:
StatsCan." *CBC News*, January 25. https://www.cbc.ca/news/canada/nova-scotia/halifax-leads
-minorities-core-housing-need-1.6724680.

Macdonald Notebook. 2024. "Exclusive: MacPolitics: An Insider's Account of Andy Fillmore's Winning
Campaign – The Real Worry About Anti-Trudeau Sentiment; Meet the Mayoral Candidates Who Took
Anti-Trudeau Votes." *The Macdonald Notebook*, October 21. https://www.themacdonaldnotebook
.ca/2024/10/21/macpolitics-an-insiders-account-of-andy-fillmores-campaign-the-real-worry-ab
out-anti-trudeau-sentiment-the-mayoral-candidates-who-took-anti-trudeau-votes/.

MacLean, Alexa. 2020. "Halifax Construction Leaves Some Intersections 'Impassable' For Wheelchairs
and Elderly, Advocate Says. *Global News*, September 16. https://globalnews.ca/news/7339119
/halifax-construction-disabled-advocate/.

MacInnis, Jonathan. 2022. "'People Rely on Those Services': Continued Halifax Transit Disruptions
Becoming a Concern for Users, Councillors." *CTV Atlantic*, December 6. https://atlantic.ctvnews
.ca/people-rely-on-those-services-continued-halifax-transit-disruptions-becoming-a-concern
-for-users-councillors-1.6182449.

Mavroyannis, Maria. 2002. "Is Bigger Better: Recent Experiences in Municipal Restructuring." *Canadian
Tax Journal* 50 (3): 1011–18. https://www.ctf.ca/common/Uploaded%20files/Documents/PDF
/2002ctj/2002ctj3_mavroyannis.pdf.

McDavid, James. 2002. "The Impacts of Amalgamation on Police Services in the Halifax Regional
Municipality." *Canadian Public Administration* 45 (4): 538–65. https://doi.org/10.1111/j.1754-7121
.2002.tb01858.x.

Milligan, Shelly. 2009. *2006 Aboriginal Population Profile for Halifax*. Statistics Canada. https:
//canadacommons.ca/artifacts/1214754/2006-aboriginal-population-profile-for-halifax
/1767854/.

Millward, Hugh. 2002. "Peri-Urban Residential Development in the Halifax Region 1960–2000: Magnets, Constraints, and Planning Policies." *Canadian Geographer* 46 (1): 33–47. https://doi.org/10.1111/j.1541-0064.2002.tb00729.x.

Murray, Karen Bridgett. 2007. "From Africville to Globalville: Race, Poverty, and Urban Governance in Halifax, Nova Scotia." In *Race, Neighborhoods, and the Misuse of Social Capital*, edited by James Jennings. New York: Palgrave Macmillan. 133–44.

Murray, Karen B., and Michael Caverhill. 2008. "The Patterning of Political Representation in Halifax." In *Electing a Diverse Canada: The Representation of Immigrants, Minorities, and Women*, edited by Caroline Andrew, John Biles, Myer Siemiatycki, and Erin Tolley, 150–204. Vancouver: UBC Press. https://doi.org/10.59962/9780774814874-011.

Nelson, Jennifer. 2008. *Razing Africville: A Geography of Racism*. Toronto: University of Toronto Press. https://doi.org/10.3138/9781442686274.

Nova Scotia. 1991. Report of the Nova Scotia Advisory Group on Race Relations.

Nova Scotia. 2014. *The Report of The Nova Scotia Commission on Building Our New Economy*. Halifax: Nova Scotia Commission on Building Our New Economy.

Nova Scotia. 2018. "New Name, Mandate for Waterfront Development." News release, May 11. https://novascotia.ca/news/release/?id=20180712002.

Nova Scotia. 2019. "More Youth Choose Nova Scotia." News release, May 11. https://novascotia.ca/news/release/?id=20190129002.

Nova Scotia. 2021. *Wortley Report Update: Summary of Department of Justice-Led Recommendations*. Department of Justice.

Nova Scotia. 2022. "Executive Panel on Housing in the Halifax Regional Municipality." https://novascotia.ca/housing-panel/.

Nova Scotia. 2022a. *HRM Housing Barrier Review*. https://novascotia.ca/housing-panel/docs/housing-development-barrier-review.pdf.

Nova Scotia Legislature. 2022. *Nova Scotia Legislature Assembly 64, Session 1*, Friday, October 28, 2022. https://nslegislature.ca/legislative-business/hansard-debates/assembly-64-session-1/house_22oct28.

Nova Scotia Legislature. 2023. "Halifax Regional Municipality Charter." https://nslegislature.ca/sites/default/files/legc/statutes/halifax%20regional%20municipality%20charter.pdf.

O'Kane, Josh. 2017. "The Casualties of a Downtown Development: The $500-Million Nova Centre Promises to Be a Centrepiece for Halifax. But Local Businesses in Its Shadow Claim the Disturbance from its Long-Running Construction Is Hurting Their Bottom Lines." *The Globe and Mail*, March 6. https://www.theglobeandmail.com/report-on-business/industry-news/property-report/little-guys-hurt-in-halifax-megaproject/article34224060/.

Outhit, Tim. 2010. Comments on the article "Appalling Democracy" by Tim Bousquet, August 12. https://www.thecoast.ca/halifax/timouthit/Profile?oid=1547362.

Pammett, Jon H., and Nicole Goodman. 2013. *Consultation and Evaluation Practices in the Implementation of Internet Voting in Canada and Europe*. Elections Canada. https://www.elections.ca/res/rec/tech/consult/pdf/consult_e.pdf.

Patil, Anjuli. 2020. "Halifax Regional Council Endorses 2 Rapid Transit Projects." *CBC News*, May 27. https://www.cbc.ca/news/canada/nova-scotia/halifax-regional-council-endorses-2-rapid-transit-projects-1.5585834.

Patil, Anjuli. 2023. "Halifax Among Fastest Growing Cities in Canada." *CBC News*, January 11. https://www.cbc.ca/news/canada/nova-scotia/halifax-among-fastest-growing-urban-regions-in-canada-1.6710481.

Poel. Dale. 2000. "Amalgamation Perspectives: Citizen Responses to Municipal Consolidation." *The Canadian Journal of Regional Science* 23 (1): 31–48.

Port of Halifax. 2022 "Deep Water Berth Extension at Port of Halifax Fully Operational." https://www
.portofhalifax.ca/deep-water-berth-extension-at-port-of-halifax-fully-operational/.

Public Services and Procurement Canada. 2022. "2021 Annual Report: *Canada's National Shipbuilding
Strategy*. https://www.canada.ca/en/public-services-procurement/services/acquisitions/defence
-marine/national-shipbuilding-strategy/reports/2021-annual.html.

Raddall, Thomas H. 1950. *Halifax, Warden of the North*. Toronto: J.M. Dent.

Rutland, Ted. 2018. *Displacing Blackness: Planning, Power, and Race in Twentieth-Century Halifax*. Toronto:
University of Toronto Press. https://doi.org/10.3138/9781487518233.

Ryan, Haley. 2022. "Proposed Halifax Boundary Changes Would Reunite Black Neighbourhoods."
CBC News, December 21. https://www.cbc.ca/news/canada/nova-scotia/proposed-halifax
-boundary-changes-reunite-black-communities-neighbourhoods-1.6692236.

Ryan, Haley. 2023a. "Halifax Council "Railroaded" into Selling Parkland for School, Councillor Says."
CBC News, February 8. https://www.cbc.ca/news/canada/nova-scotia/halifax-council-park
-west-school-parkland-sale-1.6741456.

Ryan, Haley. 2023b. "Halifax Agrees to Most Federal Tweaks for Housing Money – But Not Height."
CBC News, September 27. https://www.cbc.ca/news/canada/nova-scotia/halifax-will-agree
-to-most-federal-tweaks-for-housing-money-but-not-height-1.6979299.

Sancton Andrew. 2001. Canadian Cities and the New Regionalism. *Journal of Urban Affairs* 23 (5):
543–55. https://doi.org/10.1111/0735-2166.00105.

Sancton, Andrew. 2000. "Amalgamations in the 1990s." In *Merger Mania: The Assault on Local Government*,
83. Montreal: McGill-Queen's University Press. https://doi.org/10.1515/9780773568914-007.

Savage, Mike. 2015. *State of the City 2015*. https://www.halifax.ca/sites/default/files/documents/city
-hall/mayor/Chamber2015FinalforWeb.pdf.

Seguin, Nicola. 2022. "Cornwallis Street in Halifax to Be Renamed to Nora Bernard Street." *CBC
News*, December 12. https://www.cbc.ca/news/canada/nova-scotia/nora-bernard-halifax
-street-rename-1.6682681.

Sehatzadeh, Adrienne L. 2008. "A Retrospective on the Strengths of African Nova Scotian
Communities: Closing Ranks to Survive." *Journal of Black Studies* 38 (3): 407–12. https://doi.org
/10.1177/0021934707306574.

Sewell, John. 2021. "Toward City Charters in Canada." *Journal of Law and Social Policy* 34: 134–64.
https://doi.org/10.60082/0829-3929.1412.

Sivak, Ben. 2021. "Halifax Green Network Plan." In *Implementing Connectivity Conservation in Canada*,
edited by Lemieux, Christopher J., Aerin L. Jacob, and Paul A. Gray. Canadian Council on Ecological
Areas (CCEA) Occasional Paper No. 22. Canadian Council on Ecological Areas, Wilfrid Laurier
University, Waterloo, Ontario, Canada. 161–173. https://www.researchgate.net/profile/Karen
-Beazley/publication/350096167_Implementing_Connectivity_Conservation_in_Canada/links
/6050ac80458515e8344cf23a/Implementing-Connectivity-Conservation-in-Canada.pdf
#page=177.

Smith, Alison, and Zachary Spicer. 2018. "The Local Autonomy of Canada's Largest Cities." *Urban Affairs
Review* 54 (5): 931–61. https://doi.org/10.1177/1078087416684380.

Smith, Emma, and Haley Ryan. 2020. "1 Year After Wortley Report, Author Says Police Reform in
N.S. Has Long Way to Go." *CBC News*, June 9, 2020. https://www.cbc.ca/news/canada/nova
-scotia/scot-wortley-police-street-check-report-police-brutality-anti-black-racism-defund
-1.5604890.

Spencer, Gregory, Tara Vinodrai, Meric S. Gertler, and David A. Wolfe. 2010. "Do Clusters Make a Difference? Defining and Assessing their Economic Performance." *Regional Studies* 44 (6): 697–715. https://doi.org/10.1080/00343400903107736.

Spencer, Gregory M. 2009. *The Creative Advantage of Diverse City-Regions: Local Context and Social Networks*. PhD dissertation, University of Toronto. https://www.collectionscanada.gc.ca/obj /thesescanada/vol2/002/NR67745.PDF?is_thesis=1&oclc_number=796917569.

Statistics Canada. 2023. "Residential Real Estate Investors and Investment Properties In 2020." February 3. https://www150.statcan.gc.ca/n1/pub/46-28-0001/2023001/article/00001-eng.htm.

Taplin, Jen. 2022. "Nova Scotia Government Being Led Astray by Developers, Halifax Councillors Say." *SaltWire*, November 8. https://www.saltwire.com/atlantic-canada/news/nova-scotia-government -being-led-astray-by-developers-halifax-councillors-say-100792730/?utm_medium=Social&utm _source=Twitter.

Taplin, Jen. 2023. "It's All About Positioning for Growth: NEW YEAR Q&A: Halifax Mayor Mike Savage and New CAO Cathie O'Toole on Plans for Housing, Homelessness, Transit." *Chronicle-Herald*, January 11.

Taylor, Zack. 2019. *Shaping the Metropolis: Institutions and Urbanization in the United States and Canada*. Montreal: McGill-Queen's University Press. https://doi.org/10.1515/9780773558427.

Taylor, Zack. 2020. *Theme and Variations: Metropolitan Governance in Canada*. Toronto: Institute on Municipal Finance and Governance, Munk School of Global Affairs.

Taylor, Zack, and Alec Dobson. 2020. *Power and Purpose: Canadian Municipal Law in Transition*. Toronto: Institute on Municipal Finance and Governance. https://tspace.library.utoronto.ca/handle /1807/99400.

Tetreault. 2004. Halifax Economy Stronger through Diversity. *Trade & Commerce* 99 (3): B5.

Thomas, Ren, and Adriane Salah. 2022. "Supporting Non-Profit and Co-Operative Housing in Halifax, Nova Scotia." *Housing and Society* 49 (3): 271–300. https://doi.org/10.1080/08882746.2021.2005316.

Thomas, Jesse. 2022. "N.S. Premier Defends Appointing Business Friends as Crown Executive Chairs." *CTV Atlantic*, July 29. https://atlantic.ctvnews.ca/n-s-premier-defends-appointing-business -friends-as-crown-executive-chairs-1.6007967.

Transport Canada. 2022. "Government of Canada Invests in a New Port of Halifax Container Facility to Improve Trade." News release, March 7. https://www.canada.ca/en/transport-canada/news/2022/03 /government-of-canada-invests-in-a-new-port-of-halifax-container-facility-to-improve-trade.html.

Vojnovic, Igor. 1999. "The Fiscal Distribution of the Provincial-Municipal Service Exchange in Nova Scotia." *Canadian Public Administration* 42 (4): 512–41. https://doi.org/10.1111/j.1754-7121.1999 .tb02038.x.

Vojnovic, Igor. 2000. "The Transitional Impacts of Municipal Amalgamations." *Journal of Urban Affairs* 22 (4): 385–417. https://doi.org/10.1111/0735-2166.00063.

Ward, Rachel. 2015. "How Much Campaign Money Your Councillor Got from Developers." *CBC News*, April 21. https://www.cbc.ca/news/canada/nova-scotia/halifax-councillors-got-one-third-of -donations-from-development-community-1.3040934.

Weaver, John. 2016. *A Tale of Two Cities – Halifax and Victoria*. Amalgamation Yes. http://www .amalgamationyes.ca/uploads/2/7/4/7/27470281/_a_tale_of_two_cities_-_final_version.pdf.

Williams, Martyn. 2021. "Dangerous Sidewalk Closure On Young Street: Halifax Hostility to Sidewalk Users Must End." *Nova Scotia Advocate*, July 13. https://nsadvocate.org/2021/07/13/dangerous -sidewalk-closure-on-young-street-halifax-hostility-to-sidewalk-users-must-end/.

Withers, Paul. 2018. "Inside the Failed Fight for Federal Funding for Port of Halifax." *CBC News*, October 11. https://www.cbc.ca/news/canada/nova-scotia/inside-failed-fight-federal-funding-for-port-of-halifax-1.4857943.

Wolfe, David A. 2009. *21st Century Cities in Canada: The Geography of Innovation*. Ottawa: Conference Board of Canada. https://www.researchgate.net/publication/270685196_21st_Century_Cities_in_Canada_The_Geography_of_Innovation.

Woodford, Zane. 2018. "Halifax Council Gets CFL Update Behind Closed Doors." *The Star*, June 19. https://www.thestar.com/halifax/2018/06/19/halifax-council-gets-cfl-update-behind-closed-doors.html.

Woodford, Zane. 2019. "Halifax Mayor Mike Savage Says He Hasn't Decided Whether He'll Seek a Third Term." *The Star*, November 15. https://www.thestar.com/halifax/2019/11/15/halifax-mayor-mike-savage-says-he-hasnt-decided-whether-hell-seek-a-third-term.html.

Woodford, Zane. 2020. "Halifax Mayor Mike Savage Launches Re-Election Campaign, Releases Platform." *Halifax Examiner*, September 15. http://www.halifaxexaminer.ca/government/city-hall/halifax-mayor-mike-savage-launches-re-election-campaign-releases-platform/.

Woodford, Zane. 2021. "Halifax Council Candidates Blithely Broke the New Campaign Contribution Rules, and The Municipality Didn't Do Anything About It." *Halifax Examiner*, January 14. http://www.halifaxexaminer.ca/government/city-hall/halifax-council-candidates-blithely-broke-the-new-campaign-contribution-rules-and-the-municipality-didnt-do-anything-about-it/.

Woodford, Zane. 2022a. "Nova Scotia Housing Minister Approves Controversial Eisner Cove Development." *Halifax Examiner*, November 10. http://www.halifaxexaminer.ca/government/province-house/nova-scotia-housing-minister-approves-controversial-eisner-cove-development/.

Woodford, Zane. 2022b. "Councillors Take Halifax Police Budget Debate in Camera After Residents Call for Cuts." *Halifax Examiner*, February 23. http://www.halifaxexaminer.ca/government/city-hall/councillors-take-halifax-police-budget-debate-in-camera-after-residents-call-for-cuts/.

Woodford, Zane. 2022c. "Halifax Council Seeks Staff Report on Committee to Consider Recommendations to Defund Police." *Halifax Examiner*, November 23. http://www.halifaxexaminer.ca/policing/halifax-council-seeks-staff-report-on-committee-to-consider-recommendations-to-defund-police/.

Wortley, Scot. 2019. *Halifax, Nova Scotia: Street Checks Report*. https://humanrights.novascotia.ca/sites/default/files/editor-uploads/halifax_street_checks_report_march_2019_0.pdf.

Youthful Cities. 2022. *Real Affordability Index*. https://youthfulcities.com/urban-indexes/rai-2022/.

Zaccagna, Remo. 2015. "Butts Quits as Halifax CAO After 4 Years; Top Bureaucrat Makes Surprise Move, Heads to Clayton Developments as President." *Chronicle-Herald*, December 10.

5

Ottawa

Luc Turgeon

INTRODUCTION

In her overview of Ottawa municipal politics published in the first edition of *City Politics*, Caroline Andrew argued that the city's leading political divide was between political forces representing *consumption* and *production* orientations towards municipal government. By *consumption orientation*, Andrew (1983, 141) referred to the tendency to "look at local government primarily in terms of the services it will provide to residential neighbourhoods and to residents of the city as a whole." This perspective gives priority to consumption expenditures, be they on parks, social housing, recreational amenities, or public transit. By *production orientation*, Andrew (1983) referred to municipal government "as a means of support for economic development, and hence the maximization of profits." Such interests tend to push for lower taxes and deregulation, especially when it comes to planning, as well as high-profile infrastructure developments that might be of interest for the business community. Andrew argued that, while present in all large Canadian cities, the consumption–production divide took a unique form in Ottawa, as consumption forces were especially strong in the federal capital, a reflection of the city's large middle-class and highly educated population, itself a by-product of the city's sizeable federal public service workforce.[1]

In this chapter, I argue that the contemporary era of city politics in Ottawa has been characterized by the predominance of a new cleavage, the territorial cleavage, that divides urban, suburban, and rural interests. The importance of this cleavage can be seen in the coalitions that have developed at city hall, the voting record of councillors and of citizens, and the predominance of certain issues over others. Moreover, the difficulties in reconciling these territorial interests have frequently led to a relatively dysfunctional city council, which in turn has forced upper levels of government to intervene at times in city affairs during crises.

Figure 5.1: Municipal wards, City of Ottawa

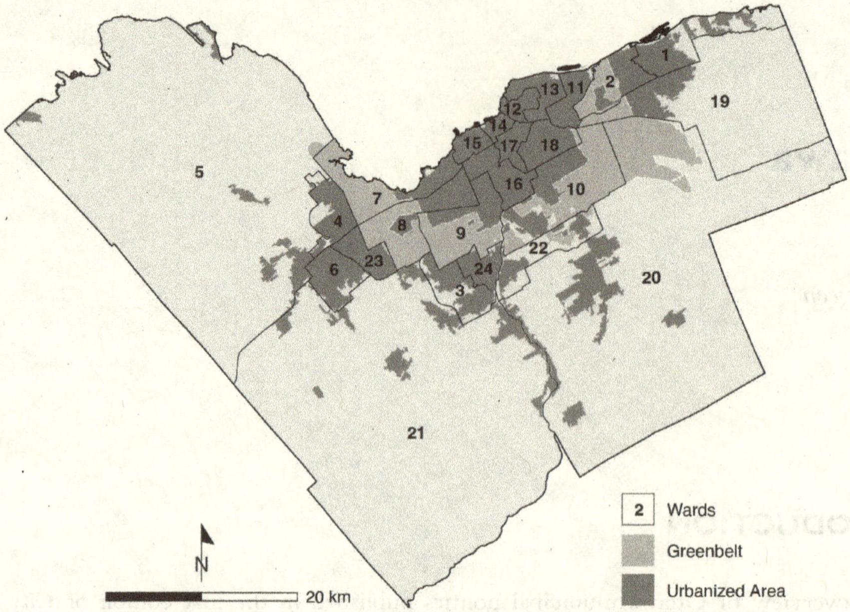

Notes: The City of Ottawa contains not only an urban core within its protected Greenbelt but also suburban and rural areas beyond it. Source: Wards – City of Ottawa; Contiguous Settlement Area – Statistics Canada; Greenbelt – National Capital Commission.

The predominance of this new cleavage is undoubtedly the result of the process of amalgamation that occurred in the 1990s and early 2000s. Post-amalgamation, the new City of Ottawa is one of Canada's largest municipalities, one that combines a relatively distinctive mix of rural, suburban, and urban interests.[2] To present a vivid portrait of such reality, the City of Ottawa (2,790.3 km^2) is larger than the cities of Montreal (431.5 km^2), Toronto (630.2 km^2), Calgary (825.3 km^2), Edmonton (684.4 km^2), and Vancouver (115.8 km^2) combined. As shown in Figure 5.1 and in Table 5.1, the city is divided in 12 urban, 9 suburban, and 3 very large rural wards. The dividing line between urban and suburban wards is primarily the extent to which they are the Greenbelt, a protected area of green spaces created in the 1950s and 1960s that divides the historical core of the city from suburban and rural wards. In reality though, several of the 12 urban wards have a built environment and a socio-demographic profile that often have more in common with suburban wards than with inner-city urban wards.

The predominance of the territorial cleavage does not mean that the production–consumption cleavage identified by Andrew has totally disappeared. Instead, the former has, to an extent, been superposed on the latter. To somewhat simplify, inner-city consumption forces have increasingly, since amalgamation, been at odds with the priorities of other parts of the city.

Table 5.1: List of Ottawa wards

1. Orléans East-Cumberland	Suburban	13. Rideau-Rockcliffe	Urban
2. Orléans West-Innes	Suburban	14. Somerset	Urban
3. Barrhaven West	Suburban	15. Kitchissipi	Urban
4. Kanata North	Suburban	16. River	Urban
5. West-Carleton March	Rural	17. Capital	Urban
6. Stittsville	Suburban	18. Alta Vista	Urban
7. Bay	Urban	19. Orléans South Navan	Suburban
8. College	Urban	20. Osgoode	Rural
9. Knoxdale-Merivale	Urban	21. Rideau-Jock	Rural
10. Gloucester-Southgate	Urban	22. Riverside South-Findlay Creek	Suburban
11. Beacon Hill – Cyrville	Urban	23. Kanata South	Suburban
12. Rideau-Vanier	Urban	24. Barrhaven East	Suburban

Moreover, political forces representing urban (and especially inner-city) interests have often been excluded from key spheres of decisions at city hall, especially during Jim Watson's decade in power, as the latter largely governed over time through a coalition dominated by suburban and rural councillors and became increasingly opposed by progressive councillors representing inner-city wards. As such, the territorial cleavage is in part superposed not only on a consumption–production cleavage but also a progressive–conservative (or left–right) one. Moreover, this transformation of the dominant political cleavage in the city has occurred in a context of growing inequalities, challenging the image of Ottawa as a "middle-class" city focusing solely on "middle-class issues." Indeed, issues such as racism, poverty, homelessness, police reform, social housing, and safe injection sites, to name but a few issues, have been the object of intense debates at city hall and in the city.

Other relatively distinctive aspects of Ottawa city politics that Caroline Andrew identified in the early 1980s are also still very much present today. These include the significant role played by upper levels of government in the development of the city. The federal government influences city politics in Ottawa first through its role as the region's most important employer, which has a significant impact on the socio-economic composition of the city. Through the National Capital Commission, which owns significant amount of land within the city border, it also plays a key role in planning and economic development. For Andrew, the key role of the federal government in Ottawa was also associated with another key cleavage, that is the French and English divide, as official bilingualism contributed, among other factors, to the maintenance of a significant francophone minority. This cleavage remains important, as francophones have mobilized to ensure greater representation in city institutions and for better services. However, they have not been the only minority groups mobilizing for greater recognition. Immigrant, racialized, and Indigenous residents of the city have also increasingly organized to ensure that their voices are heard and their rights respected.

Ottawa has thus changed significantly since Caroline Andrew took stock, in the early 1980s, of the city dividing political lines. In this chapter, I explore the evolution of Ottawa since the early 1980s, with a focus on the gradual emergence of the dominant territorial cleavage. The amalgamation process and its aftermath loom large in this transformation, as well as Jim Watson's twelve years in power. Like the other chapters in this volume, I first document key social, economic, and institutional changes in Ottawa over the past three decades. In the second section, I explore the practice of politics in Ottawa since the late 1980s, focusing on key debates and issues as well as emerging alliances across the city.

CHANGES IN A MIDDLE-CLASS CITY

Ottawa's image has long been one of a relatively sleepy, middle-class, and, with the exception of the French–English divide, relatively homogeneous city, hence perhaps the moniker "the city that fun forgot." Writing about the difficulties of recruiting francophone civil servants during the post-war period, James Mallory (1971, 175–6) argued that, among a range of factors, one should not neglect the "sacrifice imposed on a French Canadian of civilized taste condemned to live in Ottawa." However, the City of Ottawa has changed significantly since the 1980s. While readers can decide whether fun can now be associated with Ottawa, it remains true that the city has gone through important social and demographic changes over the past forty years.

Social Change

Ottawa is Canada's fourth-largest city, with a population of 1,017,449 residents, according to the 2021 census (Statistics Canada 2022a, 8). Its downtown population of 67,169 residents is the sixth largest in Canada (Statistics Canada 2022b, 16). The increase in the population of the downtown core in Ottawa (7.7 per cent) between 2016 and 2021 has been fairly small in comparison to cities like Toronto (16.1 per cent) and Montreal (24.2 per cent) (Statistics Canada 2022b, 26). In fact, the bulk of the population growth has occurred in the "intermediate suburban area" (21.4 per cent) and to a lesser extent the "distant suburbs" (10.1 per cent) (Ibid., 24).[3]

The image of Ottawa as a middle-class city is generally, from a comparative standpoint, still fairly accurate. Among the cities explored in this volume, Ottawa ($88,000) had the highest median household income in 2020 (Statistics Canada 2022c), followed by Calgary ($85,000). Ottawa is also the Canadian city explored in this volume with the highest proportion of the adult population aged 25–64 with a bachelor's degree or higher (49.5 per cent) (Statistics Canada 2022c). The high number of university graduates in the city is undoubtedly related to the importance of the federal government as an employer in the region.

The relative prosperity of Ottawa hides, however, significant variations across wards and neighbourhoods, as well as important pockets of poverty in the city, most of them concentrated in the downtown core. When it comes to median household income, the Ottawa Neighbourhood Study showed significant variation in 2015, from an after-tax income of $34,677 in Lowertown to $189,059 in Rockliffe Park. As for poverty, if the rate of low-income prevalence is rarely above 10 per cent, and even 5 per cent in suburban and rural neighbourhoods (the city average is 12.6 per cent), it is around 30 per cent in urban neighbourhoods such as Sandy Hill, Overbrook-McArthur, Riverview, and Vanier South, while just above 40 per cent in Ledbury-Herongate-Ridgemont (Ottawa Neighborhood Study 2023).

With Montreal, Ottawa is the other large city discussed in this volume with a significant linguistic divide between French and English. The proportion of citizens who have French as their mother tongue has decreased significantly since the early 1980s, from 19.3 per cent in 1981 to 15.7 per cent in 2021 (Statistics Canada 2022d). This decline reflects, first and foremost, the significant increase in people whose first language is neither French nor English, from 12.1 per cent in 1981 to 27 per cent in 2021. However, while a number of newcomers, especially from Africa, have contributed to revitalizing the institutions of the francophone community in the city, ultimately the vast majority of allophones in Ottawa have English has their first official language. In fact, the percentage of Ottawa residents who have only French has their first official language has declined from 19.5 per cent in 1981 to 14.1 per cent in 2021 (Statistics Canada 2022c). Here again, there are wide variations in the city, with a higher proportion of francophone in a few east end suburban wards (Orléans East-Cumberland, Orléans West-Innes) and urban wards in the downtown core (Rideau-Vanier and Rideau-Rockliffe) and a lower francophone presence in the western part of the city.

An important social change alluded to in the previous paragraph is related to the growing presence of immigrants in the city. In 1981, 18.3 per cent of the old City of Ottawa's population was foreign-born, compared to 25.9 per cent in 2021 (Statistics Canada 2022c). While the increase might seem relatively small, it reflects the fact that, post-amalgamation, the City of Ottawa incorporated rural and suburban wards with historically fewer immigrants. As such, these numbers underestimate the growing presence of newcomers in the city. The three most frequently reported countries of birth of foreign-born residents of Ottawa are China, India, and the United Kingdom (Statistics Canada 2022c).

Finally, the City of Ottawa has witnessed an important increase in the number of Indigenous peoples, from 1.1 per cent in 2001 to 2.6 per cent in 2021. The increase has been more important in the downtown core, especially in the neighbourhood of Vanier, which has a very large Inuit presence outside Inuit Nunangat and has become a hub for Indigenous organizations in the city. In fact, Indigenous peoples represented around 7 per cent of the population in Vanier-North in 2016 (Ottawa Neighborhood Study 2023). The growing number of both Indigenous peoples and Indigenous organizations in Ottawa, as well as, more broadly, the country's reconciliation agenda, has led the City of Ottawa to create in 2007 an Aboriginal Working

Committee and to adopt in 2018 a reconciliation action plan following the Truth and Reconciliation Commission Calls to Action.

Economic Change

The City of Ottawa's economy is highly dependent on the presence of governmental activities. In fact, since amalgamation, there has been an increase in the proportion of residents employed in "public administration," from 17.7 per cent in 2001 to 23.5 per cent in 2021 (Statistics Canada 2022c). This increase also reflects a sharp decrease in the manufacturing sector, from 8.2 per cent in 2001 to 3.1 per cent in 2021 (Statistics Canada 2022c). According to a 2015 study by the Conference Board of Canada, "With the diversity of Canada's economy as a reference point, Ottawa's economy has a diversity level of 0.37," by far the lowest among Canadian largest cities (Conference Board of Canada 2016). As such, the city's economy is highly dependent on governmental activities. At the same time, because public administration is a significantly less volatile sector of activity, the city is relatively immune from the economic downturns that often affect other cities more severely. Moreover, the weakness of its manufacturing sector means that its economy has not suffered as much from the supply-chain malfunctions that have negatively impacted the economy of other North American cities in recent years.

Nevertheless, attempts have been made to diversify the economic basis of the city over the past few decades. These efforts have tended to focus on the high-technology sector, leading in the 1990s to talks of Ottawa as "Silicon Valley North." As a result, in the late 1990s, employment in the high-tech sector was higher than employment in the public sector. However, the technology sector never fully recuperated from the problems of Nortel that emerged in the late 1990s and early 2000s, as well as from its bankruptcy in 2009. Nortel, a multinational telecommunication firm and data networking equipment manufacturer, whose head office was in Ottawa, employed in the early 2000s around 16,000 employees in the National Capital Region (including contractors). Most of these jobs were lost over the following decade, a significant blow to the city's economy at the time.[4] And while many former Nortel employees have gone on to start or to manage other technology firms, employment in the tech sector today is less than it was in 2008 (58,000 versus 44,000 in 2018 in Ottawa-Gatineau) (Bagnall 2019).

Institutional Change

Before 2001, the Ottawa region had a two-tier system of government that included eleven municipalities (see Figure 5.2). The largest city in the regional municipality of Ottawa-Carleton was the City of Ottawa, with 45 per cent of the region's overall population (Andrew 2006, 76). Created originally in 1969, the regional council saw its powers and function gradually increase over time. The expenditures of the regional council rose from around 20 per cent of

Figure 5.2: Municipalities in the regional municipality of Ottawa-Carleton prior to the 2001 amalgamation

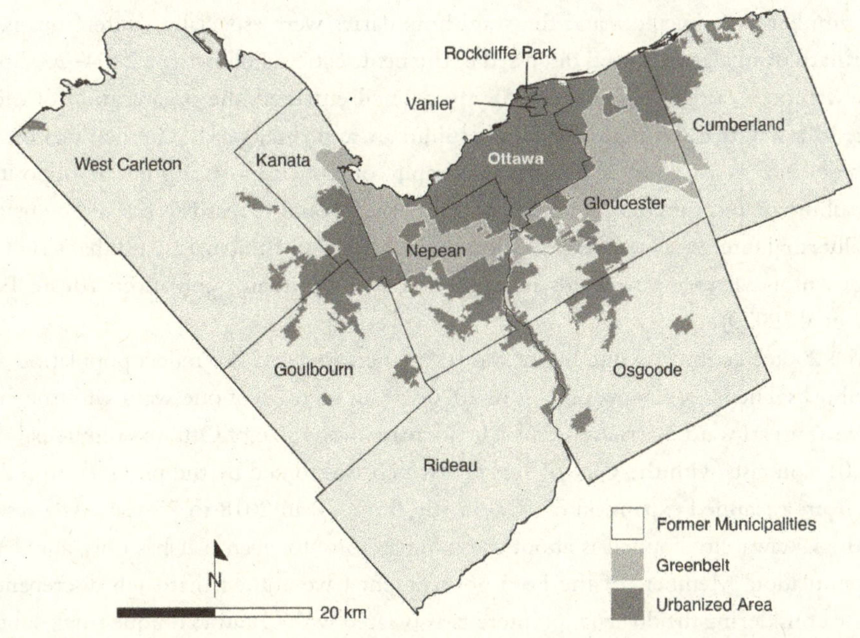

Source: 1996 Municipal Boundaries and Contiguous Settlement Area – Statistics Canada; Greenbelt – National Capital Commission.

total upper- and lower-tier expenditures at the time of its creation to over 70 per cent when it was disbanded (Graham et al. 2001, 253). Moreover, the regional council gained democratic legitimacy over time. Originally, regional councillors were selected by each municipality and were thus not directly elected. As for the regional chairs, they were selected by other members of the regional council. In the 1990s, that system was modified. The regional chair became an elected position in 1991, and regional councillors also started to be directly elected in 1994. However, over time, the existence of a two-tier system with increasing powers created frustration, especially among urban residents who viewed the growing regionalization as a form of urban subsidies for suburban development. Such frustration led to growing pressures to reform the regional authority, as explored in greater detail in the next section.

The eventual result was the adoption in 1999 by the provincial government of Mike Harris of the *Fewer Municipal Politicians Act*, which led to the creation of single-tier amalgamated cities in Hamilton, Ottawa, and Sudbury. These changes followed in the footsteps of the amalgamation of the former Municipality of Metropolitan Toronto in 1998. Contrary to Toronto and Montreal, post-amalgamation, community or borough councils were not created in Ottawa to deal with local matters. Ottawa emerged with a municipal council composed of 21 councillors and a mayor, down from the 84 municipal politicians that had been previously elected at the local and regional levels (Andrew 2006, 82).

The number of councillors and the ward boundaries were especially contentious issues in the aftermath of amalgamation, as discussed in the next section. Following a 2004–2005 boundary review process, a defining element of city council emerged: the organization of the ward structure in a way that distinguished urban, suburban, and rural wards. Council was organized along these lines as a way to reflect the geography of the city, a strong public opposition to mixing suburban and rural communities in a way that would jeopardize the representation of specifically rural interest, as well as a decision of the Ontario Municipal Board that struck down a previous proposal to create wards mixing rural and suburban populations (Beate Bowron Etcetera et al., 2020).

For the 2006 election, the number of wards was increased to 23 to reflect population growth in the suburbs. The latest review process in 2020 saw an increase of one ward with the creation of a new suburban ward, Barrhaven East. The increase in the size of Ottawa's municipal council since 2001 contrasts with the City of Toronto, which was forced by the province to reduce its council from a planned expansion of 47 seats (up from 44) in 2018 to 25 seats. As a result, the size of the Ottawa city council is about the same as Toronto, even if it has only about a third of the population. Members of the Ford government have argued that such discrepancy was legitimate considering the diversity of interests associated with Ottawa's unique rural–suburban–urban structure (Reevely 2018).

Until provincial legislative changes in 2022, the City of Ottawa had a "weak mayor" system, and as such mayoral priorities can only be adopted if the mayor obtains the support of a majority of councillors. One way for mayors to influence the working of city hall is indirectly through nominations to key city committees and boards and through the development of more or less formal coalitions. Before 2010, committee members used to choose their own chair. However, that practice was changed by Jim Watson, who started making final recommendations on the composition and chairs of committees (Chianello 2018). While such recommendations ultimately have to be approved by council, this change in the selection of committee chairs has nonetheless constituted a form of centralization of power over the past decade. The City of Ottawa currently has twelve committees and subcommittees, five advisory committees, and nine boards. The most powerful of these committees is the Finance and Economic Development Committee (FEDCO). It is chaired by the mayor and includes the Standing Committee Chairs, the Chair of the Transit Commission, the Deputy Mayors, and one member at large. The main responsibility of FEDCO, which is often presented as the mayor or the council's cabinet, is to set the city's financial framework and its strategic plan to guide the council's policy decisions.

As shown in the next section, the "weak mayor" system in place in Ontario has not always been a significant obstacle on a mayor's ability to see their priority adopted. Nevertheless, following a request by the mayor of Toronto, John Tory, to increase his ability to adopt decisions around housing, the province of Ontario adopted in November 2022 the *Better Municipal*

Governance Act, 2022. It grants the mayors of Toronto and Ottawa the ability to propose municipal bylaws related to provincial priorities and to see them adopted by just one-third of council. However, both during the electoral campaign and following his election, the new mayor of Ottawa, Mark Sutcliffe, announced that he did not intend to use these powers, to the dismay of the province's premier Doug Ford (Canadian Press 2022).

When discussing the institutional framework around municipal governance in Ottawa, one would be remiss not to mention two other key actors: the National Capital Commission (NCC) and the significant infrastructure of community associations, which are especially important considering the absence of boroughs or local councils. The NCC, which is charged with the development of the National Capital Region (NCR), was created in 1959. It replaced the Federal District Commission, which had been in place since the late 1920s. The mandate of the NCC, as stipulated in the National Capital Act, is to "prepare plans and to assist in the development, conservation and improvement of the National Capital Region in order that the nature and character of the seat of the government of Canada may be in accordance with its national significance" (*National Capital Act,* RSC, 1985, c. N-4). The Board of Directors of the Commission is composed of 15 members: a chair and a chief executive officer; three residents of the NCR from local municipalities of Ontario (two of which from Ottawa); two residents of the NCR from local municipalities in Quebec (one of which is from the section of the city of Gatineau that is west of the Gatineau River); and finally, eight from everywhere in Canada outside the NCR. Since 2016, the mayors of Ottawa and Gatineau are non-voting ex officio participants. The key role of the NCC in the development of Ottawa is a by-product of the fact that it is the largest property owner in the NCR. In fact, it owns 11 per cent of all lands in the capital region, in addition to 1,600 buildings (National Capital Commission 2023). It has been the object of significant resentment and criticisms in Ottawa, either because of its perceived lack of vision or its failure at times to actively cooperate with the City of Ottawa on development projects.

As for community associations, there are over 150 in Ottawa, representing the interests of residents of different neighbourhoods or communities. An organization, the Federation of Citizen's Associations of Ottawa, constitutes a forum that allows neighbourhood associations to share information and coordinate actions. Citizen associations are often called upon to present the perspective of residents of a neighbourhood on important files, especially with regard to planning, to organize candidate meetings and debates during elections, and, in some cases, to provide programs and amenities. Their role in planning is in part the by-product of the city's Official Plan, which stipulates that public consultation must occur when a development requires "proposed Official Plan and Zoning By-law amendments and community improvement plans" (Ottawa 2021, 242). They also play a role in the development of local plans geared towards specific areas of the city. Consultations also occur in the development of different strategies of the city, from the fight against discrimination and racism to the revitalization of certain neighbourhoods.

In the past, the City of Ottawa was criticized by experts for its lack of consistency in its public engagement, for the limited availability to the public of key information and documents, and for a lack of clarity about how public engagement impact the city decisions (Krawchenko and Stoney 2011). In 2015, the city officially released a new engagement strategy, which aimed to provide city staff with a clear approach to public consultation across the different departments of the city. In the aftermath of the adoption of this new strategy, the city launched *Engage Ottawa*, a platform that allows citizens to directly submit feedback on projects and initiatives through comments, surveys, and polls, as well as having access to relevant documents. Finally, citizens are also called upon to participate in city politics by putting their name forward to act as representatives of the public on a range of committees and boards, from the Ottawa police service board to the Environmental Stewardship Advisory Committee.

THE PRACTICE OF POLITICS

Perhaps not surprisingly for a capital city, Ottawa never has a shortage of candidates running for office at the municipal level. Since amalgamation in 2001, the number of candidates for the position of mayor has been 8 in 2003, 7 in 2007, 20 in 2010, 8 in 2014, 12 in 2018, and 14 in 2022. The high number of candidates also reflects the fact that it is relatively easy to run for mayor in Ontario. All that is needed is to pay a refundable fee of $200 and the submission of 25 original signatures endorsing one's nomination. As for political participation, it has tended to be higher in Ottawa than in other Ontario municipalities, even if it undoubtedly remains low. In the 2022 municipal election, for example, turnout was 44 per cent in Ottawa compared to 35 per cent in Hamilton and just under 30 per cent in Toronto. The city also does not lack in important political debates. Over the past thirty years, numerous issues have punctually led to a significant mobilization of citizens, from the question of amalgamation to debates about the linguistic status of the city and numerous controversial development projects.

In this section, I explore the practice of politics in Ottawa from the late 1980s to the early 2020s, focusing on not only the main political cleavages during that period but also the key debates and controversies. I focus on three distinct periods: from the late 1980s to amalgamation, the tumultuous decade that followed amalgamation, and the long period in power of Jim Watson.

The Road to Amalgamation

Ottawa politics in the late 1980s and 1990s had much in common with the middle-class politics of the 1970s and 1980s described by Caroline Andrew in the previous edition of City Politics, in which the main cleavage was between proponents of consumption and production orientations to local politics. Local debates focused especially on questions of tax and spending.

The 1991 Ottawa election was a three-way fight between a conservative councillor Jacquelin Holzman, progressive councillor Nancy Smith, and the interim mayor Marc Laviolette, who had replaced mayor Jim Durell following his resignation. Durell publicly supported Holzman, who ran on a campaign to freeze taxes and promote economic development. She was re-elected in 1994 but chose not to run again in 1997, paving the way for the election of a young Jim Watson, who had developed a reputation as a "fiscal conservative who has favored privatizing city operations and reducing government regulations and procedures that hinder businesses" (Hume 1996, C3).

The most significant debates during that decade focused on the future of the regional level of governance. The regional municipality of Ottawa-Carleton, created in 1969, originally brought together 16 lower-tier municipalities, which were reduced to 11 in 1973. Over time, a number of commission reports were issued to address the division of powers between the local and regional level, which contributed to a significant increase in the responsibilities, powers, and spending of the upper-tier regional government. However, concomitant to the rise of regional power was a growing dissatisfaction with the two-tier governance structure. First, as mentioned previously, perceptions of a growing democratic deficit led to the direct election of the regional chair in 1991 and the regional councillors in 1994. Second, considering that the City of Ottawa constituted close to half of the population of the regional municipality, the growing responsibilities granted to the regional level led to accusations that those living in the urban core were effectively subsidizing the development of suburban infrastructure and services (Graham et al. 2001, 253). Finally, there were growing calls to reduce duplications, especially among members of the business community, who denounced not just the costs and delay associated with planning but also a range of municipal services in a region with two levels of government (Graham et al. 2001, 259).

As a result, in the late 1980s, a growing number of actors started voicing their support for a significant reform of the regional governance structure. Among some proposals was the creation of a single-tier government for the region, which tended to be more popular among the local business elite, the Ottawa city council, and a majority of regional councillors (Graham et al. 2001). Moreover, the province increasingly pressured the region to reform its structure, and it appointed Gardner Church as a facilitator to find common ground among the different local councils (Dilkens 2014, 247). Church ultimately failed to secure a consensus, leading the province to warn that it was ready to appoint a commissioner to determine the best option for reform (Dilkens 2014).

The creation of a single-tier municipal government was especially opposed by suburban and rural councils for two main reasons. First, it was argued that amalgamation would contribute to the disappearance of unique local identities and traditions, which would be subsumed in an amalgamated City of Ottawa. Second, amalgamation was viewed as potentially detrimental to the interest of suburban and rural municipalities. More specifically, even if Ottawa was governed by fiscally conservative mayors from the mid-1980s on, its spending, taxation, and debt levels were often higher than in other municipalities that were part of the regional municipality of Ottawa-Carleton. As such, amalgamation was viewed as potentially contributing to higher levels of taxation as well as levels of services less in tune with the interest of their constituents (Graham et al. 2001, 259).

Those different perspectives clashed especially during consultations organized by a Citizen's Panel on Local Governance in Ottawa, which had been created in August 1997 to recommend a new governance structure for the region (Andrew 2006, 77). Several municipalities mobilized their citizens behind the status quo or other preferred options, leading at times to raucous meetings (Dilkens 2014, 247–8). The panel was disbanded in March 1998 before it finished its work. One of the panel's co-chairs, Grete Hale, stated that significant political interference had caused the "well [to be] poisoned," thus making it "impossible to continue our work" (Mohammad & Boswell, 1998, p. B1, cited in Dilkens 2014, 248).

The failure of the Citizen's Panel made it almost inevitable that the province would step in to impose a solution. And indeed, in September 1999, the Conservative government appointed four special advisers to recommend governance reform in Ottawa-Carleton, Hamilton-Wentworth, Sudbury, and Haldimand-Norfolk. The special adviser for Ottawa was Glen Shortliffe, a former clerk of the Privy Council who was viewed as closely aligned philosophically to the Conservative government. In any case, the provincial guidelines to the special advisers seemed to preclude the status quo and the maintenance of a two-tier form of regional governance, as it stated that reform should promote fewer municipal politicians, lower taxes, more efficiency in the delivery of services, less bureaucracy, and clearer lines of responsibility and accountability (Sancton 2000, 145).

Despite protests by suburban and rural councils, Mr. Shortliffe ultimately proposed the amalgamation of the different municipalities that had constituted the Regional Municipality of Ottawa-Carleton. He rejected, as such, the proposal of a "Rural Alliance" that had been formed to lobby in favour of the separation of rural municipalities from the region and the creation of a common "Service Management Board" (Sancton 2000, 148). In particular, Mr. Shortliffe believed that the rural municipalities did not possess the tax base necessary to maintain services without subsidies by the rest of the city. Among the many other proposals of the special adviser for Ottawa were the adoption of a form of "area rating," whereby different tax rates would apply to different parts of the new city depending on services available in that region. It would ultimately lead to the creation of two distinct tax rates in Ottawa: a (lower) tax rate for rural residents and another one for residents of the rest of the city. However, the adviser rejected another proposal, which was that tax rates be adjusted to account for the assets and liabilities of the former municipalities, an issue that was sensitive in light of the debt level of the former City of Ottawa and the significant capital reserves of Nepean (Sancton 2000, 147). He also rejected the creation of local boroughs while proposing that the city be recognized as officially bilingual.

In December 1999, following the adviser's report, the province adopted *the Fewer Municipal Politicians Act*. It dissolved 12 previously existing municipalities: the Regional Municipality of

Ottawa-Carleton, the City of Cumberland, the City of Gloucester, the Township of Goulbourn, the City of Kanata, the City of Nepean, the Township of Osgoode, the City of Ottawa, the Township of Rideau, the Village of Rockliffe Park, the City of Vanier, and the Township of West Carleton. It also stated that, on 1 January 2001, "the inhabitants of the municipal area are constituted as a body corporate under the name 'City of Ottawa' in English and 'ville d'Ottawa' in French."

Troubles in Post-Amalgamation Ottawa

Elections for the new City of Ottawa were organized in the fall of 2000. The outgoing mayor of the Regional Municipality of Ottawa-Carleton and proponent of amalgamation, Bob Chiarelli, was elected mayor of the new city. In the immediate decade following amalgamation, four issues were especially controversial and, in many cases, showed the continued importance of the territorial cleavage: the representation of rural interests at council, the linguistic status of the city, the city transit plan, and the city budget. Moreover, the period saw clear meddling by the province and even the federal government in the affairs of the city.

The question of representation of different territorial interests was a key issue that the new city council had to deal with. Post-amalgamation, rural interests were significantly over-represented on council. Of the original 21-member council, 5 councillors represented predominantly rural wards, and they tended to represent significantly fewer constituents than urban and especially suburban councillors. These five wards had 13 per cent of registered voters, but elected 23 per cent of the city's 21 councillors in the 2003 election that saw the re-election as mayor of Bob Chiarelli (Ottawa Citizen 2004, F6). As argued by Andrew Sancton (2000, 156), such initial discrepancy constituted a significant violation of the "representation-by-population" principle.

That five predominantly rural wards were still in place during the 2003 elections was very controversial and the product of a significant political fight in the first few years of the new city. Shortly after the creation of the new amalgamated city, a task force was created to deal with significant disparities in the population of the different wards. To promote a more equitable electoral map and adjust to the faster growing suburban population, the task force proposed to redraw the ward boundaries. Their proposal would have effectively reduced the number of predominantly rural wards from five to four. With the incorporation of suburban areas, the rural character to these wards would also have been diluted.

The proposal of the task force was ultimately adopted by city council. This decision created a significant backlash from rural residents and organizations, who challenged the decision to the Ontario Municipal Board (OMB). However, as the OMB was about to start its audience in the fall of 2002, the Ontario Municipal Affairs Minister, Chris Hodgson, chose to use some of his available executive powers to delay the OMB proceedings, effectively preventing the city from redrawing its ward boundaries in time for the 2003 election (Singer 2002, E1).[5] The result was that the 2003 election, as mentioned before, did not see any change to the number and boundaries of wards. Nevertheless, after a few months of delay, the OMB proceeded with audiences and ultimately ruled against the proposed boundary change. In a controversial and contested decision, the board argued that the task force that had adopted the new ward boundaries had put "too much emphasis on the principle of representation by population over the principle of effective representation" (cited in Ottawa Citizen 2003, B4). The board ultimately faulted the city for not exploring the possibility of adding two new wards to deal with population growth, a strange statement, as noted in an editorial of the Ottawa Citizen published in the wake of the decision, because the purpose of the 1999 *Fewer Municipal Politician Act* that created the amalgamated City of Ottawa was to reduce the number of politicians (Ottawa Citizen 2003). Ultimately, the city adopted a new ward structure that saw the addition of 2 new councillors and the creation of three suburban wards, as well as the elimination of one rural ward. The decision was again appealed at the OMB, but this time it ruled in favour of the city.

The official linguistic status of the city was another contentious subject of debate during the immediate period after amalgamation. The Ontario government of Mike Harris had rejected the special adviser's plea to declare the City of Ottawa officially bilingual. It was therefore up to the new city council to adopt a specific language policy. At first sight, besides the reluctance of the province to enshrine in law the bilingualism status of the city, several other factors made the adoption of a policy of "official bilingualism" difficult. First, of the previous eleven municipalities that had been amalgamated, only four municipalities previously had a bilingualism policy in place: Ottawa, Vanier, Gloucester, and Cumberland (Charbonneau and Coeytaux 2013, 140). Second, none of the candidates for mayor campaigned on a platform to officially recognize the bilingual status of the city (Andrew 2006b, 86).

Nevertheless, a linguistic policy had to be adopted to clarify the rights of the francophone minority, and a public debate was organized at city hall in May 2001. The acrimonious debate pitted groups, mostly representing the francophone minority, that argued that the adoption of an "officially bilingual" status was necessary both as a matter of right and of national unity, and those, mostly anglophone, who rejected the necessity of such policy, while lamenting its potential negative impact on the work prospects of unilingual Anglophones and the costs associated with such policy. Opposition to the proposed policy of bilingualism was especially strong in rural wards and western part of the cities, which had low francophone populations (Charbonneau and Coeytaux 2013, 140). Ultimately, city council decided first to maintain the bilingualism policy of the previous City of Ottawa and second, to ask the province to officially recognize the official bilingualism status of the city, a request that would again be rejected by the province.

Another key controversy in the aftermath of amalgamation was with regard to a light-rail project. It came in the wake of a pilot project, the "O-Train," an eight-kilometre, five-station system that used existing track and diesel-powered three-car unit trains along a north–south axis (Hilton and Stoney 2007, 6). Three years after the beginning of the pilot project, it was declared a success and the mayor secured in 2004 $200 million each by the provincial and the federal government. A memorandum of understanding with both government partners was signed in 2005 to finance a 27-km, double-track electrified light-rail transit (LRT) system running again predominantly along a north–south axis, which would require the shutting down of the existing O-Train for a three-year period (Drake 2007). In July 2006, city council approved, by 14 to 7, a $725 million bid by a consortium led by Siemens to build the proposed project (Drake 2007).

Nevertheless, in the following months, the project would slowly unravel. The contract was signed with Siemens in September 2006, a few months before the municipal election. However, the contract was not made public, leading to accusations of secrecy and concern about its real cost. As such, it quickly became one of the main objects of debates during the municipal election. That election pitted three candidates. The first candidate was Bob Chiarelli, the outgoing mayor of the city. The second was Alex Munter, a suburban councillor who was, nevertheless, associated with the progressive wing of council and who, while supporting the LRT, believed that a plan for an east–west line should be developed sooner than later. The third candidate was Larry O'Brien, a conservative businessman who had decided to run at the last minute and proposed to postpone the actual project for six months. O'Brien also asked the Federal Treasury Board to conduct an audit of the contract. The president of the Treasury Board, John Baird, was an Ottawa MP, one with a history of infighting with the outgoing mayor Bob Chiarelli dating to their days in provincial politics. Ultimately, in the middle of the municipal campaign, Baird announced that he was withholding federal money until after the election and that he would make the money available only after the new council would vote in favour of the project. Baird's intervention was denounced by many commentators as political and partisan muddling in municipal affairs (Benali and Bernier 2014, 11).

While Munter was leading in the polls for most of the campaign, O'Brien was able to ultimately prevail, earning 47 per cent of the votes against 36 per cent for Alex Munter and 16 per cent for the outgoing mayor Bob Chiarelli. There was a clear territorial dimension in support for Munter and O'Brien. While Munter won in the two suburban wards of Kanata, where he was a counsellor, most of his support was in the downtown core. As for O'Brien, he managed to win a plurality of vote in almost all of the suburban wards as well as in all of the rural wards. Following the election of O'Brien, council had a tight deadline to adopt the contract with Siemens, which came with a potential penalty if delayed or cancelled. A first vote led to the resurrection of a modified project, which would have seen the abandonment of the downtown portion of the project, by a vote off 12 to 11 (Drake 2007). The decision to abandon the downtown part of the project was denounced by both the provincial and federal governments, who argued that such a decision was jeopardizing their commitment to fund the project. Ultimately, the federal government asked for more time to study the project, and the city was put in a position to either move forward without federal funding or pay a significant fee. Council then decided to cancel the project by a vote of 13 to 11 (Drake 2007). Only in 2008 would the project be relaunched as the mayor and council would finally agree the launch of a first east–west line that would go through a downtown tunnel. However, doubts about the financial risk associated with the project and with the financial capacity of the city contributed to the delay of the project until 2011 (see below).

Besides the LRT, the mandate of Larry O'Brien was plagued by two other significant controversial issues. The first was the fact that the mayor was charged in a bribery probe, which forced him to temporarily step down for a two-month period in 2009 while he was on trial. Ultimately, O'Brien was acquitted. The second issue was the question of property taxes, which also had been a significant problem for previous Ottawa mayors. Indeed, while Chiarelli and council had managed to implement tax freezes during the mayor's first mandate (Adam 2003, D1), taxes were ultimately increased by a significant amount during his second mandate (Denley 2006). It is in that context that Larry O'Brien had promised, during the electoral campaign of 2006, to freeze property taxes during his mandate. However, the mayor ultimately broke his promise, as freezing taxes in the following years would have required significant cuts to services, which council ultimately rejected. Indeed, O'Brien's mandate was marked by significant conflict with council (Rupert 2008, B1)

The Reign of Jim Watson and Its Aftermath

Despite the scandals and infighting at council that rocked his mandate, Larry O'Brien decided to run again in 2010. He faced two main opponents. The first was Jim Watson, who had been mayor of the pre-amalgamated City of Ottawa and, later, a member of the provincial Liberal government. The second was Clive Doucet, a long-time progressive urban councillor. Watson was able to present himself as the centrist and competent candidate that could bring back order to city council. He received 48.7 per cent of the votes, easily defeating O'Brien, who obtained 24.1 per cent of the vote, and Doucet, who obtained 14.9 per cent. Watson was able to win a plurality of votes in all the wards except the three rural ones that supported O'Brien.

The election of Watson and the arrival of 9 new councillors, a relatively high turnover in Ottawa municipal elections, profoundly changed the dynamic of council. After the relative dysfunction of council under O'Brien, Watson brought back some semblance of stability and a certain measure of mayoral control. The Manning Centre council tracker for the period between 2010 and 2014 found that 96 per cent of votes and recommendations carried (Farkas 2015, 28). The city also spent significantly less time deliberating issues (180 hours) than was the case in Hamilton (227 hours) and Toronto (883 hours) (Peat 2015). Moreover, looking at divided votes (those that were not unanimous), Mayor Watson enjoyed a significant win rate (94.1 per cent) compared to the mayors of Hamilton (81.3 per cent) and Toronto (42.9 per cent).[6]

Mayor Watson was also able to build consensus on several projects. In fact, during his first mandate, a range of ventures that had been stalling over the previous decade were finally given the green light. The first and perhaps more important was the LRT project. The city announced in early December 2012 that it had accepted, subject to council approval, a $1.8 billion bid from the Rideau Transit Group to build a new east–west LRT system that would be named the

Confederation Line. Mayor Watson proudly declared: "We'll be doing it on time, on budget and with a fixed price" (CTV 2012). A few days after that announcement, council unanimously approved the city's development plan. The line was to be completed in 2018. The second issue was the final launch of the Lansdowne Park redevelopment in 2012, which had also been a contentious issue during Mayor O'Brien term in office. Finally, the redevelopment of Arts Court and the building of a new Ottawa Art Gallery were approved in 2013.

The relative stability at city hall, coupled with Watson's apparent ability to combine a certain degree of fiscal prudence with the development of new initiatives, contributed to his re-election in 2014 with 76.2 per cent of the vote. In fact, Watson won a plurality of votes in all the wards. The 2014 election also brought in 3 new downtown councillors with progressive views, Catherine McKenney (Somerset), Jeff Leiper (Kitchissipi), and Toby Nussbaum (Rideau-Rocklife). As a result, a divide gradually seemed to appear during Watson's second mandate between, on the one hand, more progressive councillors representing wards located in the downtown core and the mayor and the rest of council on the other.

One file where this divide was evident was over a controversial plan of the Salvation Army to close its shelter in the touristic Byward Market and to open a new one on Montreal Road, the main thoroughfare of the Vanier neighbourhood, one of the city's poorest neighbourhoods.[7] The proposal led to the creation of a grassroot organization, SOS Vanier, dedicated to oppose the plan and to propose alternatives. While dismissed at first as a simple case of NIMBYism, vocal opposition to the project came from academics and groups such as the Canadian Alliance to End Homelessness, who viewed the building of a large shelter as an outdated model and warned about the detrimental effects of concentrating social services in a few low-income neighbourhoods. Ultimately, the project, which was defended by the mayor, was adopted following a vote of 16 to 7 that saw rural and suburban councillors all vote in favour of the project while the more progressive urban councillors voting against it (Leblanc 2017).

Controversies also occurred on other issues, including the linguistic status of the city (see below), the opening of safe injection sites, a plan to have a downtown casino, problems with the police force, and the approval of controversial residential and commercial projects that were at odds with community plans. Nevertheless, these controversies did not seem to affect the mayor's popularity. In fact, in early 2017, a poll showed that Watson's approval rating was 79 per cent, the highest percentage among the mayors of the country's ten largest cities (Crawford 2017).

How does one explain the tremendous popularity of Watson? First, Watson undoubtedly engaged in "permanent campaigning." A master at retail politics, he attended an extraordinary number of events each week, from school cookie sales to the opening of small businesses. Second, Watson tended to defuse controversies by governing in a way that was conducive to blame avoidance or credit claiming. Watson often refused to take the lead or to make decisions regarding controversial issues. For example, with regard to the linguistic status of the city, Watson refused, following an important mobilization of the francophone community in 2012, to work towards the recognition of Ottawa as officially bilingual. He stated that, in practice, such recognition already existed because francophones were entitled to services in French since the adoption in 2001 of the city's bilingualism policy. Ultimately, partly as a result of the city's inaction, the provincial government stepped in and adopted in 2017 an amendment to the City of Ottawa Act to recognize the bilingual status of the city. Another example of blame avoidance was with regard to the question of safe injection sites. Watson had opposed such sites in the past, which required city approval. Ultimately, facing pressures to act in light of the ongoing opioid crisis but clearly worried of the potential political repercussions, Watson decided to let the final decision to the city's public health board, which ultimately approved the opening of such sites.

The strategies of permanent campaigning and blame avoidance of the mayor were largely successful, and he was easily re-elected in 2018. That election saw Watson won 71 per cent of the votes, almost 50 points above the second-place contestant, the former councillor Clive Doucet who threw, again, his hat in the ring at the last minute. Watson seemed at the time undefeatable. Following that election, Watson excluded councillors from the downtown core from all important committee positions. None of them became chair of a committee, and more importantly, none were appointed to the important FEDCO, while all three rural councillors were. The exclusion was denounced by urban councillors who decried such lack of representation at key positions (Willing 2018).

This territorial divide had undoubtedly been present before 2018 on council, as shown before, but it became even more visible and acknowledged after that election and the exclusion of downtown councillors from key positions. In 2019, councillor Mathieu Fleury, from the downtown ward of Rideau-Vanier, upset with the council decision over the redevelopment of the Château Laurier, denounced on Twitter the "Watson club," an expression that would follow Watson and his allies throughout his third term (Fleury 2019). It referred to a group of councillors that tended to support the mayor's priorities, voted collectively on important and potentially controversial decisions, and often used procedural manoeuvres to limit debates. That "club" was predominantly composed of rural and suburban councillors and constituted a majority of councillors.[8] They tended to be opposed by a group of mostly progressive urban councillors.

A growing pushback by a minority of councillors against the stifling of debates at council is one of the factors that would contribute to a difficult last mandate for Mayor Watson. The most damaging file for the mayor though was the LRT file. Troubles with the LRT had plagued the project since 2012, well in advance of its planned launch in 2018.[9] A massive sinkhole in 2016 led to important delays and to the closing of Rideau Street, a main thoroughfare of the city, for six months. The LRT did not open in 2018, even though it was supposed to, and ultimately, the consortium missed a number of deadlines to hand over the system to the city. It did not prevent council to vote in favour of an LRT extension, referred to as Stage 2, in March 2019. Ultimately, the Confederation Line opened in September 2019, 456 days later than originally planned. It was quickly plagued by problems, including long delays, trains that suddenly stopped working, numerous conflicts with the consortium responsible for the construction of the LRT, and a 54-day shutdown of the entire system following a train derailment. Problems with the LRT eventually led councillor Catherine McKenney to bring to council a motion in the fall of 2021 to initiate a judicial inquiry on problems associated with the LRT. The mayor and some of his closest allies contributed to the motion's defeat. They decided instead to have the city's auditor general examine the contract with the consortium (Willing 2021: A1). As a result, the auditor general would not have been able to hold public hearings nor investigate the action of the mayor and members of councils. However, before the auditor could start her work, the provincial government of Doug Ford stepped in and announced the creation of a public inquiry headed by Justice William Hourigan.

After public hearings in the summer of 2022, a final report was published a few weeks after the 2022 municipal election. It was nothing short of a scathing indictment of the leadership of Jim Watson and the inappropriate closeness that had developed between Watson and the senior leadership of the city bureaucracy during his mandate. Judge Hourigan found that the mayor and the city manager, Steve Kanellakos, had deliberately misled council by failing to report to them that testing criteria had been lowered to ensure the launch of the LRT, and as such misrepresented to council the test results and potential problems with the launch of the LRT. Judge Hourigan argued that:

> This conduct prevented councillors from fulfilling their statutory duties to the people of Ottawa. Moreover, it is part of a concerning approach taken by senior City officials to control the narrative by the nondisclosure of vital information or outright misrepresentation. Worse, because the conduct was willful and deliberate, it leads to serious concerns about the good faith of senior City staff and raises questions about where their loyalties lie. (Ottawa Light Rail Transit Commission 2022, 28–9)

Watson, though, had by then left city hall, having announced in December 2021 that he would not seek re-election. However, the LRT debacle was not the only file that contributed to a rapid decline in the popularity of the mayor during his last mandate. Another important controversy was the mayor's management of the occupation of downtown Ottawa by a "freedom convoy" opposed to vaccine mandates in January and February 2022. Even though the intentions of the protesters were known quite in advance, the police force, the mayor's office, and city council seemed ill-prepared for the arrival of the convoy. As a result of criticisms of his management of the protest, the chief of police Peter Sloly resigned during the occupation. The councillor Diane Deans, head of the police board and a long-time foe of the mayor, rapidly announced the hiring of a successor. However, her decision to go ahead with the hiring of a new chief without consultation, something she was legally entitled to do, was criticized by the mayor and his allies at city council. As a result, with the occupation still ongoing, they voted to remove her as a chair of the police board during an especially raucous and acrimonious city council meeting that presented to the rest of the country the image of a dysfunctional city. Mayor Watson was criticized during that time not only for seemingly using the occupation to settle political scores with opponents such as Deans (Adam 2022) but also for engaging in negotiations with leaders of the convoy. In the end, the federal government stepped in and invoked the Emergencies Act to deal with blockades in Ottawa and at different border crossing points throughout the country. The blockade ended after 18 days, following the intervention of a range of police forces, from the Ottawa police to the RCMP, to displace the protesters in the city.

Ultimately, the three mandates of Jim Watson as mayor of the post-amalgamation of the City of Ottawa were marked by hopeful beginnings that ended up in scandals. Watson was a mayor who prided himself of his conservative, competent, and prudent style of management. However, the LRT fiasco, the management of the "convoy occupation," and the slow, stalled, or limited progress on numerous other important files had undoubtedly challenged that image.

The election to replace Jim Watson as mayor showed the continuing importance of the territorial cleavage in Ottawa politics and the middle-class character of the city. The campaign seemed at first to be a race between three candidates. The first, Catherine McKenney, was a progressive councillor elected for the first time, as mentioned previously, in 2014 in the downtown ward of Somerset. McKenney had long been an opponent of former mayor Jim Watson on council and had seen their profile increase significantly during the convoy occupation of downtown Ottawa, which was part of McKenney's ward. McKenney campaigned on a platform of significant investment in cycling, transit, and social service infrastructure. Although associated with the New Democratic Party (NDP), McKenney was able to earn the support of the former federal Liberal environment minister Catherine McKenna as well as the former head of the Bank of Canada and the Bank of England, Mark Carney. The second candidate was Mark Sutcliffe, a journalist and businessman who positioned himself as a centrist candidate. Sutcliffe received support from Conservative and Liberal politicians, as well as member of the city's economic elites. He ran a campaign that focused especially on the proposed spending of Catherine McKenney and on keeping taxes low. Finally, the former mayor and minister in the Liberal provincial government Bob Chiarelli ran as the candidate of experience, with a rather vague agenda, which explains perhaps why the race quickly became one between McKenney and Sutcliffe.

If polls published during the campaign placed McKenney in a small lead over Sutcliffe, the final results were not especially close and had much in common with the 2006 campaign that opposed Larry O'Brien to Alex Munter. Sutcliffe won 51.4 per cent of the vote, McKenney 37.9 per cent, and Chiarelli 5.8 per cent. Once again, there was a clear territorial divide in voting. McKenney won a plurality of votes in five wards out of twenty-four, all located in the downtown core, including 73.3 per cent in their own ward of Somerset and 62 per cent in the neighbouring Kitchissippi ward (Laucius 2022, A3). They were also only a few points behind Sutcliffe in other urban wards located close to the downtown core. However, they ultimately did not manage to attract any major support in the suburban and rural wards. In fact, Sutcliffe won over 70 per cent of the votes in the three rural wards and around 60 per cent in most suburban wards. Those results were perhaps not surprising in hindsight considering that Sutcliffe had explicitly politicized the territorial divide throughout his campaign. McKenney's announcement, at the beginning of their campaign, of an ambitious and controversial plan to invest $250 million in the development of bike lanes, became central to Sutcliffe's campaign. It allowed him to present McKenney as out of touch.

The mandate of Mark Sutcliffe has so far been marked by problems that, while not unique to the City of Ottawa, had been exacerbated by the actions or inaction of the previous mayor and council, whether regarding the LRT file, the housing and homelessness crisis, or the fiscal situation of the city, and which required urgent actions. In the case of the fiscal crisis, the problem has been aggravated by Sutcliffe's electoral promise not to increase property taxes by more than 2.5 per cent during the first two years of his term. While Sutcliffe has publicly blamed the lack of federal and provincial support for the fiscal woes of the city, the reality of limited tax increases at a time of significant inflation has undoubtedly contributed to the current financial struggles of the city. In the third year of Sutcliffe's mandate, property taxes were increased by 2.9 per cent. In comparison, in Toronto, property taxes were increased by 9.5 per cent the same year.

As for the territorial cleavage, it has somewhat lessened since the election of Sutcliffe. The mayor promised early in his mandate to govern in a way that would be inclusive of all parts of the city. Indeed, a number of committee chairs were awarded downtown councillors, who had previously been excluded from key positions by the former mayor Watson. Second, beyond the territorial cleavage, Ottawa has, since 2014, experienced a growing political polarization along the left–right political axis, with the election of more progressive councillors and organizations supporting their work. A key aspect has been the creation in 2019 of Horizon Ottawa, a grassroots organization promoting a municipal government "that protects the environment, expands public ownership of infrastructure, and creates public spaces that are safe, inclusive and accepting of all of its residents" (Horizon Ottawa 2022). One of the organization's missions is the election of progressive councillors. And if the left–right cleavage has, to a certain extent, overlapped with the territorial cleavage historically in Ottawa, the last election saw two candidates endorsed by Horizon Ottawa elected in wards outside the downtown core. As such, there is some potential for a form of urban politics in Ottawa focusing more on the left–right divide than the urban–suburban–rural one. However, the territorial cleavage also momentarily seemed to recede following the election of Jim Watson before re-emerging later during his last two terms. The geographical reality of Ottawa post-amalgamation is such that the territorial cleavage is unlikely to disappear for long periods.

CONCLUSION

There has been both continuity and change in the politics of Ottawa since the 1980s. On the one hand, the politics of the city, forty years after Caroline Andrew's investigation of city politics in Ottawa in the post-war period, can still be characterized as middle-class. Ottawa remains a city with a highly educated population with significant socio-economic capital. That population remains somewhat fiscally conservative while having a significant mobilization capacity in reaction to projects that might be perceived as challenging some of its preferences when it comes to planning and urban development. On the other hand, amalgamation has undoubtedly had a significant impact on the city, political debates, and political alliances. It has limited the ability of councillors representing the downtown core to push for a more progressive agenda in the tradition of the consumption forces that Andrew had seen rising in the 1960s and 1970s. In short, territorial cleavages are especially important, and they influence public discourses and political strategies, both during elections and during council debates.

ACKNOWLEDGMENTS

This chapter is dedicated to the late Caroline Andrew, a wonderful scholar and colleague, who did so much to promote the study of urban politics in Canada, la francophonie, and the development of a more inclusive City of Ottawa. A draft version of this paper was first presented in October 2022 at the Research Centre on the Future of Cities at the University of Ottawa. I would like to thank my colleague Anne Mévellec for her generous comments and suggestions as well as the director of the centre, Vincent Mirza, for the opportunity to present this chapter. I also want to thank the editors of this volume, Stewart Prest and Katherine Graham, for their suggestions.

NOTES

1   Indeed, Andrew argued that the politics of Ottawa was best characterized as "middle-class."
2   Hamilton, Halifax, and Sudbury are other Canadian cities that combine large rural, suburban, and urban areas.
3   Statistics Canada defines intermediate suburbs as neighbourhoods located 20–30 minutes from downtown and distant suburbs as those located more than 30 minutes from downtown.
4   Corporations that bought some of the remains of Nortel in the city employed around 3,000 workers in 2019.
5   The decision of Minister Hodgson was eventually successfully challenged in court, but by then it was too late to change the electoral map.
6   Data from Ottawa and Hamilton are from Farkas (2015) and data for Toronto are from the Urban Policy Lab's Council Scorecard (Urban Policy Lab 2023).
7   Both the current shelter and the proposed one are located in the Rideau-Vanier Ward, but in distinct neighbourhoods.
8   Two exceptions were Tim Tierney and Jean Cloutier, who represented urban wards but not downtown core wards.
9   For a good timeline of some of the key problems that have plagued the LRT project, see Trick (2022).

REFERENCES

Adam, Mohammed. 2022. "Truck Protest – A City Under Siege, a Council at War with Itself." *Ottawa Citizen*, February 17. https://ottawacitizen.com/opinion/adam-truck-protest-a-city-under-siege -a-council-at-war-with-itself.

Andrew, Caroline. 1983. "Ottawa-Hull." In *City Politics in Canada*, edited by Warren Magnusson and Andrew Sancton, 140–65. Toronto: University of Toronto Press. https://doi.org /10.3138/9781487575908-005.

Andrew, Caroline. 2006. "Evaluating Municipal Reform in Ottawa-Gatineau: Building for a More Metropolitan Future?" In *Metropolitan Governing: Canadian Cases, Comparative Lessons*, edited by Eran Razin and Patrick J. Smith, 75–94. Jerusalem: Hebrew University Magnes Press.

Bagnall, James. 2019. "R.I.P., Nortel." *Ottawa Citizen*, January 19.

Benali, Kenza and Ella Bernier. 2014. "Le Train Léger sur les Rails: Les Rebonds d'un Projet de Transport Durable Conflictual." *Revue Gouvernance* 11 (1): 1–30. https://doi.org/10.7202/1038880ar.

Beate Bowron Etcetera, Hemson Consulting Ltd., The Davidson Group. 2020. Options Report. *Ward Boundary Review*. https://pub-ottawa.escribemeetings.com/filestream.ashx?documentid=33982.

Canadian Press. 2022. "Strong Mayor Powers Coming to More Large Ontario Cities in a Year, Ford Says." *CBC News*, October 17. https://www.cbc.ca/news/canada/toronto/strong-mayor-powers -expanding-ford-1.6619221.

CTV News. 2012. "Ottawa's $2.1 Billion Light Rail Transit Program Moving Forward." December 5. https://ottawa.ctvnews.ca/ottawa-s-2-1-billion-light-rail-transit-program-moving-forward -1.1066675.

Charbonneau, François, and Samuel Coeytaux. 2013. "L'Affaire Lepage et le Caractère Symbolique de la Politique de Bilinguisme de la Ville d'Ottawa (1970–2001)." *Revue d'Études Canadiennes* 47 (2): 119–49. https://doi.org/10.3138/jcs.47.2.119.

Chianello, Joanne. 2018. "Secretive Process Leaves Urban Councillors on Sidelines." *CBC News*, December 12. https://www.cbc.ca/news/canada/ottawa/ottawa-council-committee-assignments -rules-1.4941296.

Conference Board of Canada. 2016. "Economic Insights into 13 Canadian Metropolitan Economies." *Metropolitan Outlook 1, Winter 2016*. https://www.conferenceboard.ca/product/metropolitan -outlook-1-economic-insights-into-13-canadian-metropolitan-economies-winter-2016/.

Crawford, Blair. 2017. "Meet Mr. Popular." *Ottawa Sun*, January 16.

Dilkens, Drew. 2014. *A Comparative Analysis of Municipal Government in Canada: A Pre-Amalgamation and Post-Amalgamation View of Windsor-Essex, Ottawa and the Greater Toronto Area*, Unpublished PhD Dissertation. Paris: International School of Management.

Drake, Laura. 2007. "John Baird, Light Rail Killer." *The Thyee*, January 23. https://thetyee.ca/News /2007/01/23/LightRail/.

Farkas, Jeromy. 2015. *Council Tracker. Summary Report. Ottawa, 2010–2014*. Calgary: Manning Centre.

Fleury, Mathieu. 2019. "The reporting this morning on the #ChâteauLaurier vote this week at Council misses the fact that my motion was NOT revoted on yesterday, the reconsideration procedural tool had not been used at @ottawacity Council since 2009; NOT democratic by the #WatsonClub." Twitter, July 12. https://twitter.com/mathieufleury/status /1149650268738297859.

Graham, Katherine A., Allan M. Maslove, and Susan D. Phillips. 2001. "Learning from Experience? Ottawa as a Cautionary Tale of Reforming Urban Government." *Journal of Comparative Policy Analysis: Research and Practice* 3 (3): 251–69. https://doi.org/10.1080/13876980108412662.

Hilton, Robert, and Christopher Stoney. 2007. "Dreams, Deception and Delusion: The Derailing of Ottawa's Light Rail Transit Plans." *Governance Review* 4 (1): 1–26. https://doi.org/10.7202/1039114ar.

Horizon Ottawa. 2022. *About Us*. https://www.horizonottawa.ca/about.

Hume, Peter. 1996. "Mister Maybe." *Ottawa Citizen*, April 27.

Krawchenko, Tamara, and Christopher Stoney. 2011. Public Private Partnerships and the Public Interest: A Case Study of Ottawa's Lansdowne Park Development." *Canadian Journal of Nonprofit and Social Economy Research* 2 (2011): 74–90. https://doi.org/10.22230/cjnser.2011v2n2a92.

Laucius, Joanne. 2022. "Sutcliffe Support Outside City Core was Solid." *Ottawa Citizen*, October 31.

Leblanc, Daniel. 2017. "Armée du Salut à Vanier: Le Conseil Municipal Dit Oui." *Le Droit*, November 23. https://www.ledroit.com/2017/11/22/armee-du-salut-a-vanier-le-conseil-municipal-dit-oui-1332e77b9b4e3368026427c619b10b68.

Mallory, James. 1971. *The Structure of Canadian Government*. Toronto: Macmillan.

National Capital Commission. 2023. "About Us." https://ncc-ccn.gc.ca/about-us.

Ottawa. 2021. *Official Plan*, Section 11: Implementation. https://documents.ottawa.ca/sites/documents/files/section11_op_en.pdf.

Ottawa Citizen. 2003. "Not the Last Word on Wards." *Ottawa Citizen*, May 13.

Ottawa Citizen. 2004. "Rural Citizens Belong in City." *Ottawa Citizen*, March 26.

Ottawa Neighborhood Study. 2023. "Neighbourhood Maps." https://www.neighbourhoodstudy.ca/maps-2/.

Ottawa Light Rail Transit Commission. 2022. *Report of the Ottawa Light Rail Transit Public Inquiry: Final Report*. Toronto: Queen's Printer.

Peat, Don. 2015. "Rob Ford Says 'No' the Most: Report Tracks Voting Habits of Council." *Toronto Sun*, October 28.

Reevely, David. 2018. "Doug Ford Listens to 'the People' in Toronto, but Not in Ottawa." *Ottawa Citizen*, September 11.

Rupert, Jake. 2008. "Change of Heart." *Ottawa Citizen*, January 15.

Sancton, Andrew. 2000. *Merger Mania: The Assault on Local Government*. Montreal: McGill-Queen's University Press. https://doi.org/10.1515/9780773568914.

Singer, Zev. 2002. "Minister Vetoes New Ward." *Ottawa Citizen*, October 18.

Statististics Canada. 2022a. *Census in Brief: Canada's Fastest Growing and Decreasing Municipalities from 2016 to 2021*. Ottawa: Minister of Industry.

Statistics Canada. 2022b. "Canada's Large Urban Centres Continue to Grow and Spread." *The Daily*, February 9.

Statistics Canada. 2022c. *Focus on Geography Series, 2021 Census Population*. https://www12.statcan.gc.ca/census-recensement/2021/as-sa/fogs-spg/Index.cfm?Lang=E.

Statistics Canada. 2022d. Mother Tongue by Geography, 2021 Census. https://www12.statcan.gc.ca/census-recensement/2021/dp-pd/dv-vd/language-langue/index-en.html.

Trick, Sarah. 2022. "Ottawa's Colossal LRT Debacle: A Brief-ish History." January 14. https://www.tvo.org/article/ottawas-colossal-lrt-debacle-a-brief-ish-history.

Urban Policy Lab. 2023. *Urban Policy Lab Scorecard*. https://urbanpolicylab.ca/projects/councilscorecard/.

Willing, Jon. 2018. "Mayor Watson's New Cabinet Appears to Sideline Critics, Strengthen Allies." *Ottawa Citizen,* December 18.

Willing, Jon. 2021. "Council Opts for AG Probe of LRT, not Judicial Inquiry." *Ottawa Citizen*, October 14.

6

Winnipeg

Ursula M. Stelman and Aaron A. Moore

INTRODUCTION

For students of Canadian urban politics, Winnipeg has served as an important case study for urban reform efforts in Canada. Over the decades, institutional reforms spearheaded by provincial governments, along with innovative tripartite agreements between city, provincial, and federal governments, have attempted to address serious urban challenges. At the time of publication of the first edition of *City Politics in Canada* (1983), Kiernan and Walker argued that the much-touted amalgamation of the City of Winnipeg and its adjoining municipalities, known as Unicity, had failed to meet its goal of politicizing the local government in the form of political debate and policy formation at the city level. They argued that structural reforms were not enough to combat the "anti partisan ethos," lack of true participatory motivations, and the importance of internal dynamics in local politics (Kiernan and Walker 1983).

Our analysis shows how subsequent reforms of local government in Winnipeg gave rise to powerful executive mayors who, in some cases, undermined the legal accountability and status of the public administration. This was in part justified by an ideology and rhetoric of the need to prioritize pro-business, managerialist efficiency and to "de-politicize" local politics by establishing a non-partisan model. Structural reforms refocused the mayor and council on fiscal constraints and the efficient delivery of services over citizen engagement. The erosion of democratic representation by decreasing the numbers of wards for elected councillors and the redrawing of electoral boundaries to over-represent suburban middle-class areas and property owners contributed to the disenfranchisement of the poor non-property owners in the inner city. Vulnerable inner-city citizens were often poorly serviced or neglected as available monies were diverted to development and infrastructure in the downtown. These structural reforms were consistently authored and pushed by partisan provincial governments, and not by the electorate of Winnipeg as a whole.

Today, Winnipeggers bristle at the image of a city that in recent years has had the highest crime and child poverty rates in the country, growing homelessness, and a drug dependency epidemic, all highly visible on city streets. In addition, Winnipeg has unique environmental and fiscal realities that would pose special challenges for any city government: highly segmented communities with many conflicting needs and interests; extreme weather conditions; and poor soil conditions, combined with an extensive road and infrastructure network, that create ongoing and costly demands for maintenance and repair of the built infrastructure.

Arguably, some of the pressing issues the city faces are beyond the capacity of the municipal government to address. However, the political cleavages between the historic core (the City of Winnipeg as it was before the 1971 amalgamation) and suburban areas within today's City of Winnipeg, the focus on managerialism, the persistent tension between elected officials and administration, the growing fiscal deficit and continuous downsizing of the organization, have resulted in a government that often seems incapable or unwilling to substantially act on many growing social and economic problems impacting its citizenry.

This chapter first examines the major socio-economic and demographic changes that Winnipeg has undergone in the past four decades, and then discusses both the resilience of, and shifts in, the city's economy. We periodize political dynamics and administrative reforms by mayoralty, providing an in-depth account of how institutional changes and internal power dynamics have shaped the politics of Winnipeg since the 1980s. As we will see, Winnipeg's post-1989 institutional reforms marked the beginning of a consistent pattern of strengthening the power of the mayor and reducing democratic accountability, one that continues to this day.

SOCIAL AND DEMOGRAPHIC CHANGES SINCE 1980

By 1982, Winnipeg had long since declined in importance in Canada. Once the third-largest city in the country and the gateway to the west, the city took a back seat as the city regions of Calgary, Edmonton, and Vancouver overtook it and began to dominate the economic development of western Canada. While Winnipeg experienced moderate population growth through the 1980s (the city's population grew by 9.2 per cent during that decade), it experienced a long period of stagnant population growth through the 1990s (growing by 1 per cent over the entire decade; Statistics Canada 1981, 1986, 1991, 1996, 2002). Whereas suburban flight accounted for periods of slow growth and decline in cities like Toronto and Montreal, Winnipeg's stagnation encompassed the entire urban region.

For a time, population stagnation appeared to be the new norm for the city (Leo and Brown 2000; Leo and Anderson 2006). By around 2005, the city population was growing again, but this growth remained much slower than Canada's other major cities experienced, particularly the other large prairie cities of Edmonton and Calgary. As with the rest of Canada, Winnipeg and Manitoba experienced declining birth rates (Bougas 2015). While other areas of the country compensated for this through immigration, Winnipeg's immigrant population declined throughout the 1980s and 1990s.

As Manitoba's largest city, accounting for roughly 60 per cent of the province's population, Winnipeg was and remains strongly affected by decades of population loss to other provinces. Every year between 1980 and 2022, more people left for other parts of Canada than moved to Manitoba. Over this forty-year span, the province of Manitoba lost 52,000 residents within the 20–29 age cohort to interprovincial migration (Statistics Canada n.d.a.) – though it experienced a gain in university-aged residents (20–24 years old) between 2020 and 2022 (Statistics Canada n.d.b.).

Given interprovincial losses and the steady or declining immigrant population through the 1990s, Winnipeg's population may have continued to stagnate throughout the new millennium, if not for a substantial change in provincial policy towards immigration. In 1996, the provincial Progressive Conservatives (PCs) under Premier Gary Filmon penned the first ever agreement with the federal government to allow the provincial government to actively target immigrants (Carter 2009; Leo and Brown 2000).

This new deal, the first Provincial Nominee Program (PNP) in Canada, came into effect in 1998 and ushered in a new period of population growth in Manitoba, and particularly in the City of Winnipeg (Carter 2009; Statistics Canada 2022b). After a decade of losing immigrants to other parts of Canada, Winnipeg quickly became an important secondary hub for immigration. In the first decade of the program (1999–2009), the province increased immigration by almost 250 per cent, with a disproportionate number of immigrants moving to Winnipeg.

Largely driven by immigration, the City of Winnipeg added 80,000 new residents between 2011 and 2021. Where Winnipeg once lagged other large Canadian cities, it became a leader. From 2016 to 2021, Winnipeg experienced the fourth fastest growth of any of Canada's major cities after Halifax, Edmonton, and Ottawa (Statistics Canada 2022a). By 2016, the City of Winnipeg, as a proportion of its census metropolitan area (CMA), had grown to over 90 per cent (Statistics Canada 2017). Winnipeg, as the central city in the larger region, likely bucked the trend of central city decline experienced in some other Canadian CMAs (Statistics Canada 2022a).

With significant growth in immigration into the city, Winnipeg's population has diversified, though it still lacks the diversity of the country's traditional immigrant magnet cities of Toronto, Vancouver, and Montreal. Residents of European origin still account for most of the city's population. Currently, English and Scottish are the most common ethnic or cultural identities. However, even Winnipeggers of European descent differ in composition than in many other cities. Ukrainians are the third-largest group in the city, followed by Germans, many of whom come from the province's large Mennonite community. Filipinos are the next largest cultural group, accounting for over 10 per cent of the population. Other large visible minority groups

Figure 6.1: Change in the foreign-born and Indigenous identity populations by location, 1981–2021

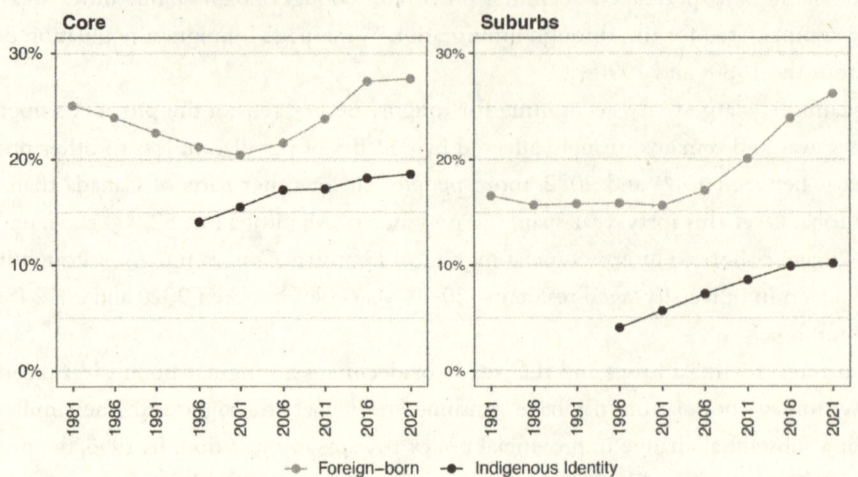

Notes: "Core" indicates the area of the City of Winnipeg before the 1971 amalgamation; "suburbs" indicates neighbourhoods within former Metropolitan Winnipeg suburban municipalities.

include South Asians (8.3 per cent), Black (5.3 per cent), and Chinese (3.0 per cent) (Statistics Canada 2022a). Beyond immigrants, other groups have accounted for a significant portion of the city's growth in the past two decades.

Winnipeg's Indigenous population almost doubled between 1996 and 2021. In the 1996 census, 7 per cent of the city's population identified as Indigenous (Statistics Canada 1996). By the most recent census (2021), 12.4 per cent of Winnipeggers identified as Indigenous (Statistics Canada 2022a). Today, Winnipeg has the highest proportion and greatest number of Indigenous peoples of all major cities in Canada. Not surprisingly, given Manitoba's status as the home of the Métis Nation, residents identifying as Métis account for a slight majority of residents that identify as Indigenous. Of the remaining residents identifying as Indigenous, over 40 per cent identify as First Nation, while the remainder comprise a small Inuk population and individuals identifying with multiple responses (Statistics Canada 2022a). While immigration largely accounts for the change in the city's population trajectory, Winnipeg's Indigenous population is today the fastest growing demographic in the city (see Figure 6.1).

While Winnipeg's population is now growing as fast, or faster than many other large cities in Canada, the city has also experienced a considerable increase in youth poverty. Overall, the poverty rate in Winnipeg (measured using Statistics Canada's low-income cut-off) is lower than in Toronto, Montreal, and Vancouver (Statistics Canada 2022a). However, according to the 2021 census, a little over 10 per cent of children between the age of 0 and 5 and just under 10 per cent of children between 0 and 17 lived in poverty in Winnipeg, compared to 8.6 per cent and 8.4 per cent in Montreal, the city with the second highest levels of poverty among children and youth in Canada (Statistics Canada 2022a). The high level of youth poverty in Winnipeg is, in part, a reflection of the relative youth and continuing high poverty rates of its Indigenous residents.

Based on 2016 census data, the median age of Indigenous Winnipeggers (termed Aboriginal identity in 2016) was only 27, compared to 39.8 for non-Indigenous Winnipeggers. Among Winnipeggers that identified as First Nation, the median age was 22.9. In 2015, 44 per cent of residents identifying as First Nation lived below the poverty line, followed by 27.3 per cent of Inuit people, and 19.7 per cent of Métis residents. Today, high rates of poverty among youth, Indigenous, and Black residents remain a pressing issue. However, among these groups, levels of poverty declined significantly between 2016 and 2021. As of 2021, 23.2 per cent of Winnipeg First Nation residents lived in poverty, as did 14.4 per cent of Inuit residents and 10.5 per cent of Métis residents (Statistics Canada 2022c). However, this apparent decline, which was noted across Canada, is likely due in part to the temporary rollout of additional income replacement supports during the COVID-19 pandemic.

As shown in Figure 6.1, Winnipeg's urban core is more diverse than its suburban neighbourhoods. A significant increase in immigration to suburban locations has resulted in similar proportions of the foreign-born population in the two zones. The proportion of the core and suburban populations that identifies as Indigenous has increased, but remains almost twice as high in the core.

ECONOMIC CHANGE SINCE 1980

The City of Winnipeg is situated in the Red River Basin in southeastern Manitoba. The heart of the city lies at the confluence of the Red and Assiniboia Rivers. This site, now known as the Forks, was important for trade among the Indigenous peoples of the area and eventually became a major trading post and centre with the arrival of French fur traders and emergence of the Métis Nation. Following the decline of the fur trade and the arrival of European settlers, the Forks – and Winnipeg – became the main rail depot for a growing agricultural region and the gateway to the west. Perhaps because of this early role as entrepôt and gateway, Winnipeg over time has developed a highly diverse economy, with limited reliance on volatile industries like resource extraction (Bougas 2015; Leo and Anderson 2000).

Like every other city in Canada, Winnipeg's economy was subject to the many upheavals over the past forty years, with certain sectors, notably manufacturing, ebbing over time. Nevertheless, while the city's economy has changed over the past four decades, the composition of the economy has remained stable, with no single industry dominating. The city's economy has continued to grow over this period, even during the decade of population stagnation in the 1990s (and the brief period of high unemployment for three years in the early 1990s).

By the beginning of the 1980s, Winnipeg's economy was performing well. Unemployment was low, with five industries accounting for most of the labour force (community, business, and personal services; trade; manufacturing; transportation, communication, and other utilities; and public administration and defence). This broad pattern of diverse services is consistent through to the most recent census, with the slow decline of manufacturing as a proportion of the economy being the most notable change (Statistics Canada 1981). As of 1997, manufacturing remained the second largest employment sector in Winnipeg, the largest one being transportation and communications. However, while some areas of manufacturing, such as aerospace, continued to grow, by the beginning of the new millennium, manufacturing as a proportion of the economy, and in total number of jobs, was clearly in decline. Despite this, Winnipeg's unemployment rate fell to 5.7 per cent by 2000 (Statistics Canada 2002).

Through the first decade of the 2000s, the province's new immigration policy not only bolstered the city's population growth but also contributed to the city's growing economic diversity. This diversity, in turn, played a substantial role in minimizing the shock of the 2008–2009 recession. Winnipeg lost only 200 jobs in 2009 at the height of the recession. While manufacturing jobs declined in 2008, they quickly rebounded in subsequent years (Bougas 2015). Following the recession, Winnipeg enjoyed low unemployment and strong economic growth led by construction, professional, scientific, and technical services, agriculture, and utilities (Manitoba n.d.; EDW 2015). By 2015, the decline of manufacturing had slowed, and the sector began to further diversify (Bougas 2015). In 2021, the city added over 1,000 new manufacturing jobs (Statistics Canada 2022).

Today, although Winnipeg continues to struggle with the after-effects of the COVID-19 pandemic, with unemployment at 8.9 per cent in 2020, the city's economy remains robust. Nevertheless, there are some areas of economic weakness looming. Well before the current staffing shortages resulting from COVID-19, and despite the province's aggressive immigration program, economists identified the potential for labour shortages in the province (Prism n.d.). The continued loss of young residents to other provinces is likely to only exacerbate this problem.

Institutional Change in Winnipeg Before 1980

While Winnipeg's economy proved resilient over the last decades of the twentieth century and into the new millennium, the city did experience significant economic stress throughout much of the twentieth century, as its role as entrepôt declined and wealth shifted west to Saskatchewan and Alberta. This period of decline resulted in a long history of municipal fiscal imbalances and financial pressures (Hum et al. 1986). As a result, the fiscal realities of the city played an important role in instigating institutional reforms.

Unicity Amalgamation

In 1969, Manitobans elected a New Democratic Party (NDP) government following a decade of PC rule. The newly elected NDP government inherited a two-tier form of metro government for the region of Winnipeg that the PCs had introduced a decade earlier. Initially, the NDP government was cautious in its approach to regional governance. Unlike the PC government,

Figure 6.2: The 1971 Unicity amalgamation

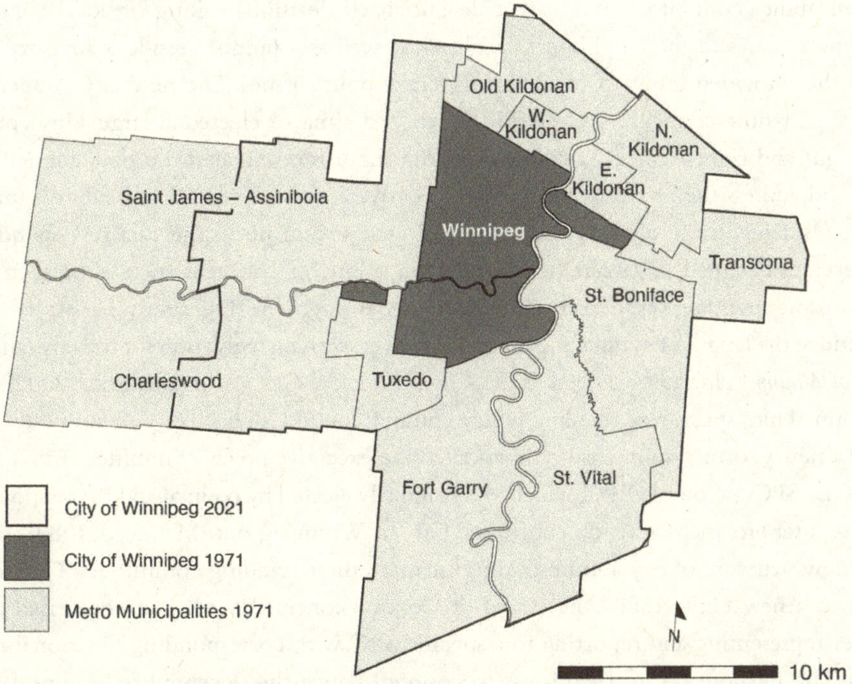

Notes: The 1971 Unicity amalgamation merged the existing municipalities in the two-tier Metropolitan Winnipeg system into a single-tier municipality.

the NDP wished to address socio-economic and budgetary inequalities within the region and favoured amalgamation as means to force the wealthy suburbs to address the decline of the inner city and downtown. However, the party was concerned about public backlash among right-leaning suburban voters. As a result, they initially planned a slow progression towards a unified city region. This approach changed in 1971, after the NDP handily won a by-election in the Conservative stronghold of St. Vital, a city to the south of Winnipeg that had a mayor vehemently opposed to amalgamation. This victory emboldened Premier Ed Schreyer to initiate a second round of major institutional changes just ten years after the creation of Metro Winnipeg. On 27 July 1971, the NDP amalgamated all the municipalities in the Winnipeg region into a single new City of Winnipeg, colloquially known as "Unicity" (Lightbody 1978; see Figure 6.2).

The new Unicity government reflected the NDP's belief in decentralized participatory government even as it introduced a centralized administrative structure (Axworthy 1978; Kiernan and Walker 1983). Axworthy (1978, 6) postulates that "this reorganization was aimed not only at providing a more effective form of urban regional government, but it also sought to decentralize the operation of government and give more access to the people." He further notes that the centralized bureaucracy and decentralized political structure were problematic from the start as the province overestimated the capacity of structural reform to alter decision-making dynamics at city hall.

The NDP aimed to increase democratic participation in the Unicity by introducing thirteen community committees (CCs) that decentralized decision-making on local zoning and development issues to the neighbourhood level, as well as a number resident advisory groups (RAGs) that provided input to council on specific policy issues. The new city council comprised 50 part-time councillors, elected in wards, and a mayor elected at large. Unsurprisingly, centre-right and right-wing councillors from the suburbs dominated the new council. While Unicity did address the property tax imbalance between former central city suburbs (property taxes for residents in the old city dropped, while taxes went up in the suburbs), spending on social services declined in favour of spending on suburban concerns such as protection and recreation, and spending on infrastructure for new development (Lightbody 1978).

To reduce the length of council meetings, the new governing legislation for the city (an updated *City of Winnipeg Act*, later changed to the *City of Winnipeg Charter Act* (Manitoba, 2002)) subdivided council into specialized standing policy committees (SPCs). Each SPC reported to council through a newly formed municipal cabinet called the executive policy committee (EPC) and the chairs of the SPCs sat on the EPC, which was originally elected by council and later appointed by the mayor after provincial amendments to the City of Winnipeg Act (Manitoba, 1989).

The new structure of city administration mirrored these standing committees. The province introduced a new entity called the Board of Commissioners (Board) that comprised a commissioner representing and reporting to a specific SPC with corresponding functional responsibilities. The purpose of the Board was to counterbalance the decentralized political process with a centralized bureaucracy under the leadership of the commissioners (Cassidy 1971). The introduction of a group of commissioners also prevented the consolidation of administrative power into one individual.

The Board initiated policy and administrative changes and improvements, led civic responses to crisis situations, provided mentoring to department heads, pursued federal or provincial funding (or both) for civic works, and coordinated service delivery. During its existence, the Board was made up of senior bureaucrats who were educated as engineers, planners, or public administrators. They worked closely with the mayor and the EPC to steer the policy, strategic, and financial direction of the city.

In 1977, the new Conservative provincial government weakened the role of the mayor by removing the mayor as chair of the EPC and prevented the mayor from attending meetings of the Board of Commissioners (Axworthy 1978, 19). This change was likely a response to the increasing profile and authority of mayor Stephen Juba. Juba had his own independent power base and could intimidate politicians who did not want to be seen in his public disfavour.

In addition, the province reduced the size of council from 50 to 29 part-time councillors (1977). This change negatively impacted political and community representation (via CCs and RAGs) as the number of wards shrunk while constituency size grew. These changes were again efforts to decrease the power of the populist Mayor Juba and to address the dysfunction in decision making and policy formation associated with a large council.

WINNIPEG GOVERNMENT AND POLITICS SINCE 1980

For most of Winnipeg's history, including the Metro era and the first decade of Unicity, city council functioned along highly politicized, pseudo-partisan lines. Dating back to the General Strike in 1919, council was split between labour and socialist councillors – who represented the north and west ends of the city, areas with lower-income residents and a high proportion of immigrants – and liberal–conservative coalitions representing the city's wealthy south.

Following the creation of Metro and the subsequent amalgamation of the city region, socialist and labour forces coalesced under the NDP banner, while the coalition of liberals and conservatives would eventually run under the banner of various citizen election committees. The purpose of the citizen election committees was to prevent labour and the "socialists" from taking over city hall. Members of these committees considered the socialist and labour candidates as an existential threat to business and a disaster for the city, should they ever govern.

During the Metro era, the liberal–conservative bloc ran as the Greater Winnipeg Election Committee at the Metro level. Following the creation of Unicity, the Independent Citizens Elections Committee (ICEC) emerged as the main liberal–conservative bloc on council. As conservatives dominated much of the former suburbs of Winnipeg, the ICEC became the de facto governing bloc in Winnipeg, with NDP councillors still largely confined to the old north end, the west end, and downtown.

While Winnipeg councillors today still hold varying political ideologies, beginning in the 1980s, ongoing changes to city institutions slowly eroded the pseudo-partisan nature of council and significantly narrowed the areas of political dispute on council. As a result, Winnipeg's government is far less politicized today than it was at the beginning of the 1980s. Over the past forty years, the politics and policy direction of the city has consistently combined promises and programs to address poverty in the inner city and of revitalization of the city's downtown with austerity measures and infrastructure spending geared towards the conservative suburbs. The result is a city still struggling with poverty and homelessness in the core, and increasingly throughout the rest of the city, as well as a growing structural deficit that tests Winnipeg's ability to provide basic services to its residents.

The Bill Norrie Era: 1979–1992

The reduction of city council to 29 councillors in 1977 heralded the gradual weakening of politicization in the city, while further cementing the dominance of suburban interests. Bill Norrie first became mayor in a by-election in 1979 (Winnipeg Free Press, 1979). In the following regular municipal election in 1980, he distanced himself from his association with ICEC and successfully ran for mayor as an independent. The 1980 election brought significant loses for the ICEC, resulting from its pro-development stance and perceived conflicts of interest (Fayerman

1980). On the eve of the 1983 election, in which Norrie won re-election, the ICEC formally disbanded and its former members ran as independents. In its place, the so-called Gang of 18, which comprised former ICEC members and additional councillors, continued to meet with Norrie and the EPC to informally determine the agenda for city council. As with the former ICEC, this group actively supported the depoliticization of city council. By 1987, the NDP stopped running candidates in local elections. However, many council candidates (NDP, PC, Liberal) retained ties with provincial and federal parties.

Contrary to what one might expect, the declining politicization and dominance of suburban conservative councillors did not result in the city abandoning the downtown core and traditional inner-city neighbourhoods (the north and west ends). Rather, Bill Norrie's mayoralty ushered in a new era of spending and programming for urban renewal. Accounts of the development of Winnipeg's Core Area Initiative suggest that the then-federal minister of employment and immigration, and local MP for the Fort Rouge riding, Lloyd Axworthy, provided the political impetus for the creation of the new program. Before entering federal politics, Axworthy was a key figure in the creation of the Institute of Urban Studies at the University of Winnipeg and had devoted his own time and research to studying Winnipeg's core area and the social problems associated with it (Layne 2000).

This first Core Area Initiative (CAI I) was a five-year, $96 million tripartite federal-provincial-municipal intergovernmental revitalization agreement focusing on approximately twenty-six square kilometres of Winnipeg's inner city. The area encompassed approximately 100,000 residents. The mandate of CAI I combined economic development, employment and training, and physical revitalization of inner-city neighbourhoods. The initiative included thirteen cost-shared programs and over 1,000 projects. CAI I led to the creation of the North Portage Development Corporation (NPDC), a tripartite corporation responsible for the re-development of the area just north of Portage Avenue in the city's downtown (Layne 2000). NPDC's projects included the construction of the Portage Place mall, a failed retail venture that is a current focus of redevelopment and revitalization discussions.

In 1986, the three levels of government extended the initiative (CAI II) for an additional five-year term with a $100 million budget. CAI II focused more heavily on revitalizing the city's commercial and tourist industries than CAI I. The second phase of the CAI involved the creation of yet another tripartite corporation, the Forks Renewal Corporation (FRC). The FRC focused on the renewal and revitalization of old east railway yards at the confluence of the Red and Assiniboine Rivers.

In redeveloping "the Forks," the FRC focused on building parks and amenities, included a food hall and farmer's market, intended to attract suburbanites to the downtown. For years, the Forks struggled to turn a profit, resulting in the merger of FNC and NPDC into the Forks North Portage Partnership in 1994. Although Portage Place failed as a retail mall shortly after opening, the mall's parking garage turned a sizeable profit, which the new partnership used for over a decade to subsidize the development of the Forks before it became profitable.

Towards the end of the CAI, the downtown experienced some new private investment as well. In 1989, TD Bank undertook to develop a thirty-four-storey building on the northwest corner of Portage and Main. This was the first large-scale private investment downtown since the Trizec building (1970–78) and Lakeview Square (1973).

Although the first phase of the CAI focused heavily on renewal of existing residential neighbourhoods and providing jobs and skills training for inner-city residents, CAI II suggests that, from the perspective of most city councillors at least, the point of inner-city renewal was to bolster the city's image while attracting suburban residents and businesses downtown to generate more revenue for the city, rather than addressing inequities. While renewing the downtown economy and improving the lives of inner-city residents are not mutually exclusive policies, the tension between a focus on image building and addressing issues of poverty in the inner city remains to this day, and underlines most of the subsequent attempts to re-envision and revitalize the downtown.

The Bill Norrie era also highlighted a continuing contradiction in the direction of development and growth in Winnipeg. While all three levels of government directed significant funds to renew the downtown, city council continued to support large-scale suburban expansion in the form of the Linden Woods development in the southwest corner of the city. The predilection towards sprawling development would continue even during the period of population stagnation in the 1990s.

While the city pursued downtown renewal with funding from senior levels of government, residents and businesses began to challenge the continued property taxes increases championed by the Board of Commissioners. In the 1979 election, Bill Norrie said that the city was in "good financial shape with a strong tax base" (Winnipeg Free Press, 15 June 1979, 2). However, to meet operating and capital requirements, Winnipeg had been raising annual property tax rates throughout the 1970s and 1980s at rates significantly higher than other prairie cities. Population and economic growth throughout the 1980s limited increases, but the 1990–92 recession and the period of population stagnation that followed severely strained the city's finances. Towards the end of Norrie's time in office, the city was facing a cash crunch due to shrinking civic revenues, soaring debt servicing costs, and an infrastructure that was deteriorating in much of the historic city and older suburban neighbourhoods.

The battle of the 1991 budget set the tone for future directions in the city's taxation and budgeting policies. It also highlighted some of the problems with the existing institutional configuration at city hall. Bill Norrie had succeeded as mayor by adopting a collaborative and consultative style of leadership. He relied heavily on the expertise of city staff and the Board of Commissioners, while relying on his coalition of liberal–conservative councillors. This came to an end when the 1991 budget debate revealed the institutional weakness of the mayor and growing division between the elected council and appointed Board of Commissioners.

During the 1991 budget deliberations, the deputy mayor and three members of the EPC would not support the chief commissioner's proposed draft budget, which included an 11.8 per cent tax increase that Norrie initially supported. The right wing of the council, which included many former ICEC members, was adamant that the budget must be brought into line with revenues, and proposed a zero per cent tax increase for 1991. Council rejected this. In response, council members of "Winnipeg into the Nineties" (WIN) – a short-lived pseudo-party that took the place of the NDP in the 1989 and 1992 elections – proposed an alternative budget that called for a 5 per cent increase in property tax. Council again rejected this proposal.

Ultimately, Norrie bypassed the EPC, by working with the deputy mayor and four standing committee chairs to create an alternate budget that proposed a 4.78 per cent property tax increase and restored $1.3 million in funding for library and health services. To pass this budget, Norrie relied on support from WIN councillors instead of his traditional allies on council (Thampi 1991).

The Norrie era in Winnipeg came to end when the mayor announced he would not seek re-election in 1992. While the end of Norrie's mayoralty revealed some of the inherent problems with the institutional structure of city government, the practice of politics during his time in office differed little from previous eras. Political parties, and a highly politicized council, were in decline but remained throughout his tenure, as did the relationship between elected officials and city staff. The election of Winnipeg's first, and so far only, woman mayor, Susan Thompson, would usher in significant change and flip Winnipeg governance and politics on its head.

The Susan Thompson Era: 1992–1998

Before the 1992 election, the provincial government once again changed the City of Winnipeg Act. Council was reduced from 29 to 15 councillors, and councillors became full-time. The reduction in council seats ended the Gang of 18 era, as it pitted former allies against each other in some instances. WIN was the sole remaining organization running a slate of candidates for council. Although WIN performed well, winning seven seats, right-leaning, suburbanite councillors continued to dominate council.

Susan Thompson entered the mayor's race in 1992 as a political neophyte. Norrie was leaving behind a troubled political reality that included council infighting between WIN and remaining Gang of 18 councillors, escalating property taxes, demands to keep services at status quo, growing budget deficits, high debt costs, decreasing revenues and provincial grants, and approximately 5 per cent inflation exacerbated by a slow growth economy. The citizenry were actively expressing their unhappiness with the steady property tax increases.

Susan Thompson had returned to Winnipeg to take over her father's western and leather goods business on Main Street near city hall. The economic difficulty of the times (1989–1994) had an impact on small businesses in Winnipeg. Thompson was a personal witness to the frustration of the small business community with increasing property and business taxes.

Thompson argued for a new approach to city politics. Her interests in the local economy, marketing, management, and women's issues were reflected in her "it's time for change" campaign, which promised: (1) to freeze property taxes in the next three years, (2) to streamline the bureaucracy at city hall, (3) to promote and market economic development, (4) to implement employment equity in the civic service, (5) to set performance standards and goals for bureaucrats, (6) to end duplication of service, and (7) to join the province in marketing Winnipeg (Thompson 2016).

The outcome of the 1992 election was dramatic: Susan Thompson won the battle for mayor, and the experienced councillors that ran against her were suddenly out of council and out of the important power portfolios. Winnipeggers had chosen an inexperienced non-politician to lead city hall, expressing a desire for change and to get rid of existing political gangs (Lett 1992).

Once elected, Thompson needed to form an EPC from a smaller council dominated by independents. Ultimately, she chose three of the remaining Gang of 18 and assigned them new responsibilities as deputy mayor, finance chair, and protection, parks, and culture chair. She included one WIN candidate as planning chair and an independent as works and operation chair.

In her first term, Thompson ran into fiscal problems as she and her EPC faced large unbudgeted property tax refund demands (due to reassessment) that immediately compromised her goal of a tax freeze. This began an intense period of tension and conflicts with the powerful Board of Commissioners. As Thompson recalls in her autobiography, "I was advised by the senior administration that the business of the city had to be approved by nine votes on council and as a new mayor, I only had one, I was also advised that as a new mayor, I most certainly did not have enough influence to garner nine votes on council, but the Board of Commissioners were most certain that THEY did, in other words, they ran things" (Thompson, 2016, 110).

During the mayor's first years in office, the political and administrative climate was tense and polarized. In 1993, she removed two capable members of her EPC and replaced them with ones committed to her philosophy of freezing property taxes (Lett 1993).

Thompson ran and won in 1995 with a similar popular vote share: 39 per cent. The 1995 election marked the last time council candidates ran under a banner, as WIN would dissolve in 1996. While council maintained a right–left divide, the high-stakes partisan politics of past decades came to an end, replaced with councillors focused on shoring up their support in their own wards. Although the ICEC dissolved early in the Norrie era, provincial changes to the structure of municipal government in Winnipeg largely realized the ICEC's goal of a depoliticized city council.

The full-time councillors wanted to exercise their power more directly in administrative matters and in running the city. However, many politicians were unclear about their new role as full-time councillors and were frustrated at being "out of the loop" on many important issues under the purview of the powerful mayor/EPC and the Board. The bureaucracy, meanwhile, was generally unsure how to cope with the ongoing attacks and cuts to operating budgets, negatively impacting staff morale. The mayor and her executive believed that the Board and bureaucracy were responsible for escalating municipal budgets, perceived red tape, and inefficiencies in general. The Board represented the old (bureaucratic) way of running things while Thompson represented a business approach.

In response to the pressure from the new mayor and council for a low tax environment, efficiency, and effectiveness, and to increase innovation and decrease costs, in 1994 the Chief Commissioner received approval from council for "A New Direction for Civic Administration." The organization was reduced from twenty-two to seventeen departments with a corresponding decrease in middle management employees. This was the first significant cut to city management and the introduction of departmental amalgamations.

Conflicts between the mayor, the commissioners, senior administrators, and politicians escalated. Thompson complained that the commissioners were too powerful and rigid – controlling policy, budgets, and implementation mechanisms. At one point, she put the chair of the SPC on Finance to work directly inside administration overseeing the commissioners' role in holding departmental budgets (line by line) under control. This was a major sign of "lack of confidence" in the commissioners.

Susan Thompson ultimately received the support of council to hire an external consultant to review the city's operations with a mandate to recommend changes to the political and administrative system. The "City of Winnipeg Organizational Review and Performance Assessment" in 1997 (the Cuff Report) made the following observations and recommendations for reform:

> It is our view that council and committees have either been encouraged or condoned in their involvement in what we term 'administrivia.' This has occurred as a result of a less than rigorous approach to what issues are placed on the committee agendas and the lack of a policy framework around those issues. As a result, and not necessarily because of any intended design, Council members have been increasingly drawn into administrative decision-making. (Cuff 1997, 54–5)

The report did not address the real power struggles playing out daily between the political actors and administrators at the apex of city government. Cuff instead commented on management accountability issues. He noted the absence of performance appraisal for management, which, he argued, was detrimental to the organization. In addition, he questioned the need for continuous improvement when the city needed fundamental change (Cuff 1997).

Due to the ongoing conflict between the Board and the mayor and EPC, the provincial government agreed to change the structure of the city's bureaucracy and increase the authority of the mayor. The PC government amended the *City of Winnipeg Act* in 1998 to replace the Board of Commissioners with a chief administrative officer (CAO) model (Manitoba, 1998).

While the new reforms concentrated authority in a single top bureaucrat, the new position of CAO was directly accountable to the mayor through the EPC. The mayor retained authority to choose standing policy committee chairs and to appoint members of council to the powerful EPC. Under the new CAO model, the commissioners were gone, along with many long-serving directors. This substantial shift in bureaucratic leadership fundamentally destabilized the city's administration. Under the amended Act, the CAO gained unprecedented power over the bureaucracy; the mayor and her powerful EPC maintained control over the CAO and policy influence over the full council.

With a new CAO aligned with her agenda, Thompson was able to usher in her promised tax freezes. After a tax increase of 5.8 per cent in 1993 and 3.2 per cent in 1994, Thompson succeeded in freezing taxes in 1995 and 1996 and limited increases at or below 2 per cent in her last two terms of office. Thompson's commitment to reduce taxation instigated an era of tax freezes and cuts that would extend through Glen Murray's mayoralty and most of Sam Katz's time in office. The mentality of zero tax increases would later stretch the capacity of the city to meet basic levels of service.

While Thompson spent much of her time as mayor trying to address the city's budgetary issues and making the city operate as a business, the city did not abandon attempts to address the continued decline of the city's downtown and inner city. Despite the headway the city made with the CAI in the 1980s, residents continued to leave the downtown and surrounding inner city, resulting in further loses in tax revenue. In addition to this exodus, the city's image, which was already in decline, took a major hit in the 1990s as the crime rate skyrocketed, particularly in the downtown and the north and west ends. In 1996, for the first time, Winnipeg earned the unwanted title of Canada's homicide capital. The city would continue to trade this title with other prairie cities and Thunder Bay, Ontario, going forward (Statistics Canada n.d.c.).

That same year, Winnipeg also lost its NHL team, the Winnipeg Jets, to relocation and new owners in Phoenix, Arizona, dealing a major blow to the city's national image and its citizenry. Mayor Susan Thompson, the business community, and ordinary citizens tried desperately to "Save the Jets." Winnipeg was seen as too small of a market and lacking enough business headquarters to make a team financially viable.

Real concerns about crime and poverty in the city and image concerns led the city into another tripartite agreement with the federal and provincial government in 1994, titled the Winnipeg Development Agreement (WDA). This agreement involved an investment of $75 million over five years towards community development and security, including a new focus on the area of Main Street, just north of the city's central business district, labour force development, and strategic and special investments (WDA News Release, 1994). At this point in time, north Main Street, a highly symbolic area close to city hall, was Winnipeg's "skid row."[1]

In 1997, Thompson created a task force, headed by two prominent Winnipeggers, a businessperson and an Indigenous leader, to guide $10 million in funding for the North Main Street Development program. The program was to define a new role for Main Street and support for an Indigenous community facility. The inclusion of Indigenous leadership reflected the demographic shift that was occurring in the inner city, particularly in the city centre and the north end. From 1986 to 1996, the number of residents identifying as Indigenous (Aboriginal) quadrupled (from 12,000 to just under 50,000) (Statistics Canada 1996). As noted earlier, a disproportionate number of Indigenous residents in the city were living in poverty. As a result, many Indigenous residents and families migrated to the north end, where housing was most affordable.

In addition, the city and the Institute of Urban Studies at the University of Winnipeg began a series of consultations within the inner city. These consultations, and subsequent reports, identified several issues that plagued the inner city. While physical and economic decline remained an important focus for the city, particularly as they affected the city's image, the city began focusing on social problems such as street culture, crime, poverty, discrimination, and mental illness (Stelman 1998).

Despite these interventions, Thompson's focus on budget freezes and municipal restructuring severely limited the capacity of the city to tackle the underlining social problems that maligned the inner city and downtown. At the federal and provincial level, the 1990s ushered in an era of austerity, which resulted in cuts to support for social services. Once again, as under Norrie, the city continued to expand outward, focusing its capital investment on supporting the new development on the city's fringe.

The Glen Murray Era: 1998–2004

Leading up to the 1998 municipal election, mayor Susan Thompson chose not to run, leaving the next mayor to take over the new model for a downsized city administration that she had shaped. At the time of the election, Winnipeg's decade of population stagnation saw it drop to Canada's eighth largest city. Along with the city's declining importance federally, aging infrastructure, and escalating socio-economic problems in the inner city led one observer to note that "Winnipeg talks of itself as a city that was once going to be great and now tries not to slip below mediocre" (Moore 1998, 12). Winnipeg's media suggested that the city needed a new mayor, in fact a "messiah," who could change the direction of the city and revitalize its image (Cole, 1998).

Glen Murray, a former AIDS activist turned WIN city councillor, and an openly gay candidate, was that hope. Despite his public perception as a left-of-centre and pro-union candidate, Murray ran on an election platform of reducing property taxes by 10 per cent ($50 million), trimming business tax, rationalizing civic facilities, and expanding alternative service delivery. His election promises were more conservative than those of his closest competitor (Cole 1998).

Despite his fiscally conservative platform, Murray managed to galvanize the public with grand visions for the city and promises to strengthen the city's role in the Canadian federation. Ultimately, his somewhat contradictory plan for austerity and promise to lift Winnipeg out of a decade of stagnation and decline won him the election. In his victory, Murray became the first "left-wing" mayor of Winnipeg since the 1940s and the first openly gay mayor elected in Canada (WFP, 1998).

In his first year of office, Murray froze property taxes. In the following three years, he reduced property taxes by two per cent a year. While reducing taxes, he also managed to reduce the city's debt, replenish its reserve funds, and tackle some of the city's infrastructure deficit. Murray achieved many of these feats in part due to the uploading of social services and public health to the province between 1999 and 2002 and the sale of Winnipeg Hydro to Manitoba Hydro in 2002. Beginning in 1999, the province took over social housing from the municipalities, ending Winnipeg's involvement as a housing provider. Murray's time in office also began a long trajectory of reductions in city staff that would continue under his successor Sam Katz.

Although the public and media perceived Murray as an advocate for the centre-left and unions, in practice he instructed the CAO to continue with a downsizing agenda that would deliver on his election promises of significant tax cuts and alternative service delivery options. His key interests were external to the city organization, focusing on productive partnerships with the provincial government on high-profile urban development projects, while the CAO/CFO were carrying out the downsizing and efficiency agenda.

Murray, to an extent, delivered on a promise to work with all of council and to build a balanced EPC (Rollason 1998). As one councillor noted, Murray distributed important tasks to non-EPC council members and treated them as "valued decision makers" (O'Brien 1999, A6). Despite his early focus on working with council as whole, Murray further centralized decision making in the mayor's office and EPC. Under Murray, the city developed a new unit in the administration called the EPC Secretariat, which aimed to facilitate the policy influence of the political leadership and introduced a new layer of accountability for the administration. The mayor recruited a veteran of city government and his transition team leader into the position of secretary of the EPC, a new and powerful administrative role.

Under Murray, the informal meetings begun by Thompson were strengthened, and the EPC began meeting "informally" as well as "formally." The EPC began shaping the agenda for both administration and council. The mayor and the EPC used "informal" EPC meetings to direct the administration to engage in EPC priorities and projects, instructing administration to write specific reports. Ordinarily in the past, such requests for reports would come officially and directly via formal SPC, EPC, or council meetings. Now EPC used the power of the CAO to direct bureaucracy in the informal EPC settings.

The political and administrative expertise of the head of the EPC Secretariat instantly made him a very influential and powerful force over both administration and policy formation. He was able to actualize Murray's vison by whipping up EPC or council support and keeping one eye on administrative compliance. Following his successful re-election in 2002, Murray further strengthened his hold over council and the administration by using the EPC to recruit a new CAO more aligned with his ideologies and working philosophies.

Despite, or perhaps because of, his cuts to the city's largest revenue source – property taxes – Murray spent a good part of his tenure as mayor lobbying both the provincial and federal governments for new revenue sources. At the provincial level, Murray pushed for the transfer of one per cent of the provincial sales tax to the city. Federally, Murray joined urbanists Jane Jacobs and Alan Broadbent, and the mayors of Vancouver, Toronto, Calgary, and Montreal (known as C5), to lobby the federal government for a new deal for big cities, including new powers for large central cities and a municipal gasoline tax. Murray failed in his bid to capture a portion of the provincial sales tax, and while the federal government's initial response to the C5 was positive, ultimately all the lobbying efforts achieved was the federal government setting aside one per cent of its gas tax for all municipalities (not just the big cities).

Back in Winnipeg, the days of tripartite agreements between the city and senior levels of government for inner-city renewal ended. Austerity policies at the federal level ushered a period of declining federal involvement. With a decline in federal appetite for urban renewal, the city, under Murray, focused increasingly on growth and new development in the core and less on social planning for existing residents.

The creation of the Centre Venture Development Corporation (styled CentreVenture) in 1999 reflects this shift in the city's approaches to downtown renewal and image building. Increasingly, the city would turn to the private sector for its involvement. Though a creation of the province, CentreVenture became an arms-length agency of the City of Winnipeg, whose mandate was to provide financial incentives and support for development projects and activities in downtown (McArthur 1999). With the support of CentreVenture, several small-scale private sector residential projects were initiated.

Murray and his council brought several large new projects to downtown. He worked with a newly created entertainment and sports venture, True North Sports and Entertainment (TNSE), to replace the old Eaton's building with a new downtown arena (a vital piece in the city's hope to reclaim an NHL team). During Murray's tenure, Red River College, a publicly funded post-secondary institution, built a new downtown campus and the city twinned the Provencher Bridge, facilitating easier movement of drivers from eastern Winnipeg into downtown and creating the now-iconic pedestrian bridge symbolizing architectural dynamism. The city also began the redevelopment of Riverfront Drive (Moore 2002).

While many of these projects altered the built form of downtown, they did little to address the underlying issues of poverty, homelessness, and crime that plagued the inner city and downtown. Although Murray was touted as a left-wing progressive mayor, the city achieved little during his tenure in addressing Winnipeg's significant social issues. Nonetheless, there were a few small successes. In Murray's second term in office, the city introduced its first policy for addressing Indigenous issues, titled "First Steps: Municipal Urban Aboriginal Pathways." This new policy document facilitated the introduction of urban reserves in the city and introduced plans for a north end recreational centre catering specifically to Indigenous peoples.

Glen Murray's tenure as mayor ended abruptly when he chose to run (unsuccessfully) as a federal Liberal candidate in the 2004 election. Nevertheless, despite his shortened tenure as mayor, the changes he introduced in the city – the EPC secretariat, tax cuts and freezes, cuts in city staffing, and the increasing involvement of the private sector in downtown and inner-city revitalization – had a profound and long-lasting impact.

The Sam Katz Era: 2004–2014

Following Murray's abrupt departure from office, Winnipeggers elected Sam Katz, owner of the local baseball team, the Goldeyes, as mayor in a by-election (Welch, 2004). Like Susan Thompson, Katz was a political neophyte. As an infant, Katz had immigrated with his family from Israel to Winnipeg. Like many other new immigrants, they settled in the city's north end (Bellan 2004). Having grown-up in the north end, Katz was particularly troubled by the escalating crime, gang violence, and drug houses he saw popping up in what he fondly remembered as a vibrant immigrant community growing up.

As mayor, Katz promised to carry on the property tax freezes begun under Thompson, to invest in more policing, to reduce "red tape" at city hall, and provide tax incentives for people to replace or renovate existing homes in older parts of the city. As with Thompson and Murray, Katz promised downtown revitalization while catering to the fiscal conservatism of the majority suburbs.

During his three-term tenure, he delivered on his promises to freeze property taxes and to increase investment in policing and public safety. Katz froze the property tax for seven years. Combined with Glen Murray's reduction and freeze, the City of Winnipeg did not raise property taxes for thirteen years in a row (from 1999 to 2011).

While freezing taxes, Katz and council substantially raised spending on police, fire, and paramedic services. From 2002 to 2012, police, fire, and paramedic services experienced a budget increase of 44 per cent *above* inflation and population growth. In total, their combined budgets increased by 73 per cent over a decade. To cover these increases and support the property tax freeze, Katz, following the approach of Thompson and Murray, continued to make cuts to the city's administration. By 2012, the city was spending 15 per cent less on organizational support, corporate, and governance compared to 2002 (a 44 per cent cut when accounting for population growth and inflation) (City of Winnipeg 2013).

Katz also continued with Murray's approach to downtown and inner-city redevelopment. While the city did provide funding for some investments, it relied increasingly on the private sector to deliver new development. Instead of directly funding projects, the city and province began offering property tax abatements to developers willing to build affordable housing in downtown, with a particular focus on the city's Exchange District. The province enabled the tax abatements by introducing the Community Revitalization Tax Increment Financing Act in 2009 (Manitoba, 2009).

Despite being billed as a form of tax increment financing (TIF), in practice the tool differed little from traditional forms of tax abatement used in the past to attract manufacturing to cities. In practice, the city and province would agree to hold property taxes at the original assessed amount before redevelopment for 25 years if developers built new housing (i.e., the developers retained any incremental increases to the property tax generated from the new development for that period) (Manitoba n.d.). In contrast, the prototypical TIF program involves the municipality using tax increments to pay for capital investments intended to stimulate new development and increases in property values within a defined area.

For many Winnipeggers, however, the greatest hope for renewal and image building in the city was not new affordable housing in downtown and inner city. Although it was largely the product of a partnership between prominent Winnipeg businessman Mark Chipman (chair of True North Sports and Entertainment) and Toronto billionaire David Thompson, for Sam Katz, the return of an NHL team to Winnipeg (the Winnipeg Jets 2.0) was a major coup (ESPN 2011). The return of the Jets has been the most unifying action of government and business in Winnipeg's recent history, re-establishing symbolic value, a positive image and pride in the city for rich and poor, young and old, and political left and right. While the new arena, started under Murray and completed under Katz, and the return of the NHL did little to address poverty in downtown, they did attract the interest and support of the city's suburbanites.

Despite such successes and continued popularity with the electorate through three elections, Sam Katz did little to stabilize the decline of the city's administration. Under Katz's tenure, a complicated, tense, and difficult time for political–administrative relationships re-emerged, this time leading to questioning of the ethical conduct of both the mayor and senior city administrators. Katz's tenure resulted in ongoing court challenges, growing scrutiny of municipal actions and decisions by the news media, and a growing revolt against corruption at city hall among Winnipeg residents.

A major problem that Sam Katz experienced almost immediately on assuming office was the scrutiny, bureaucracy, and due process that came with political office. He saw due diligence by administration as an obstruction to his vision of running the city like a small business without being challenged. His relationships with many long-serving councillors began to deteriorate. A long-term councillor complained that councillors outside the mayor's inner circle did not have the same access to information and that "there is an atmosphere at city hall where it's a no-no to speak up with a different point of view" (Edgar 2005, 18).

As with previous mayors and EPCs, Sam Katz and his EPC eventually recruited their own candidates for CAO. In 2008, the city hired, with Katz's approval, the former CAO of Kingston, Ontario. Although chosen by Katz and the EPC, the new CAO's tenure was short. Two years later (2011), Katz's friend and business partner Phil Sheegl became CAO. The hiring was controversial. With a background in land development, Sheegl had no experience in government administration and had only joined city staff in 2008 when hired as the director of planning and property development. Sheegl resigned from the city in 2013 under a cloud of suspicious land deals, pending audits, RCMP investigations, and court cases (CBC News 2013).

The relationship between Katz and Sheegl lacked many of the expected checks and balances between the most senior administrator and the mayor. Perceptions of conflict of interest plagued the two from the start, having a negative impact on public trust because personal, public, and private interests were intertwined. Repeated concerns about their dealings eroded public support for Katz. In response to allegations of corruption and low approval ratings, Katz chose not to run again in 2014.

Following his departure, Winnipeg's media has continued to uncover a string of questionable practices and deal making during Katz and Sheegl's tenure. Among these, the construction of the new police headquarters stands out. Under Katz, the city decided to embark on converting the old Canada Post mail sorting warehouse into a new police headquarters downtown. The project was plagued by time and costs overruns. The initial price tag that city staff presented to council was $136.6 million, including the purchase of an attached office building, which the city expected to sell. Today the estimated cost of the renovation is over $212 million, and the city has yet to sell the office tower (KPMG 2014). The cost overruns and administrative failures of the project are only part of the story, however. Early in Sam Katz's successor's first term in office, an RCMP investigation into the police headquarters and possible criminal corruption was initiated at the city's request. While RCMP chose not to charge anyone – a controversial decision to this day – the investigation raised questions about the choice of construction company for the project and issues of over-billing and kickbacks (Thorpe 2022).

The relationship between Sam Katz, Phil Sheegl, and various private interests brought into clear focus the continued problems surrounding the relationship between a strong mayor, who controls the EPC, and the CAO, who ultimately controls the conduct of the bureaucracy. Legislated powers and functioning informally behind closed doors prevented elected officials and city staff from challenging the conduct of the mayor and the EPC. After decades of continuous changes to strengthen the powers at the apex, the city council (as a whole) lacked a coherent structure for transparency and oversight of city decision making.

The Brian Bowman Era: 2014–2022

Following Katz's scandal-plagued tenure as mayor, Winnipeggers again chose a political neophyte, Brian Bowman, over experienced and notable politicians. In an election with a voter turnout of over 50 per cent, Bowman, a lawyer and former head of the Winnipeg Chamber of Commerce, became Winnipeg's first mayor of Métis descent. Elected on an ambitious platform of transit expansion and infrastructure investment, Bowman also promised to conduct business differently than his predecessor. Though he spoke little of changes to the city's administration during the 2014 election, he did promise to have council choose the members of the EPC and to reduce the power of the mayor (Kives 2016).

Few of Bowman's most ambitious proposals came to fruition during his two terms in office. While the city did complete the city's sole bus rapid transit (BRT) line, he was unable to deliver on his promise to build three more lines, and his ambitious agenda for new infrastructure investments for roads was curtailed. Although Bowman defended the city's finances through most of his two terms in office, the limited resources available to the city after over a decade of property tax freezes and cuts became increasingly apparent. While he raised property taxes by 2.33 per cent each year in office, he and council devoted all the additional revenue to covering the cost of the BRT (0.33 per cent) and to address the city's crumbling road infrastructure (2 per cent).

As noted in the summary of the city's financial trends in the 2020 budget, Bowman's tax increases did little to address what the report referred to as a "structural operating budget deficit" (City of Winnipeg 2019, 4-1). To balance the budget over the past decades, the city had repeatedly deferred infrastructure spending. While the tax increases did partially address the city's infrastructure deficit, the city continued to make cuts elsewhere to address yearly funding shortfalls (City of Winnipeg 2019). During Bowman's time in office, the city repeatedly cut transfers from the operating budget to capital (a 75 per cent cut in five years), increased the frontage levy,[2] and drew on dividends from the city's water and sewer utilities. By 2020, the city had largely exhausted these methods for balancing the budget (City of Winnipeg 2019). Given the city's restrictive finances, Bowman's decision to run a far more conservative campaign in 2018, with few big promises beyond continuing to fix the city's roads, was not surprising.

Bowman also struggled to change the city's relationship with traditional powers-that-be in the city. Bowman was successful in distancing himself from the various power players, returning some sense of ethics in the mayor's office. However, critics continued to complain about the opaqueness of city hall during his tenure (Kives 2015b).

Bowman often struggled when publicly challenging embedded and influential groups. Early in his tenure, he publicly criticized a deal made between CentureVenture and TNSE, where TNSE agreed to purchase, for $2 million, a parcel of land that CentureVenture had bought for $6.6 million. This criticism led to a public rebuke from TNSE chair Mark Chipman. For many Winnipeggers, Chipman was an unassailable hero for bringing the NHL back to city, and Bowman ultimately backed down following sustained criticism from the news media (Kives 2015a). Ultimately, along with the city's newly expanded Convention Centre, the new True North Square became a focus for downtown renewal, an area in the west of downtown advertised as the Sports, Hospitality, and Entertainment District (SHED).

Under Bowman, the city's approach to downtown and inner-city renewal differed little from that of Katz. While the city continued to discuss the need for more programs to address homelessness and poverty in downtown, its main investments continued to be tax abatements for new development, including TIF funds for True North Square (despite the absence of an affordable housing component).

The city did continue to work on its relationship with the Indigenous population, which had grown to over 90,000 residents as of 2021 (Statistics Canada 2022). Bowman and city staff focused on policies to make Winnipeg more accessible and welcoming to Indigenous residents, while First Nations established more urban reserves in the city. Nevertheless, the city's limited coffers prevented substantive funding for social programs to address homelessness and poverty in the inner city and downtown.

One of Bowman's priorities for downtown renewal in his first term in office involved the reopening of Portage and Main, the heart of the city's downtown, to pedestrians. Though lauded by urbanists, downtown residents and businesses, the reopening of Portage and Main became a major point of contention in the city, which reinforced the existing division between suburbs and inner-city residents. Under pressure from critics on council, Bowman agreed to add a question on reopening Portage and Main to the ballot in the 2018 election. Though Bowman would handily win re-election, over 70 per cent of residents voted against reopening the intersection, with a clear distinction between suburban and inner-city residents. Although billed by the media as one of the main issues going into the election, the question of Portage and Main did not resonate with most of the city's residents. Instead, residents were primarily concerned with rising crime in the city (Moore 2021).

Bowman's approach to governing also differed little from his predecessor. Bowman did return an element of trust to the mayor's office, as he actively distanced himself from many of the city's major movers and shakers and worked with the city administration to institute changes to city processes to provide stronger oversight. However, during his time in office, he strengthened the role of mayor and the EPC. Early on in his tenure as mayor, Bowman used the EPC's authority to suspend acting CAO Deepak Joshi, Phil Sheegl's replacement. Bowman indicated a lack of confidence in the CAO. At the time, questions about further deals hidden from council emerged in the media (CTV Winnipeg 2015). Following his suspension, Joshi chose to step down. This ushered in a period of revolving CAOs in Winnipeg, many of whom were restricted to "acting" status.

Instead of giving council the authority to elect members of the standing committees and the EPC (a promise requiring changes to Winnipeg's Charter Act), Bowman increased the number of members of the EPC from 7 to 9, despite a provision in the Winnipeg Charter limiting the number of members. Known as the EPC+2, the expanded executive body included the mayor, the six chairs of the standing committees, and deputy and acting deputy mayors, both positions appointed by the mayor. While the latter two roles were often held by chairs of standing committees, Bowman used them to expand his executive to a majority on council. Further, despite his promises to address the lack of transparency under the previous mayor, Bowman continued to hold informal meetings with his EPC+2 and city staff (Pursuga, 2021).

Bowman did bring some positive change to the city. Nevertheless, councillors outside of the EPC+2 raised many of the same concerns expressed by those left on the outside under Katz. And while the city has begun implementing measures to prevent the scandals that occurred under Katz and Sheegl, the compromised relationship between elected officials and city staff – the strong mayor/EPC with a subordinate CAO – remains in place in Winnipeg (Pursuga, 2021).

The Scott Gillingham Era: 2022–Present

Following Bowman's departure as mayor, Winnipeg experienced one of the most tightly contested mayoral elections in its history. Eleven candidates ran, including former leader of the provincial Liberals, Rona Bokhari, former MP for Winnipeg Centre, Robert-Falcon Ouellette, and high-profile city councillor, Kevin Klein. In the end, former city councillor Scott Gillingham edged out former mayor Glen Murray for victory.

Gillingham's tenure as mayor has coincided with a growing list of Indigenous-led projects in the inner city and downtown. Chief among these is a joint venture between the Southern Chiefs Organization and the real estate arm of TNSE to revitalize the former Hudson Bay building and Polo Park Mall. These projects include the new Indigenous-focused space, affordable rental development, and new space for social services (Bernhardt 2023). As mayor, Gillingham has delivered on his promise to increase transparency and the role of all councillors in city governance. One of his first acts was to reduce the number of standing committees from six to five, thus reducing the size of the EPC, and he ended Bowman's practice of including deputy and acting deputy mayors as members of the EPC. He also asked the province to give authority to appoint committee chairs back to council. Mayor Gillingham's biggest challenge, however, is to address a fiscal crisis that has been decades in the making. No longer able to cut transfers to capital or administration to balance the budget, the city is now scrambling to address an expected $9.1 million shortfall in 2024, *after* draining the city's reserve fund (MacLean 2024).

DISCUSSION AND CONCLUSION

In the first edition of *City Politics in Canada*, Kiernan and Walker (1983) argued that class conflict, made most explicit in the 1919 General Strike, ceased to be a factor in Winnipeg city politics and that the subsequent structural reforms failed to strengthen the urban polity despite their democracy and participation agenda. Successive reforms through the 1980s and 1990s would only exacerbate these problems.

The provincial government, through legislation and over time, strengthened the powers of the mayor such that they were effectively in control of the EPC, the CAO (administration), and the policy and planning agenda of council. This level of mayoral or EPC power without oversight and accountability of political/bureaucratic behaviour has compromised democratic representation and participation by citizens. In certain cases, private interests can override public interests.

In the 1990s, after the depoliticization of city council was complete, elected representatives and administrators embraced a managerial ethos that de-emphasized the role of citizen engagement and politics and emphasized the business principles of outputs, contracting out, and fiscal management. Since Susan Thompson, successive mayors have constrained civic finances through tax freezes and cuts, severely restricting the city's capacity to engage in substantive programming, while council's ability to address the most pressing needs of the city depended on the whim of mayor or the EPC. Mayor Gillingham has demonstrated a willingness to empower council at the expense of his own office, but it remains to be seen whether a more transparent and collegial council can address the problems resulting from prior mayors' fiscal decisions.

From 1998 to 2016, Winnipeg property taxes only increased by 9 per cent in nominal terms. The average of other major western Canadian cities was 96 per cent (City of Winnipeg 2017). While most cities effectively doubled their revenue from property taxes over time, Winnipeg's revenue stayed relatively flat (and indeed declined in real terms, accounting for inflation). Moreover, the 9 per cent increase occurred towards the end of this period.

By the end of mayor Katz's last term and the beginning of Bowman's tenure as mayor, the erosion of the city's infrastructure and services became apparent. This dissatisfaction culminated in the election of Scott Gillingham as mayor in 2022. Gillingham, the former finance chair under Bowman, was one of only two mayoral election candidates out of eleven to explicitly promise to raise property taxes to tackle the city's fiscal problems and infrastructure needs. Even Gillingham's promised increases of 3.5 per cent a year may be inadequate to make up for over a decade of freezes.

The state of the city's finances enabled successive mayors to run on efficiency platforms and enforce austerity measures focused on keeping property tax rates low, eliminating "non-core" services, and relegating social and health needs to other levels of government and community-based organizations.

In the past, fiscally conservative suburban politicians, who made up the majority of the council and the powerful EPC, influenced approaches to city revitalization by prioritizing suburban development and downtown revitalization over inner-city renewal. Dealing with inner-city social needs often took a back seat to investments in new development and infrastructure intended to improve the city's image and lure tourists and suburbanites back into the downtown core. Today, in some cases, the city provides inner-city CBOs with small grants, coordination, and facilitates access to other sources for long-term funding (e.g., homelessness initiatives).

While the five mayors who have held office since 1980 all recognized the issues of poverty and homelessness in the inner city, their response has been mixed and has depended on tripartite agreements and the availability of private funding. The scale and complexity of the social, economic, and health challenges and a lack of financial resources have been serious impediments to revitalization efforts and to addressing poverty and homelessness.

Beginning with Susan Thompson, the mayors and councils increasingly recognized both the importance of the growing Indigenous population in the city and the fact that a disproportionate number of Indigenous residents lived in poverty in the inner city. However, programming to help vulnerable inner-city residents often gave way to investments in new development and infrastructure.

Although the city's dominant focus has been on urban revitalization schemes supported by tripartite agreements for financing, Winnipeg's future is also linked to the health and vitality of its most vulnerable citizens. Winnipeg needs to find appropriate strategies and partnerships to deal with the social and economic needs of all its citizens in all their complexity, contradictions, and conflicts.

NOTES

1 Skid Row refers to a neighbourhood in Los Angeles known for its high concentration of homeless people, many struggling with health issues and addiction, and the hotels, temporary shelters, and encampments that "house" them.
2 The frontage levy is a form of tax based on the length of a property where it fronts a municipal street. In Winnipeg, the levy only applies if the street contains a sewer or water main.

REFERENCES

Axworthy, Lloyd. 1978. *The Best Laid Plans Oft Go Astray: The Case of Winnipeg.* The Institute of Urban Studies. https://winnspace.uwinnipeg.ca/bitstream/handle/10680/909/029-1978-Axworthy -BestLaidPlansOftGoAstray-WEB.pdf?sequence=1.

Bellan, Matt. 2004. "Sam Katz, Winnipeg's First Jewish Mayor, Expressed Mixed Feelings on Election Night." *The Jewish Post and News,* June 23. https://jewishpostandnews.com/samkatz.html.

Bougas, Constantinos. 2015. *Final Report: Long-Term Population, Housing and Economic Forecast for Winnipeg.* The Conference Board of Canada. https://legacy.winnipeg.ca/cao/pdfs/ConfBoardCanFinalRpt _LTPop-Housg-Ec_Forecast_Wpg.pdf.

Bernhardt, Darren. 2023. "True North, Southern Chiefs Sign Partnership they Say Will Be Catalyst to Revitalize Downtown Winnipeg." *CBC News,* December 12. https://www.cbc.ca/news/canada /manitoba/redevelop-plans-portage-place-bay-true-north-sco-winnipeg-1.7056507.

Carter, Tom. 2009. *An Evaluation of the Manitoba Provincial Nominee Program.* Manitoba Labour and Immigration Division. https://www.immigratemanitoba.com/wp-content/uploads/2017/11/pnp -manitoba-provincial-nominee-program-tom-carter-report-2009.pdf.

Cassidy, J. 1971. "The Policy Paper in Brief." In *The Future City: A Selection of Views on the Reorganization of Government in Greater Winnipeg,* edited by Lloyd Axworthy, 5–8. Winnipeg: University of Winnipeg Press. https://winnspace.uwinnipeg.ca/bitstream/handle/10680/829/1971-Axworthy-FutureCity1 -SelectionViewsReorgGovt%20%28split%29.pdf.

CBC News. 2013. "Police HQ cost overruns final straw forcing Sheegl out." *CBC News,* October 24. https://www.cbc.ca/lite/story/1.2223829.

City of Winnipeg. 2013. *2013 Adopted Operating Budget: Adopted by Council on January 29, 2013.*

City of Winnipeg. 2017. *2018 Budget – Volume 1: Community Trends and Performance Report.* July.

City of Winnipeg. 2019. *Community Trends and Performance Report: Volume 1 for 2020 Budget.* July.

Cole, Brian. 1998. "Editorial: Election Promises." *Winnipeg Free Press,* September 19. https://access -newspaperarchive-com.uwinnipeg.idm.oclc.org/ca/manitoba/winnipeg/winnipeg-free-press /1998/09-19/.

CTV Winnipeg. 2015. "Mayor Bowman Suspends Acting City CAO Deepak Joshi." *CTV News,* January 16.

Cuff, George B. 1997. *The City of Winnipeg Organizational Review and Performance Assessment: A Corporate Review, Draft Report.* A Report by George B. Cuff & Associates.

Edgar, Patti. 2005. "Mayor Katz's Year in Review and a Look Ahead to Whether He'll Run Again." *Winnipeg Free Press,* December 28. https://access-newspaperarchive-com.uwinnipeg.idm.oclc.org/ca /manitoba/winnipeg/winnipeg-free-press/2005/12-28/page-18.

Economic Development Winnipeg (EDW). 2015. *Economic Performance Indicators.*

ESPN. 2011. "Winnipeg Mayor: Thrashers Coming." ESPN, May 20. https://www.espn.com/nhl/news /story?id=6571287.

Fayerman, Pamel. 1980. "Fresh Faces Cite Anti-ICEC Vote." *Winnipeg Free Press*, October 23. https:
//access-newspaperarchive-com.uwinnipeg.idm.oclc.org/ca/manitoba/winnipeg/winnipeg-free
-press/1980/10-23/page-18.

Hum, Derek, Frank Strain, and Michelle Strain. 1986. "Fiscal Imbalance and Winnipeg: A Century of
Response." *Urban History Review* 15 (2): 137–50. https://doi.org/10.7202/1018619ar.

Kiernan, Mathew J., and David C. Walker. 1983. "Winnipeg." In *City Politics in Canada*, edited by
Warren Magnusson and Andrew Sancton, 222–54. Toronto: University of Toronto Press. https://doi
.org/10.3138/9781487575908-008.

Kives, Bartley. 2015. "CentreVenture Fires Back." *Winnipeg Free Press*, February 3. https://www.proquest
.com/newspapers/centreventure-fires-back/docview/1650361461/se-2?accountid=15067.

Kives, Bartley. 2015. "Still Waiting for the Mayor's Promised Accountability. *Winnipeg Free Press*, July 3.
https://www.proquest.com/newspapers/still-waiting-mayors-promised-accountability/docview
/1692901220/se-2?accountid=15067.

Kives, Bartley. 2016. "Bowman Learns Some Promises Aren't Worth Keeping." CBC News, October 30.
https://www.cbc.ca/news/canada/manitoba/epc-shuffle-analysis-1.3827739.

KPMG. 2014. *Draft report: Winnipeg Police Service Headquarters Construction Project Audit*.

Layne, Judy. 2000. "Marked for Success??? The Winnipeg Core Area Initiative's Approach to Urban
Revitalization." *Canadian Journal of Regional Science* 23 (2): 249–78. https://idjs.ca/images/rcsr
/archives/V23N2-Layne.pdf.

Leo, Christopher, and Kathryn Anderson. 2006. "Being Realistic About Urban Growth." *Journal of Urban
Affairs* 28 (2): 169–89. https://doi.org/10.1111/j.0735-2166.2006.00266.x.

Leo, Christopher, and Wilson Brown. 2000. "Slow Growth and Urban Development Policy." *Journal of
Urban Affairs* 22 (2): 193–213. https://doi.org/10.1111/0735-2166.00050.

Lett, Dan. 1992. "Thompson Was 'Best Man for the Job'." *Winnipeg Fee Press*, October 29. https:
//access-newspaperarchive-com.uwinnipeg.idm.oclc.org/ca/manitoba/winnipeg/winnipeg-free
-press/1992/10-29.

Lett, Dan. 1993. Wielding Knife, Thompson Ends Up Wounding Herself. *Winnipeg Free Press*, October
16. https://access-newspaperarchive-com.uwinnipeg.idm.oclc.org/ca/manitoba/winnipeg/winnipeg
-free-press/1993/10-16/page-14.

Lightbody, James. 1978. "Electoral Reform in Local Government: The Case of Winnipeg." *Canadian
Journal of Political Science* 11 (2): 307–32. https://doi.org/10.1017/S0008423900041111.

MacLean, Cameron. 2024. "Projected $23M Deficit, Depleted Reserve Fund Leave City of Winnipeg
in 'New Territory': Finance Chair." *CBC News*, November 21. https://www.cbc.ca/news/canada
/manitoba/city-in-new-territory-without-reserve-fund-to-cover-expected-deficit-1.7389949.

Macpherson, L.G. 1940. "Report of the Royal Commission on the Municipal Finances and
Administration of the City of Winnipeg, 1939." *The Canadian Journal of Economics and Political Science*
6 (1), 68–72. https://doi.org/10.2307/137056.

Manitoba. n.d. Backgrounder: Tax Increment Financing Questions and Answers.

Manitoba. 1989. *The City of Winnipeg Amendment Act, SM 1989-90, c 8*. https://web2.gov.mb.ca/laws
/statutes/1989-90/c00889-90e.php.

Manitoba. 1998. *The City of Winnipeg Amendment and Consequential Amendments Act, SM 1998, c 37*.
https://canlii.ca/t/b4vn.

Manitoba. 2002. *The City of Winnipeg Charter, SM 2002, c 39*. https://www.canlii.org/en/mb/laws/stat
/sm-2002-c-39/latest/sm-2002-c-39.html. https://doi.org/10.5860/CHOICE.39-5568.

Manitoba. 2009. *The Community Revitalization Tax Increment Financing Act, SM 2009, c 29*. https://web2
.gov.mb.ca/laws/statutes/2009/c02909e.php#.

McArthur, Keith. 1999. "People, People, People: Thriving Downtown Key to How Entire City Is Viewed." *Winnipeg Free Press*, May 9. https://access-newspaperarchive-com.uwinnipeg.idm.oclc.org/ca/manitoba/winnipeg/winnipeg-free-press/1999/05-09/page-115.

Moore, Aaron A. 2021. "Winnipeg." In *Big City Elections in Canada*, edited by Jack. Lucas & R. Michael McGregor, 193–211. Toronto: University of Toronto Press.

Moore, Terry. 1998. "Fresh promises." *Winnipeg Free Press*, October 30. https://access-newspaperarchive-com.uwinnipeg.idm.oclc.org/ca/manitoba/winnipeg/winnipeg-free-press/1998/10-30/.

Moore, Terry. 2002. "Second Term for Murray." *Winnipeg Free Press*, October 21. https://access-newspaperarchive-com.uwinnipeg.idm.oclc.org/ca/manitoba/winnipeg/winnipeg-free-press/2002/10-21/page-14.

O'Brien, David. 1999. "Glen Murray – The First Six Months: Mayor Wants to Prove a Few Points." *Winnipeg Free Press*, May 11. https://access-newspaperarchive-com.uwinnipeg.idm.oclc.org/ca/manitoba/winnipeg/winnipeg-free-press/1999/05-11/page-2.

Prism Economics and Analysis. *Regional Manufacturing Profile: Winnipeg Region*. Prepared for Canadian Manufacturers & Exporters and the Canadian Skills Training & Employment Coalition. https://static1.squarespace.com/static/5c5b05dd9d41495d4f344e70/t/5cbde4528165f58c1db00174/1555948648204/MANUFACTURING-PROFILE-Winnipeg.pdf.

Rollason, Kevin. 1998. "Murray Won Inner City, Was Competitive in Suburbs." *Winnipeg Free Press*, October 30. https://access-newspaperarchive-com.uwinnipeg.idm.oclc.org/ca/manitoba/winnipeg/winnipeg-free-press/1998/10-30/page-7.

Statistics Canada. n.d.a. *Table 17-10-0021-01 Estimates of the Components of Interprovincial Migration, Annual*. https://www150.statcan.gc.ca/t1/tbl1/en/tv.action?pid=1710002101.

Statistics Canada. n.d.b. *Table 17-10-0015-01 Estimates of the Components of Interprovincial Migration, By Age and Sex, Annual*. https://www150.statcan.gc.ca/t1/tbl1/en/tv.action?pid=1710001501.

Statistics Canada. n.d.c. *Table 35-10-0189-01 Crime Severity Index and Weighted Clearance Rates, Police Services in Manitoba*. https://www150.statcan.gc.ca/t1/tbl1/en/tv.action?pid=3510018901.

Statistics Canada. 1981. *1981 Census of Population, Statistics Canada Catalogue no. 97-570-X1981005*.

Statistics Canada. 1986. *1986 Census of Population, Statistics Canada Catalogue no. 97-570-1986003*.

Statistics Canada. 1991. *1991 Census of Population, Statistics Canada Catalogue no. 95F0168X*.

Statistics Canada. 1996. *1996 Census of Population, Statistics Canada Catalogue no. 95F0181XDB96001*.

Statistics Canada. 2002. *2001 Community Profiles. Released June 27, 2002. Last modified: 2005-11-30. Statistics Canada Catalogue no. 93F0053XIE*. Ottawa. https://www12.statcan.gc.ca/english/Profil01/CP01/Index.cfm?Lang=E.

Statistics Canada. 2007. *Winnipeg, Manitoba (Code4611040) (table). 2006 Community Profiles. 2006 Census. Statistics Canada Catalogue no. 92-591-XWE*. Ottawa. https://www12.statcan.gc.ca/census-recensement/2006/dp-pd/prof/92-591/.

Statistics Canada. 2012. *Winnipeg, Manitoba (Code 4611040) and Canada (Code 01) (table). Census Profile. 2011 Census. Statistics Canada Catalogue no. 98-316-XWE*. Ottawa. https://www12.statcan.gc.ca/census-recensement/2011/dp-pd/prof/index.cfm?Lang=E.

Statistics Canada. 2017. *Winnipeg, CY [Census subdivision], Manitoba and Division No. 11, CDR [Census division], Manitoba (table). Census Profile. 2016 Census. Statistics Canada Catalogue no. 98-316-X2016001*. Ottawa. https://www150.statcan.gc.ca/n1/en/catalogue/98-316-X2016001.

Statistics Canada. 2022a. *Census Profile. 2021 Census. Statistics Canada Catalogue no. 98-316-X2021001*. Ottawa. https://www150.statcan.gc.ca/n1/en/catalogue/98-316-X2021001.

Statistics Canada. 2022b. Immigrants Make Up the Largest Share of the Population in Over 150 Years and Continue to Shape Who We Are as Canadians. *The Daily*, 26 October. https://www150 .statcan.gc.ca/n1/daily-quotidien/221026/dq221026a-eng.htm.

Statistics Canada. 2022c. *Census in Brief. Disaggregated Trends in Poverty from the 2021 Census of Population. Catalogue Number 98-200-X, Issue 2021009.* 9 November. https://www12.statcan.gc.ca/census -recensement/2021/as-sa/98-200-X/2021009/98-200-x2021009-eng.cfm.

Stelman, Ursula. 1998. *Winnipeg's Main Street: A Search for Meaning. Local Government Case Studies.* London, ON: University of Western Ontario.

Thampi, Radha Krishnan. 1991. "Norrie Catches Brickbats While Taking Bow for Budget." *Winnipeg Free Press*, March 21. https://access-newspaperarchive-com.uwinnipeg.idm.oclc.org/ca/manitoba /winnipeg/winnipeg-free-press/1991/03-21/page-16.

Thompson, Susan A., and Terry Létienne. 2016. *Her Worship, Moments in History, Moments in Time.* Victoria, BC: Friesen Press.

Thorpe, Ryan. 2022. "Capital Projects, Critical Issues, Clear Indifference." *Winnipeg Free Press*, June 3. https://www.proquest.com/newspapers/police-not-innocent-hq-cost-overruns-report/docview /2672551179/se-2?accountid=15067.

Welch, Mary A. 2004. Grand Slam for Sam. *Winnipeg Free Press*, June 23. https://access-newspaperarchive -com.uwinnipeg.idm.oclc.org/ca/manitoba/winnipeg/winnipeg-free-press/2004/06-23/page-1.

Winnipeg Free Press. 1979. "Varied Platforms highlight race: 12 men seeking election to mayor's post." Winnipeg Free Press, June 15, 2. https://access-newspaperarchive-com.uwinnipeg.idm.oclc.org/ca /manitoba/winnipeg/winnipeg-free-press/1979/06-15/page-4.

7

Vancouver

Ian Bushfield and Stewart Prest

INTRODUCTION

The past forty years have transformed Vancouver. What was then a comfortably second-tier western Canadian city and regional transportation hub is now a world-renowned city with a distinct global identity. Two international events nearly thirty years apart – Expo 86 and the 2010 Winter Olympics – put the city on the map globally and provided the impetus to significant urban development in the ensuing decades.

The same period saw the city inspire a new trend in urban planning and experience backlash in the inability of many residents to share in that vision. Indeed, Vancouver's story in recent decades is deeply ironic: even as Vancouver gave rise to *Vancouverism* – a byword for liveable mixed-use urban design – the city became less affordable and unliveable for many residents.

We explore two dominant themes in this chapter. First, the political dynamics of the city exhibit characteristics of a political "punctuated equilibrium" – periods of political stability marked by moments of rapid change in response to a shift in the political environment.[1] The moments of stability tend to be characterized by prolonged dominance by a single party, whether the Non-Partisan Association (NPA) in the 1990s or Vision through the 2010s. At such times, ideological commitments provide an invaluable guide for voters (Cutler and Matthews 2005). Those periods of stability are interspersed by moments of transformative change, not only of the ruling party but, at times, the party system as a whole: a pattern first documented by Tennant (1980). Governing coalitions broke down as they confronted policy challenges that pulled them outside ideological comfort zones, giving rise to periods of instability and party reformation. Notable examples include the city's hosting of Expo 86 and the 2010 Winter Olympics, events that challenged parties of the left and the right in distinct ways.

Second, when the party system does shift, issues of land use and broader disagreements over urbanization are at the heart of the fragmentation. Real estate is the unofficial pastime of the Lower

Mainland, and debates over land use and other aspects of urban development policy lie at the heart of multiple governing party crack-ups. The politics of land use accordingly shape every era of politics we talk about in this chapter, in ways that often defy ideological expectations. Vancouverites' views over where, how, and how much to urbanize the city are shaped as much by one's housing situation as by ideology. Whether one is an owner or a renter, whether one is a long-term tenant or new arrival, whether one is comfortably, uncomfortably, or precariously housed – or homeless[2] – such experiences decisively inform Vancouverites' lived experience of the city and their views of what ought to be done to improve it. Vancouver is at once a city of long-time homeowners-turned-paper-multimillionaires, house-poor holders of massive mortgages, long-term renters, and marginal tenants struggling to get by. It is a city marked by persistent challenges related to homelessness and precarious housing. The ensuing debates over housing and related issues of urbanization – from basement suites, to transit investments, to bike lanes, to municipal zoning, to drug use and other challenges facing the city's marginalized communities in the Downtown Eastside (DTES) and elsewhere – have played a significant role in splintering traditional political coalitions on both the left and right of the city's political spectrum in the last thirty years. The results of those debates have reshaped both the physical and political geography of the city.

The remainder of the chapter proceeds as follows. The next section outlines a thumbnail sketch of the city and the broader metro region. We then delve into the institutional characteristics of the city, including several salient features that distinguish Vancouver from other major centres in the country – most notably, the presence of a federated municipal governance structure, an at-large voting system, and a well-developed municipal party system. These characteristics are distinct from much of the rest of the country and have shaped Vancouver politics in significant ways.

We conclude by charting successive political eras in the city. We narrate that history in three parts, each denoting a period of stability bookended by political disruption. First, there was the era of NPA dominance from 1986 to 2002. Second, we chronicle the steady rise of the new urban left, led first by a moderate faction of the Coalition of Progressive Electors (aka "COPE Lite") and subsequently by Vision Vancouver under Gregor Robertson from 2002 to 2018. Finally, we look at the recent period of uncertainty, abruptly ended by the recent resounding victory by Ken Sim and his nascent A Better City (ABC) party in late 2022. With each era, we note significant developments precipitating the rise and fall of each governing coalition.

SHIFTING TERRAIN: SOCIAL, ECONOMIC, AND GEOGRAPHICAL REALITIES

Social Change

Vancouver is among the fastest-growing regions of Canada. From 1981 to 2021, the City of Vancouver grew from 414,000 people to over 662,000 (Statistics Canada 1981 and 2023). Its growth is made all the more remarkable given that the city's population was actually declining in the 1970s. Metro Vancouver, which includes the city and surrounding area, commonly referred to as the Lower Mainland to distinguish it from both Vancouver Island and the rest of the mainland, encompasses 21 suburban municipalities, a set of unincorporated areas, which includes the University of British Columbia, and the Tsawwassen Treaty First Nation. It grew from 1.15 million to 2.64 million over the same period. Beyond the urban core, Metro Vancouver is a complex region of criss-crossing governance structures, including vibrant Indigenous communities responsible for some of the most dynamic, innovative, and controversial land use decisions in the region in recent years. While the City of Vancouver attracts much of the world's attention – and constitutes the focus of this chapter – the politics of the wider region have continued to develop a distinct and varied character of their own.

Vancouver sits on the unceded traditional territories of the xʷməθkʷəy̓əm (Musqueam), Sḵwx̱wú7mesh (Squamish), and səlilwətaɬ (Tsleil-Waututh) Nations. Recent census data indicates that Indigenous peoples in the City of Vancouver account for about 2.4 per cent of the municipal population.[3] Between 1996 and 2016, the reported Indigenous census population nearly doubled in Metro Vancouver, growing at more than three times the region's rate. Politically, in the last forty years, the institutions of government in Vancouver have gradually become more attuned to Indigenous issues, an awareness reflected in everything from embracing the now routine land acknowledgments to more recent civic support for land development under the leadership of local First Nations. This trend is most apparent with the planned Sen̓áḵw development project on the Squamish lands surrounding the south end of the Burrard Bridge. The project, while proceeding with city support in the form of a service agreement, has been planned and approved entirely through Squamish governance processes.[4] It has also attracted significant criticism along the way.[5]

The city now welcomes newcomers from around the world, in stark contrast with the city's chequered history of racial exclusion and discrimination.[6] More than half the population identified as neither white nor Indigenous in the 2016 census (City of Vancouver 2020). Even more striking, in the 2016 census nearly as many Vancouverites reported a non-official language as their first language as English. Beyond the increase in absolute and relative numbers, the human geography of the city and region has changed as well. Many municipalities in Metro Vancouver have developed unique cultural identities associated with the arrival of immigrant populations from parts of the world in ways that shape politics at all levels. Given the importance of strong networks in municipal politics, these communities have become essential players in elections. Even so, elected representation continues to lack diversity in the region thanks to factors such as incumbent advantage and a lack of term limits (Singh 2022). The election of Ken Sim in 2022 was a watershed moment, marking the first time Vancouver has elected a mayor of Asian descent.[7] Vancouver City Council has slowly come to better reflect the city's population in other ways, with at least half the council composed of women following each of the last three elections. That said, the next woman elected mayor will be the city's first. Other political ceilings remain unbroken as well; for instance, the city has never elected a South Asian councillor.[8]

Economic Change

Of all the cleavages dividing Vancouver, the divisions created by wealth, and lack thereof, may be the most politically salient. While the sources of wealth are varied, the divisive effects are most keenly felt in property ownership and tenancy. It is a divide, moreover, that has deepened dramatically in recent decades. The City of Vancouver is, in reality, two cities: one of homeowners and another of renters.[9] As with other areas in the country, homeowners have benefited tremendously from the growth in property values in the city over the last four decades. These trends date back to at least the 1980s.[10] New homeowners have benefited less, though as prices continue to climb, even those who bought a few years ago are, at least on paper, significantly better off.

Nationwide, the net worth of families owning a home rose to $685,400 in 2019, up from $323,700 in 1999, as renting families' net worth increased from $14,600 to just $24,000 (Statistics Canada 2022a). Given that housing prices in Vancouver continually vie for the distinction of highest in the country – the average home in the city cost well over $1.1 million in January 2023 (Canadian Real Estate Association n.d.) – the gap is that much more significant in the Metro Vancouver area, comparable only to Toronto. With even starter homes costing so much, homeownership of any kind is increasingly available only to the very highest income earners, those already with a foot on the property ladder, or those with access to intergenerational wealth. Indeed, owners in Vancouver tend to have income at least twice that of renters at the median (Statistics Canada 2022b). Just over half the city's households rent, in contrast with the metro region, where 63 per cent of households own their home (City of Vancouver 2020). With a one-bedroom unit costing more than $2,000 per month in Vancouver as of mid-2022, the divide has become more visible and politically salient in the city's recent political history (Anderson, 2022).

Even so, politicians have not always successfully converted renters' fears and frustrations into an effective voting bloc. There are several potential reasons for this gap. One factor relates to the varied demographics of home ownership and renters. Renters tend to be less socio-economically advantaged, younger, and from more marginalized communities.[11] Other research makes clear that such renters (less socio-economically advantaged and younger residents) are less likely to vote than their older, wealthier counterparts (Nakhaie 2006; Blais 2000). Thus, those more likely to rent are also less likely to vote, particularly in municipal elections (Kushner and Siegel 2006; McGregor and Spicer 2016).

Second, while renters generally have a shared interest in seeing rental prices stay low, in other respects, they may differ in terms of what policies to pursue to achieve that end. Given British Columbia's limited form of rent regulation, where rent increases are limited for current tenants but not when units are listed on the market, long-time tenants may focus on policies such as enhanced rent control and minimal neighbourhood development. In contrast, those looking for a place to rent may favour candidates promising to increase the overall housing supply in general and rental housing in particular.[12] Voters, like others in the province, still filter politics through the left–right spectrum. While renters tend to skew left, polarization nonetheless further complicates efforts to build a renter-based voting coalition.[13]

Beyond the cost of housing, the wealth divergence shows up in other ways as well. Some effects are visible and politically divisive, as with the continued growth of the homeless population of the DTES neighbourhood, and the concerns of residents in adjacent neighbourhoods about issues of public safety. While indirectly related to economic issues, the continuing toll of the province's poison drugs crisis has left a further mark, fuelling calls in numerous elections for everything from drug decriminalization and the provision of safe supply to action on substandard housing to increased law enforcement. These strains are evident in the recurring conflicts between the housed and the unhoused in neighbourhoods like Strathcona in Vancouver and elsewhere in the Lower Mainland (Little and Armstrong 2020). The emergence of a "Vancouver model" of money laundering fuelled by international gambling, and the subsequent scandals and efforts to combat, constitute another intersection of wealth and politics in the city and province (Porter et al. 2022).

Returning to the licit economy, the region's economic basis has changed in other ways in the last forty years. While the port remains a crucial part of the city's economic success, other aspects of the economy have transformed dramatically. Manufacturing, for instance, accounted for less than 5 per cent of Vancouver's employment in 2021, down from more than 13 per cent in 1981 (Statistics Canada 1981, 2023). This decline has had multiple effects. There has been growth in various new industries ranging from film and television to animation and gaming, increase in support for the province's resource sector, and increased construction related to ongoing development. It has also allowed residential growth and densification, as the deindustrialization of Coal Harbour, Yaletown, and the eastern half of False Creek freed significant new urban areas for residential and mixed-use development.

INSTITUTIONS AND INSTITUTIONAL CHANGE

Regional Governance

Located at the mouth of the Fraser River in southwest British Columbia, Metro Vancouver is defined and shaped by its geography in myriad ways. It is nestled in a triangular space between the Coastal Mountains, the Salish Sea, and the American border and is in many ways an outdoor paradise. Downhill skiing is just a bus ride away from sandy beaches stretching for kilometres in and around the metro core. Regions within the metro area are defined and separated by the waterways between them. The municipality of Vancouver constitutes a regional hub and political and economic core, surrounded by a ring of smaller municipalities of varying degrees of urbanization. The North Shore communities lie to the north across the Burrard Inlet, while

Figure 7.1: Metro Vancouver

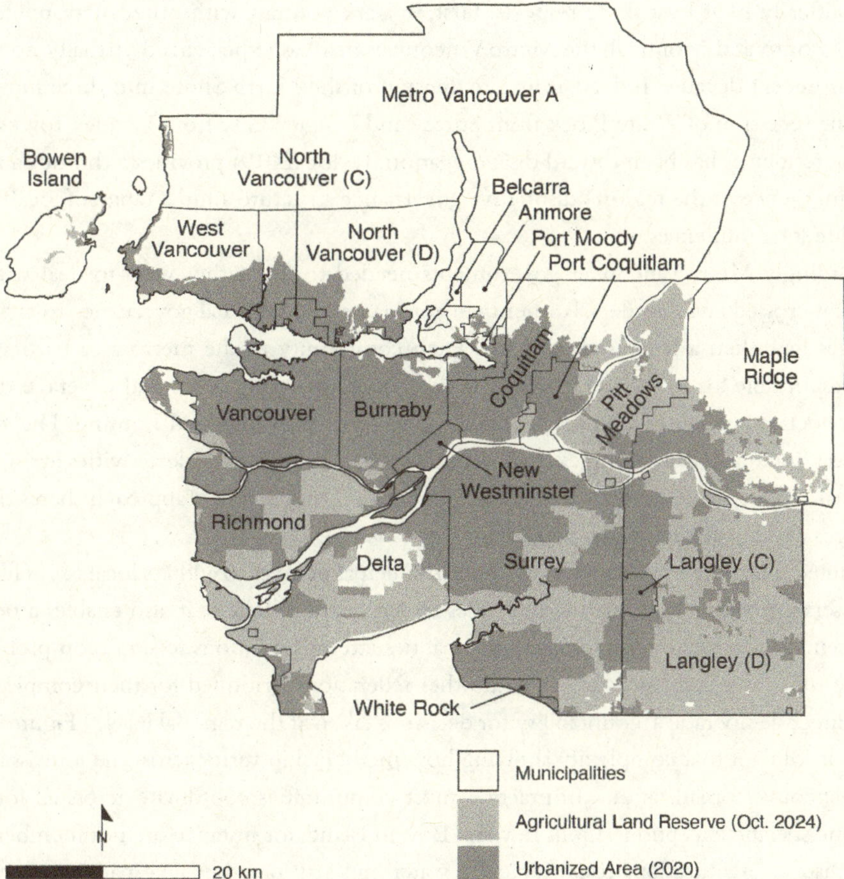

Notes: The City of Vancouver is one of 21 constituent municipalities of the Metro Vancouver Regional District, along with several Indian Reserves (not labelled) and unincorporated areas administered as Metro Vancouver Electoral Area A. Much developable land is already urbanized; most of the remainder is part of the provincial Agricultural Land Reserve (ALR). Source: ALR – https://www.alc.gov.bc.ca/alr-maps/ (Oct. 31, 2024); Contiguous Settlement Area – Statistics Canada.

Surrey and other suburban communities, known collectively as "South of the Fraser," are to the south across the Fraser River. Richmond sits on a river delta between Vancouver and Surrey. Last but not least, Burnaby, New Westminster, and other smaller municipalities stretch to the east of Vancouver proper, running along the north shore of the Fraser. Much of the land between the city and the mountains to the east remains undeveloped farmland, due to a provincial decision in the 1970s to set it aside as part of a decision to create a permanent Agricultural Land Reserve (see Figure 7.1).

Vancouver, while in many ways like other major urban centres in Canada, stands out as distinct, politically, in at least three respects. First, in stark contrast with other metropolitan areas such as Toronto and Montreal, the Metro Vancouver area has experienced virtually no amalgamation in recent decades. Indeed, from the division of the North Shore into three municipalities, to the secession of White Rock from Surrey and Langley City from Langley Township, the historical tendency has been toward disaggregation. Taylor (2019) provides a thorough account of the emergence of the region's distinctive governance structure amid a range of local, provincial, and federal influences.

Accordingly, Metro Vancouver governments needed to find other ways to deal with challenges that crossed civic borders. Rather than amalgamation, regional governance to tackle such challenges has taken a federated structure. Each community in the metro area participates in boards responsible for a range of region-wide activities, including water and sewerage, drainage, waste collection and disposal, transportation, long-term planning, and housing. The resulting federated structure is thus complex and, at times, ungainly; region-wide activities are subject to occasional restructuring, further complicating matters. This federated approach shares the same advantages and challenges as any federation.

Advantages include implementing bespoke municipal policies to address local issues like community services while coordinating responses to regional challenges. It also enables a degree of experimentation across the region as municipalities attempt solutions to common problems. At the same time, such structures are, as with other federations, critiqued for their complexity and lack of direct democratic accountability for decisions taken at the regional level.[14] Figure 7.2 tries to unpack some of that complexity, showing how membership varies across the four issue-based Metro Vancouver organizations. In practice, most communities coordinate across all four areas, with some specific exceptions. Lions Bay and Bowen Island, for instance, are not members of the Metro District organizations responsible for water and sewerage but take part in other boards responsible for regional parks, long-term planning, social housing, and other regional initiatives. The Village of Belcarra and Tsawwassen First Nation do not participate in the regional sewerage board, and White Rock (a former ward of Surrey) is not a member of the regional water board.

Cities receive voting power commensurate with their population size. Thus, larger centres – notably Vancouver and increasingly Surrey – loom large in regional decisions. A motion is passed in most instances when board members representing more than half the region's population support the measure. The defined voting structure ensures that definitive decisions may be taken even without consensus among members. Simply put, once a vote occurs, the majority (weighted by population) rules.

While at times ungainly, the federated regional structure in the Vancouver region enables a fascinating combination of local decision making for most municipal decisions, from property taxes to policing. At the same time, it still allows region-wide cooperation in areas of shared responsibility.

Figure 7.2: The complex federal structures of Metro Vancouver

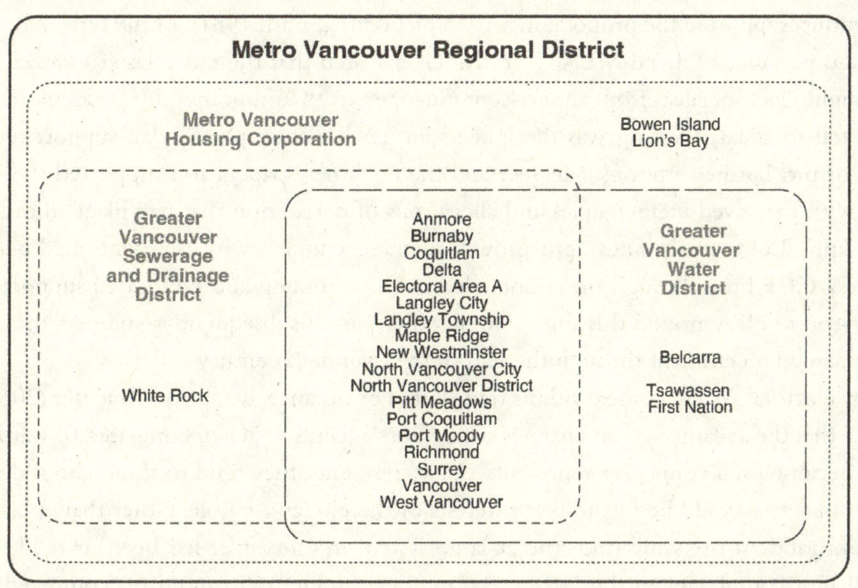

Notes: Not all constituent entities of Metro Vancouver participate in all of its services. Local representatives serve on four interlocking boards: the Metro Vancouver Regional District, the Metro Vancouver Housing Corporation, the Greater Vancouver Sewerage and Drainage District, and the Greater Vancouver Water District. See Metro Vancouver Regional District, Board Members, and Voting Strength, https://metrovancouver.org/boards/board-members.

At-Large Voting and the Role of Municipal Parties

Two further interrelated institutional practices distinguish Vancouver from most other large municipalities in Canada: the at-large voting system and the municipal party system. Unlike most other provinces in Canada, British Columbia's municipal government legislation allows for the formation of municipal "elector organizations," or civic political parties. Politics in the larger centres – Surrey, Richmond, Burnaby, and, above all, the City of Vancouver – tends to be defined by local party politics. These parties, in turn, play a major role in simplifying the at-large voting process for voters.

Effectively, each city using the system constitutes a single large multi-member ward, and the candidates with the most votes win seats on council.[15] Voters may vote for as many candidates as there are council positions; they elect the mayor on a separate citywide ballot. In the case of the City of Vancouver, the top ten candidates are elected to city council; in Surrey, the top eight candidates win seats. Following the election, each councillor represents the entire city.

The at-large system was introduced in 1936 after voters (69 per cent in favour, with 19 per cent turnout) supported the proposal in a 1935 plebiscite (Smith 1981). At the time, Vancouver's mayor and provincial Liberal MLA Gerry McGeer feared that the city's eastern working-class wards might elect socialists from the nascent Co-operative Commonwealth Federation (CCF). The switch to at-large voting was thought to entrench a solid plurality of support for more centrist or pro-business options; McGeer and the city's newspapers also supported the change to deal with perceived inefficiencies and allegations of corruption that had plagued the council. The appeal of a pro-business, pro-growth message, coupled with the continued perceived threat of a CCF breakthrough on council, led to the founding and continued support of the right-of-centre NPA around this time as well; other parties subsequently emerged, though the NPA remained preeminent through the end of the twentieth century.

Every electoral system comes with its trade-offs. For instance, Koop and Kraemer (2016) find evidence that the at-large system changes councillors' incentives when compared to ward-based voting. Because each councillor represents all city residents, they tend to think about decisions in terms of what would be in the best interests of the city as a whole, rather than a particular neighbourhood. At the same time, the at-large system in Vancouver has been marked by substantial ballot complexity. In the 2018 election, for example, 158 candidates competed for 27 positions across council, parks board, and school board (Li and Grauer 2018).[16]

Accordingly, parties play a crucial role in simplifying the choices facing voters. Even so, the lengthy ballot had other unintended effects. For instance, given that candidates were listed alphabetically, voters "without well-defined preferences" were more likely to practice a variation of "donkey voting" – in this case, simply selecting the first-listed names – as a way to manage cognitive challenges (Hagiwara 2018). To address this, the city started randomizing the order of names on the ballot in 2018. While this did not eliminate the effect, because all voters still received the same ballot order, it did randomize who benefited.

As alluded to above, the existence of not only a school board but also the Vancouver Park Board further contributes to ballot length and complexity. The park board was formed in 1888 to oversee the newly dedicated Stanley Park (The Parks of Vancouver 1972); its responsibilities now include the municipal park system. Since 1928, the board has consisted of seven elected commissioners. The only other elected park board in North America serves Cultus Lake in British Columbia's Fraser Valley ("Cultus Lake Park Governance" n.d.).

Despite its longevity, the at-large system has been repeatedly scrutinized and reconsidered. Indeed, voters in Vancouver have repeatedly reconsidered the electoral system. In all, five referenda have been held on the subject (Berger 2004). While there was little appetite for change at council during the NPA's near-total dominance over the city's governance from its inception in 1937 until 1972, things changed in the late 1960s with the founding of the left-wing Committee (later Coalition) of Progressive Electors (COPE) and the centrist The Electors' Action Movement (TEAM). Both parties favoured reforms to the voting system, although they differed in preference for pure wards versus a hybrid approach.

In a 1973 vote, 58.9 per cent supported maintaining the at-large system. In 1978, 51.7 per cent supported wards; while a majority, council deemed the result too close to proceed with the reform. In 1982, 57 per cent supported a full ward system, but the provincial government insisted a referendum meet a threshold of 60 per cent. In 1988, 56 per cent supported wards; again, the result fell short of the required threshold. In 1996, 59 per cent supported maintaining the at-large system. Finally, in the most recent referendum in 2004, 54 per cent supported retaining the at-large system (City of Vancouver n.d.-a).

Though dormant for a time, the issue emerged again after the 2018 election. Following Kennedy Stewart's election as mayor, he and council voted to request that staff prepare a report on various reform options, including employing a citizens' assembly to consider alternatives (Lekovic 2022). Staff reported back that such work should resume after the 2022 election. While the idea of reform continues to enjoy significant support in polling (Canseco 2022a), mayor Ken Sim and his ABC majority have, thus far, shown little interest in the issue.

Vancouver's Party System

The continued presence of stable and successful local parties is thus inextricably linked to the city's use of an at-large voting system. Parties allow voters to navigate the city's long and cumbersome ballots. Party candidates likewise benefit from enhanced name recognition in a very crowded field. Parties also help with fundraising and structuring the process of governing once elected. Perhaps most significantly, from a substantive standpoint, the system magnifies plurality advantage, making it easier for a party to win a majority on council with a minority of votes cast. By the same token, the system clarifies mechanisms of accountability, as voters may "throw the bums out" en masse once public opinion hardens against a given council majority party.

Armstrong and Lucas (2022) provide a useful framework within which to understand the party system in the city over time. Parties first tend to sort themselves along a left–right ideological axis. This divide has survived intact even as particular parties on the left and right have emerged, evolved, and disappeared. The parties change, but the split remains, as Vancouverites have proven adept at identifying ideological commitments of not only established parties but also the multiple upstart parties of both the left and right that have tried and, at times, succeeded in replacing the incumbents. Indeed, even independent mayoral candidates tend to be sorted on a left–right axis (Cutler and Matthews 2005; De Rooij et al. 2021).

At the same time, parties and voters alike resist reduction to a single dimension; parties have, since the emergence of TEAM in the 1960s, also distinguished themselves along a secondary "urbanist–conservationist" axis, favouring either the assertive urbanization practices in vogue at the time or resisting such efforts in favour of preservation of some form of urban status quo (Prest and Bushfield 2019; Armstrong and Lucas 2022). While during periods of relative stability the urban issue axis may seem to collapse into the traditional political divide, at moments of political flux provoked by emerging civic challenges – for example, the mid-2000s debates around the Olympics and drug policy or the debate around housing during the 2018 election – parties and individual candidates alike sort themselves along both economic and urban axes, creating a distinct two-dimensional party space. Indeed, party crack-ups have often been marked by growing divisions between relatively more urban mayors encountering resistance from sceptics within their own parties – a pattern that describes the experience of Philip Owen, Larry Campbell, and Sam Sullivan, as the next section makes clear.

More generally, while discipline varies somewhat between parties, in general, successful civic parties have maintained a significant level of party discipline.[17] This fact, combined with the plurality advantage, has allowed parties with a governing majority to leave a strong imprint on city politics in even a single term in office. Conversely, significant intraparty divides often preceded electoral failure as civic issues have exposed divides among ideological allies. The result may be collapse, schism, or the emergence of entirely new, often highly successful parties. Examples of the latter include both the emergence of Vision in the early 2000s and ABC prior to the 2022 election. This cycle of stability and change continues a pattern that Tennant (1980) traces throughout Vancouver politics dating back to 1929.

The implications of Vancouver's distinct electoral and party systems thus go beyond structuring the choices available to voters, to shaping politicians' roles and incentives, and influencing the larger political discourse and resulting political outcomes. Beyond the party accountability effect described above, it is also possible that the system somewhat weakens individual incumbent advantage by introducing more competition for each seat in comparison with ward-based voting. However, the statistical evidence for the effect is limited (Lucas 2021, 374). At the same time, the at-large system has been criticized for limiting the descriptive representation of marginalized communities, particularly those that are regionally concentrated and that might be more effectively represented through a ward system (Trounstine and Valdini 2008).

As a result of the effects described above, Vancouver's politics has been defined by the ascendance and subsequent precipitous decline of successive ruling parties, with periods of instability and evolution in between. We now tell that story in greater detail as we sketch out the evolution of politics in the city over the last forty years.

THE PRACTICE OF POLITICS

Party Systems and Evolutionary Events

While most of Vancouver's municipal parties have been formally independent of their provincial and federal counterparts, many have had informal yet widely known associations with one side of the provincial political spectrum or the other.[18] As a result, it is possible, from the 1990s to the present, to trace a kind of counterweight logic in municipal elections – one reminiscent of how Ontario voters often vote Liberal federally and Conservative provincially, or vice versa.

When a right-of-centre government held sway in Victoria, Vancouver's council and mayor have tended to veer left. Moreover, when the NDP was in control provincially, the right-of-centre NPA (or, more recently, ABC) made gains civically. The few exceptions to this rule, such as Sam Sullivan's win in 2005 and Kennedy Stewart's in 2018, tended to be close-fought multi-corner contests – in short, moments of instability and change following periods of equilibrium. Those wins also, not incidentally, tend to be sharply repudiated in a decisive subsequent election as voters returned to the pendular pattern with centre-left Vision's 2008 victory during a period of provincial Liberal dominance or centre-right ABC's win in 2022 with the provincial NDP ensconced in power. Table 7.1 tracks the political makeup of Vancouver City Council from 1980 to 2022.

Several observations may be made about this pattern. First, there are a few possible explanations for such periodicity. One may be voters' proclivity for divided power between Vancouver and Victoria. Another may relate to party organization. Links between provincial and Vancouver civic politics are informal but profound. When parties lose power provincially, top organizers may turn their attention to municipal politics.[19] A third, and possibly related, explanation is the tendency for offshoots of municipal political movements, after some time out of office, to present more moderate positions in pursuit of electoral success – as with the rise of "COPE Lite" in the early 2000s, or the sweeping success of ABC in 2022.

A second related observation is that, as the above discussion implies, the pattern of alternating governments masks a more nuanced and unpredictable dynamic, as parties and the political movements within which they are embedded struggle to build and maintain internal alliances. When such coalitions break down, new parties emerge. The party system in Vancouver has thus undergone a series of transformations as existing parties have struggled to react to changing realities facing an increasingly complex urban environment. The specific challenges were many and varied, but they were inevitably related to issues of urban life – whether housing affordability, transit policy, notions of civic "liveability," or the complex social challenges facing residents in and around the DTES. Collectively, these "urban issues" constitute a political dimension distinct from more familiar economic and social left–right spectra.

Indeed, the story of the era in question in Vancouver is, in many ways, the story of the decline of the two leading parties, the socialist COPE and the business-oriented NPA, in the face of such challenges. In their place, Vancouver has witnessed the rise of a series of more moderate parties, who, despite leaning left or right, try to draw votes from both sides of the political aisle, often through the articulation of coherent positions on urban issues that to some extent transcend older left–right political fault lines.

To summarize, while the city's politics have been defined by ideology, tensions over the urban issues of the day have at critical moments torn through the seemingly stable polarized left–right dynamic. These moments of political churn place additional burdens on voters, as they try to locate new parties on the traditional left–right spectrum. It is possible that the ensuing confusion presents a barrier to voters, overwhelmed by the abundance of choice and lack of

Table 7.1: Vancouver Election Results by Party, 1980–2022

	Party representation on city council								Mayor
Year	COPE	OneCity	Vision	Green	TEAM	NPA	ABC	Ind.	
1980	4				2	4			Mike Harcourt (ind.)
1982	6				2	2			Mike Harcourt (ind.)
1984	6				2	2			Mike Harcourt (ind.)
1986	2					7		1	Gordon Campbell (NPA)
1988	3					6		1	Gordon Campbell (NPA)
1990	5			0		5			Gordon Campbell (NPA)
1993	1			0		9			Philip Owen (NPA)
1996	0			0		10			Philip Owen (NPA)
1999	2			0		8			Philip Owen (NPA)
2002	8			0		2			Larry Campbell (COPE)
2005	1		4	0		5			Sam Sullivan (NPA)
2008	2		7	0		1			Gregor Robertson (Vision)
2011	0		7	1		2			Gregor Robertson (Vision)
2014	0	0	6	1		3			Gregor Robertson (Vision)
2018	1	1	0	3		5			Kennedy Stewart (ind.)
2022	0	1	0	2	0	0	7		Ken Sim (ABC)

reliable voting heuristics. Over time, however, some form of the dominant left–right fault line seems to re-emerge even as the party names and the specific lines of debate change, and a new, albeit inevitably temporary, consensus emerges on urban issues. In the following sections, we highlight this story of stability and punctuated change.

NPA Dominance (1986–2002)

Our story picks up with Mike Harcourt, Vancouver's first independent mayor since the emergence of the party system, who won three successive elections in 1980, 1982, and 1984. Despite his own opposition to the event, Harcourt's era was dominated by the lead-up to Expo 86. This focus would ultimately reshape the city's perception of itself and the pattern of urban growth for decades.

While his council tried to freeze rents and evictions in the lead-up to the event, between 500 and 850 low-income residents were evicted when the province overruled the city's motion

(Olds 1988). Nevertheless, many observers considered the event a success, given the investment and exposure it garnered for the city. An estimated 22 million people attended the Expo, bringing the light of the world to what had previously been a quiet community on Canada's west coast. The event left Vancouver with multiple new attractions and infrastructure, ranging from the city's first SkyTrain line to Science World, BC Place, and Canada Place.

Harcourt then jumped to provincial politics, joining the ranks of the provincial NDP. The long-serving leftist councillor Harry Rankin ran as COPE's mayoral candidate against the NPA's Gordon Campbell in 1986. The election was primarily fought over what legacy the city would assume following Expo 86: embrace the international attention the event brought to the city with additional development, or refocus on providing social services for current residents. Campbell ran on an aggressive pro-business and pro-development platform, building on the perceived success of the event. He won the election decisively. Campbell won re-election in 1988 and 1990, before becoming leader of the new BC Liberal Party in 1993. Like Harcourt before him, he would eventually become premier.

Campbell's focus, supported by an ambitious civic planning department, was on developing a series of neighbourhood megaprojects to support the new demand brought on by Expo. In this, the city sought to repurpose land made available through the deindustrialization of the lowlands around False Creek, as well as the northwest and southeast corners of downtown (Coal Harbour and Yaletown, respectively). These projects brought significant additional density into the city's core and encapsulated much of the approach to urban planning, eventually labelled as "Vancouverism." Larry Beasley, who served as the city's co-chief of planning from 1994 to 2006, described Vancouverism as an approach based on "experiential urbanism" (Beasley 2019). It was focused on creating liveable communities. The approach emphasized an architectural style of dense glass towers atop low-rise podiums of either townhomes or retail businesses. Vancouver's West End – where people could live, work, and play all within walking distance – provided essential inspiration. The approach also emphasized public and active transportation, including new infrastructure dedicated to biking, walking, and transit.

These changes, at times, came at the expense of existing car infrastructure, creating controversy that repeatedly pitted urbanist mayors against sceptics within their political movements. Ironically, this focus on liveability was necessitated, in part, by the city's rejection of a previous generation's vision of urbanization – the planned but mostly unfinished downtown freeway network.[20] If people could not live in large homes in the suburbs and commute to work in private vehicles, then there needed to be a significant increase in housing in neighbourhoods in and around the downtown peninsula.

Campbell, with a background in real estate development, also saw an opportunity to leverage these new denser, multi-use developments as a new revenue stream for capital projects in the city. Developers in most cities pay a cost proportional to the size of their proposed project to support the municipal infrastructure their project will require. Knowing that the large areas of former industrial land were to be rezoned to accommodate these new neighbourhoods,

Campbell and council entrenched a practice of discretionary rezoning for the City of Vancouver, whereby developers would negotiate with city staff to pay community amenity contributions (CACs) in exchange for increased density allowances (Lamontagne 2019).

This flexible approach evolved following feedback from the development industry. It focused on market conditions, as evaluated by city staff, as the determinant for the CAC levels for any specific project. Ultimately, it enabled the city to recover roughly three-quarters of the increased value from upzoning; in recent years, CACs have generated more than $100 million annually for capital projects (O'Donnell 2022). However, the practice attracted significant criticism over the subsequent decades due to its opacity and fears that it would incentivize the city to "sell" zoning (Condon 2014).

During this same period, austerity from the federal and provincial governments led to massive reductions in total funding for social housing construction and further downloading of costs to cities (Tremblay 2007, 75; Treff and Perry 2005). Accordingly, from the 1990s onwards, market housing dominated Vancouver's new housing construction, with little in the way of new non-market housing. In the face of continued in-migration, the lack of social housing investment contributed significantly to the decline in housing affordability in later years.

Once Campbell retired from civic politics, NPA councillor Phillip Owen ran as the party's mayoral candidate in 1993. Owen defeated Libby Davies, the COPE standard bearer and future federal NDP stalwart, decisively in a rare case of successful dynastic succession. The NPA took every seat it ran for, leaving only one COPE member on council.

Owen would continue Campbell's pro-business and pro-development path. At the same time, council could not ignore problems growing in the shadow of the city's ongoing development, particularly regarding the issues of homelessness and drug use on the DTES. In an oft-repeated pattern, emerging urban-related issues sowed division among erstwhile ideological allies. Indeed, one of Owen's significant legacies was drug reform, an area not typically considered the purview of a right-of-centre party like the NPA. The reforms, often in response to advocacy on the part of the city's drug-using community (Kerr et al. 2006), set the stage for Vancouver's progressive approach to treatment that continues to this day, emphasizing considerations such as safe use as core components of any successful strategy. At the same time, as described below, opposition among NPA councillors would hasten Owen's departure from the scene.

One other sombre episode from the period requires mention. Residents of the DTES struggled to draw attention to the disappearances of sex workers at the time. By March 1999, twenty women had disappeared. After pressure from families and advocates, Owen and the province agreed to provide a $100,000 reward for information. Nevertheless, sixteen more women disappeared in the next three years. After the Vancouver Police and RCMP formed a joint task force, they finally announced the arrest of Robert Pickton, a farmer from nearby Port Coquitlam, for the murder of two women in February 2002. Pickton would be sentenced to life in prison for killing six women but is suspected to be involved in the deaths of many more. The city and

Vancouver Police Department were the subject of intense scrutiny during and following the trial, including a provincial inquiry that concluded in 2012 (Oppal 2012).

The Rise and Fall of Vision Vancouver (2002–2018)

By 2002, even as the right-of-centre BC Liberals were rising to power provincially, Owen's support within the NPA was failing, in no small part due to his ideologically heterodox support of innovative drug reform policy. The leadership of the NPA informed him that he would face a nomination contest for the party's mayoral nomination, at which point he opted to retire instead (CBC News 2002). The party chose Jennifer Clarke, referred to by some in the media as "Lady Macbeth" (Sullivan 2002), as their standard bearer, but she struggled throughout the campaign to recapture the NPA's previous levels of support (Braid 2002).

COPE, on the other hand, had moderated, endorsing prominent former provincial chief coroner and ex-RCMP drug squad officer Larry Campbell as its mayoral candidate. Part of Campbell's fame was through the hit television series *Da Vinci's Inquest*, which was loosely based on his time as chief coroner. Supporters during the campaign even handed out "Mayor Da Vinci" buttons (Swift 2004).

Like the earlier mayor Campbell, Larry Campbell swept to power, winning 58 per cent of the vote to Clarke's 30 per cent. COPE dominated every contest, with candidates topping the polls for council, park, and school boards. They elected eight councillors, giving the party a commanding majority over the two NPA councillors elected. One of Campbell and council's first actions was decidedly "pro-urban," moving to allow legal basement suites just six months after their election. While this had been controversial in previous eras, the change was met with little protest in the city and was one of the first of such reforms across the region. This pattern of controversy later giving way to acceptance on an issue of urbanization was to recur with several other "urbanist" innovations in the era that collectively redefined the city's approach to land use. From innovative drug policy to the introduction of dedicated bike lanes to the embrace of city-wide multi-unit zoning, repeatedly, debates over innovative policies gave way to broad acceptance as a new urban consensus emerged – and the traditional left–right dynamic reasserted itself.

Despite COPE's commanding victory, Campbell's term suffered from the tensions between the more moderate and radical factions within the party. In a manner reminiscent of NPA's divisions over drug policy, COPE's factions were bitterly divided over whether the city should host the 2010 Winter Olympics. The plans had started under Owen and were supported by the provincial BC Liberal government under former mayor Gordon Campbell. Mayor Larry Campbell was moderately supportive of the initiative, hoping it would bring new investments to the city. Many of the more left-wing members of COPE, however, were opposed, citing many of the same criticisms levelled at Expo 86.

Ultimately, COPE's caucus ruptured under the strain. One subset of councillors held closer to the party's left-wing roots, while another three, dubbed "COPE Lite" by members of the media, aligned with the more moderate Campbell on issues like taxation and land development (Mickleburgh 2004). These members eventually formed Vision Vancouver, a new moderate political party (Woodward 2005). Campbell opted not to run for re-election, and councillor Jim Green ran as Vision's 2005 mayoral candidate. Vision and COPE each ran five council candidates. The NPA picked councillor Sam Sullivan over future BC Liberal premier Christy Clark as its mayoral candidate. Sullivan narrowly defeated Green, and the NPA took five seats on council, a slim majority. Vision elected four candidates, while COPE was reduced to a single councillor.

Sullivan was aware of the challenges facing the people living on the city's streets; he and his council majority pledged to buy at least one single-room occupancy (SRO) hotel per year for a decade to secure the dwindling stock of relatively more affordable – albeit increasingly decrepit – housing in the DTES. Two years after his election, activists protested the lack of action on the promise; at that point, former mayor and now premier Gordon Campbell and the BC Liberal provincial government intervened.[21] Campbell and Sullivan jointly announced provincial funding to purchase 10 SRO buildings, securing 1000 units of social housing (Hume 2007). The direct government intervention in the housing market by a right-of-centre government at both municipal and provincial levels highlighted again how urban challenges forced politicians to respond in ways that defied conventional ideological expectations.

Other initiatives during his time similarly defied partisan expectations, often echoing priorities of the nascent opposition centre-left Vision party. This put the mayor at odds with NPA orthodoxy on several issues. As an example, Sullivan championed environmental awareness, putting significant political capital into passing an EcoDensity charter at a time when many NPA party faithful were still profoundly sceptical of things like bike lanes (CBC News 2008). The principles encouraged environmentally sustainable development by spreading growth across the city, particularly replacing single-family homes with more dense built forms. In this, it was an approach years ahead of its time. Sullivan continued to support progressive approaches on issues like drug policy as well.

Between the bitter partisan battles with a strong opposition, a major civic strike, the internal party challenges generated by Sullivan's heterodox approaches, and a hard-to-quantify but widely reported dislike for the mayor's style (Mason 2008), the NPA coalition again foundered. For the second time in less than a decade, the incumbent NPA mayor lost out to an internal challenger as councillor Peter Ladner successfully challenged Sullivan for the party's mayoral nomination.

Vision Vancouver, by contrast, had used their time on council to solidify their political credentials. The party selected neophyte BC NDP MLA Gregor Robertson as its mayoral candidate over sitting caucus members. Robertson went on to trounce Ladner, and Vision elected seven of the eight candidates they ran for council. The NPA took just one seat and COPE two. Robertson would lead Vision to successful re-election in 2011 and 2014 and remains Vancouver's longest consecutively serving mayor – though as events would show, by his departure from the scene Vision's coalition too was breaking apart.

Further to the political left, COPE again struggled to gain traction after its near shutout in 2005. The party elected two councillors in 2008 and was shut out in 2011 and 2014. Frustrated by the ideological entrenchment of the party's leadership, a few prominent members who had remained with the party following the split with Vision finally abandoned the party and put their energy into OneCity Vancouver as a new vehicle for left-of-centre and broadly urbanist politics (Lee 2014). Once again, divisions over land use and other urban issues were at the heart of the intraparty divide.

A core of Vision's focus over its tenure was on environmental initiatives, including launching a Greenest City Action Team and overhauling the city's cycling infrastructure. As with the battle over secondary suites earlier in the millennium, the anti-urbanist anti-biking position attracted considerable attention politically and on radio talk shows but ultimately moved few votes, such that all competitive parties effectively adopted some version of both policies.

During Vision's time in office, the Truth and Reconciliation Commission of Canada released its final report and calls to action to address the legacy of residential schools on Indigenous peoples (Truth and Reconciliation Commission of Canada 2015). Led by councillor Andrea Reimer, the city formally redefined its relationship with the land it occupies through a formal framework for reconciliation and by opening council meetings and formal events with an acknowledgment that the city sits on the unceded traditional territories of the xʷməθkʷəy̓əm (Musqueam Indian Band), Sḵwx̱wú7mesh (Squamish Nation), and səlilwətaɬ (Tsleil-Waututh Nation) (City of Vancouver n.d.-b; Reimer 2014). Some of this work had taken shape during the city's Olympics-hosting processes, which left a legacy of new collaboration between the local government and First Nations (International Olympic Committee 2019).

The three local Nations themselves also began to transform their relationship with the land and the city. Together, they formed the MST Development Corporation, the body responsible for developing some 160 acres of land held by the Nations. Some projects are already underway, most notably the Sen̓áḵw development on land returned to the Squamish people following a successful court battle. The Squamish people have been able to proceed with developments without typical approvals from the City of Vancouver planning department, as the land is "reserve" land and not subject to the city's jurisdiction. For its part, the city has supported the development, concluding a services agreement for the development in 2022 (Skwxwú7mesh Úxwumixw (Squamish Nation) and City of Vancouver 2022).

During his campaigns, Robertson had pledged to end homelessness in the city by 2015, but the number of people experiencing homelessness in the city continued to rise throughout and after his term (Fumano and Culbert 2018). According to the Greater Vancouver Regional District, the number of people experiencing homelessness in the City of Vancouver grew from 628 in 2002 to 1,372 in 2008 (Social Planning and Research Council of BC et al. 2009). In 2010, the city launched its own count, and in 2015, it recorded 1,746 people experiencing homelessness (Thomson 2015).

Most prominently during their tenure, Vision oversaw the 2010 Winter Olympic Games in the city. Debate continues on whether the event can be considered a success (Keller 2014). Like Expo, the Olympics left Vancouver with several major infrastructure upgrades. A new community, Olympic Village, was built along the south shore of False Creek. Another SkyTrain line was built from downtown Vancouver to Richmond, with a branch to the Vancouver International Airport. New recreation facilities in Vancouver and Richmond were built for the Games and converted to more multipurpose use afterward. The Sea-to-Sky Highway from Vancouver to Whistler received significant structural and safety improvements.

Initially, the city had pledged to convert the Athletes' Village into a mixed-income neighbourhood with significant affordable housing following the Games. This became the Olympic Village neighbourhood in the southeast corner of False Creek. Ultimately, construction cost overruns led the city to nix the planned affordable units to cover the project's debts (CBC News 2010). This controversy was emblematic of Vancouver's challenge over this entire period as the fiscal needs of the city and private developers ran up against the increasing unaffordability of housing in the city. Messy as it proved, the construction of Olympic Village exhausted one of the last opportunities to redevelop the deindustrialized former "sawmills, foundries, shipbuilders, railyards, metalworks, and salt distributors" that once ringed False Creek (City of Vancouver n.d.-c). Future developments would inevitably involve displacing current residents, not just former industry. In the face of growing residential frustrations with growth plans, city staff tried a variety of innovative strategies, including the use of enhanced local deliberative democratic practices (Beauvais 2018).

Equilibrium Punctuated (2018–2022)

When the end came for Vision, it came swiftly. Remarkably, 2014 would prove the last time Vision managed to elect a municipal councillor. As with previous transitions, the fall was a function of both politics and policy. At the provincial level, the emergence of an NDP government removed a possible political foil for the centre-left Vision party, setting conditions for another pendular swing. More prosaically, as with previous mayors, Vision's style under Robertson reportedly began to grate on many (Lee 2010). While Vision's large majorities and admirable party discipline made it possible to enact its preferred policies at council, the result was a closed shop that led many outside the party to feel shut out of the political process.[22] Council decisions were often seen as predetermined, with critics and opposition left with few inroads to influence policy.

More substantively, the seeming inability of council to do anything about spiralling housing costs became a significant component of political discourse. An ominous turn for Vision came in 2017 when the party's candidate came fifth in a by-election to replace resigning Vision councillor Geoff Meggs (CityNews Vancouver 2017). NPA candidate Hector Bremner was elected on an aggressive platform that sought to massively increase the housing supply in Vancouver through radical zoning reforms.

In January 2018, Gregor Robertson announced he would not seek a fourth term as mayor – setting in motion a transformative election. That decision left Vision scrambling to find a replacement, even as other Vision councillors followed Robertson's decision to retire from municipal politics. In the end, only one incumbent Vision councillor would (unsuccessfully) seek re-election, leaving the field open to competitors. The 2018 election also ended the so-called Wild West era of unregulated municipal electoral spending, temporarily eroding incumbent advantages as parties scrambled to adapt to new disclosure requirements and fundraising and spending limits (Levin 2017).[23] Between declining fortunes and the new rules, Vision only raised $272,000 for the 2018 campaign – a fraction of the $3.4 million the party had raised in 2014.[24]

Meanwhile, Vision's opponents were experiencing their own challenges. Building off his by-election victory, Bremner set his sights on the NPA's mayoral nomination. However, the party refused his application to run and instead selected local business owner and political neophyte Ken Sim. Though the exact reasons were never specified, many observers concluded that Bremner's aggressive position on housing and other pro-urbanization issues rubbed many the wrong way in the increasingly development-sceptical NPA (Little 2018). In effect, Bremner's candidacy exposed divides within the NPA over urban vision.

Regardless, Bremner soon pivoted to launch his own vessel in cooperation with political strategist Mark Marrisen: Yes Vancouver. The party would run on Bremner's "Let's Fix Housing" platform, centred around reducing the restrictions on development across the city with the goal of increased private-sector development and greater density, all in a predominantly market-focused approach to alleviating the housing crisis – effectively, an attempt to create a distinct right-of-centre urbanist party. Alongside Bremner and Sim, two notable independents joined the race: SFU academic Shauna Sylvester and NDP MP Kennedy Stewart. Other candidates threw their hats in the ring as well, and the final ballot featured 21 candidates for mayor, which quickly became a three-way race between Stewart, Sylvester, and Sim.[25]

Competition and fragmentation also proliferated among the broadly progressive parties, Vision Vancouver, COPE, OneCity, and the Greens, with debate on who would run for mayor and how many candidates each party should run for council. These parties also sorted themselves on urban issues, with OneCity championed a variety of aggressive initiatives intended to increase density and housing supply in the city. In contrast, COPE's policy tended to be more sceptical of development that might threaten the housing of existing tenants. The Greens – as champions of local grassroots democracy – tended to emphasize the need for more extensive consultations, a requirement that inevitably slowed the increase in housing supply through neighbourhood redevelopment.

The Vancouver & District Labour Council (VDLC), an umbrella union organization, facilitated negotiations among the left-leaning parties over the creation of a coordinated slate of 10 candidates and 1 mayor. In July 2018, the VDLC announced its endorsements, which included Stewart for mayor and a rainbow coalition for council consisting of four Vision, two Greens, two OneCity, and two COPE candidates. The slate guided some voters' choices; moreover, the desire to win slate endorsement and avoid overcrowding the ballot encouraged the parties to limit the number of candidates each ran. The VDLC and other brokers were not wholly successful in imposing slate discipline on left-of-centre parties, however. Most ended up running candidates beyond the slate endorsements, with 14 candidates running for ten seats from these four parties – a step back from the coordination of Vision and COPE in previous elections and a harbinger of the chaos to come on the left in 2022.[26] Ultimately, 71 candidates from ten parties, plus many independents, put their names forward for council in 2018.

As in previous elections, voters sought ideological markers to guide their voting, even amid the changing parties and ballot complexity (De Rooij et al. 2021). The fragmentation of the left and right made that process more complex, however. The election itself was fought primarily over the issue of housing affordability, with many referring to the "housing crisis" (Balca 2018). In turn, those housing issues became a distinct axis of debate in the election, creating new divisions in both right-of-centre and left-of-centre coalitions. As old parties foundered in the face of new challenges, new parties on the right and left emerged in a realignment of the city's political system (Prest and Bushfield 2019).

In the end, Stewart took only 28.7 per cent of the vote but managed to narrowly defeat Sim by 957 votes. Sylvester finished third with 20.5 per cent. Despite the frequent coverage of Bremner's Yes Vancouver, he ultimately received fewer than 10,000 votes (5.7 per cent) and placed fifth. More significantly, voters also returned a very divided council, with the NPA taking five seats, the Greens three, and COPE and OneCity each electing one. Together with Stewart's vote, this gave the progressives a narrow 6 to 5 vote majority. No single party held a majority on the park or school boards either. Vision Vancouver did particularly poorly, electing just a single candidate – a trustee to Vancouver School Board.

The early days of the Stewart council were optimistic, with the mayor and councillors speaking highly of their efforts to collaborate both in and out of council chambers (Fumano 2018). The council sought to address the many crises facing the city, initially housing and the toxic drug supply and later COVID-19. On housing, council worked to renew leases on the city's cooperative housing units, launched programs to encourage rental housing development, and introduced an Empty Homes Tax to push vacant homes onto the rental market. Stewart launched an Overdose Emergency Task Force following his election, and council unanimously endorsed the recommendations, including a call for a regulated, safe supply of drugs. With the council's backing, Stewart would go on to lobby the federal government for an exemption to the Controlled Drugs and Substances Act, which would decriminalize the simple possession of hard drugs in the city. Ultimately, Health Canada approved a broader application from the province, which would operate for three years starting in early 2023 (Government of British Columbia 2023).

Despite these efforts, both the housing and overdose crises continued to surge throughout Stewart's term. Housing sale prices had reached record levels by December 2017 in Metro

Vancouver, with the average detached home costing over $1.6 million (Real Estate Board of Greater Vancouver 2018). Prices then declined to $1.4 million (Real Estate Board of Greater Vancouver 2020) by 2019, but prices and drug deaths both again spiralled during the ensuing COVID-19 pandemic.

By December 2021, the average single detached home in Metro Vancouver sold for $1.9 million (Real Estate Board of Greater Vancouver 2022). In a decade, already high housing prices in the city had doubled. Similarly, deaths from illicit drug toxicity declined in 2019 before increasing to new records in 2020. In 2021, 2,293 British Columbians died from an overdose. In 2022, the number increased to 2,382, with 578 of those deaths in the City of Vancouver (BC Coroners Service 2023).

The climate continued to be a significant focus during Stewart's term, with council passing a motion early declaring a climate emergency (Crawford 2019). Staff recommended critical actions for the city to meet its ambitious goals of reducing greenhouse gas emissions by 50 per cent by 2030 and reaching net zero by 2050. However, when the specifics finally reached council in 2021, the plan remained significantly underfunded. It was further undercut when council voted against a controversial proposal to implement a parking permit system with a carbon surcharge for large polluting vehicles. Notably, that specific proposal was defeated with Stewart joining the five councillors elected with the NPA. This failure captured well the frustrations and, at times, the seeming paralysis of the divided multi-party council.

During the 2018 election, the Greens pledged to establish a new citywide development plan. The last one, CityPlan, had been completed by Gordon Campbell's council in the early 1990s, and proponents wanted a new one to serve as a consensus document from which to move forward with future development plans. Green councillor Adriane Carr proposed the motion to start the planning process at city council, which was supported unanimously. That process only culminated in the final meetings before the 2022 election, where it was narrowly passed by council.

Despite its strong showing in 2018, the NPA floundered during Stewart's term. In late 2019, a few social conservatives were elected to the party's board of the NPA. This led to an increasing number of concerns among the elected council members of the party. Rebecca Bligh, the only openly LGBTQ+ council member, announced in December 2019 that she had left the NPA. She would be followed by councillors Sarah Kirby-Yung, Lisa Dominato, and Colleen Hardwick in early 2020. By the 2022 election, Melissa De Genova was the sole remaining NPA member on council; she would go on to lose her seat in that election.

A New Normal?

By late 2021, a new political vehicle began to take shape on the political right, spearheaded by former NPA mayoral candidate Ken Sim. The party was called "A Better City Vancouver" and later just ABC. Former NPA-turned-independent councillors Bligh, Kirby-Yung, and

Dominato joined the party in 2022 as it became the clear alternative to Kennedy Stewart's re-election campaign. ABC would focus its campaign on issues of public safety with a typically Vancouver twist: Sim's flagship promise was to hire 100 police officers, but also 100 public health nurses to be paired as mental health crisis responders working alongside police. At the same time, Sim largely managed to skirt funding and jurisdictional questions associated with the policing plan (McElroy 2022) and likewise avoided tangible promises on housing beyond his renewed commitment from 2018 to reduce permitting delays. Adapting to the new financing environment, the party also proved a fundraising juggernaut under the new rules (Howell 2023; Elections BC, n.d.). Very quickly, the party emerged as the dominant force on the right of the political spectrum.

As with Vision in the previous cycle, when the end came for the NPA, it came quickly. Like previous party crack-ups of both the left and right, the NPA's failure to moderate and evolve in the face of changing political circumstances presaged electoral defeat and political oblivion. The NPA initially declared long-serving Park Board commissioner John Coupar as its mayoral nominee, but he backed out of the role in the summer of 2022. In his place, the party appointed Fred Harding, who had run with the fringe party Vancouver 1st in 2018. Harding would earn fewer votes as the NPA's candidate in 2022 than he had in the previous election. It was the NPA's worst showing in the party's history, and given the rise of ABC, a return to glory for the city's long-time natural ruling party now seems difficult to imagine.

Stewart, for his part, was successful in staving off mayoral challengers from the left. He was endorsed by the VDLC almost a year out from the election, and no one opted to mount a serious independent campaign. Frustrated by a lack of coherent support on council and the weakness of the mayoral office as chair of a divided council, Stewart formed a new party called Forward Together with Kennedy Stewart, which would run candidates for council. Stewart's campaign would focus on building homes, expanding transit, and responding to climate change but was criticized for its lack of detail, particularly compared to the detailed housing plans presented by several other parties (Bushfield and Naylor 2022).

Council was a different story, however. In contrast with prior elections, left-wing parties failed to achieve even a modicum of coordination around candidates – a crucial failure given the at-large system's tendency to punish vote splitting. Exhortations by the VDLC and others to avoid vote splitting went unheeded. Forward Together nominated six candidates, the Greens five, and COPE four. Only OneCity nominated as many candidates as they had received labour endorsements (four). Vision Vancouver also ran three candidates for council, and a splinter electoral organization from COPE, VOTE Socialist, nominated a candidate. In total, left-leaning parties ran 23 candidates for ten council seats. Similar vote splitting occurred at the park and school boards.

While voters continued to cite housing as their top priority throughout the election campaign, a significant amount of attention was paid to a growing concern over crime rates in the city (Canseco 2022b). Part of this was fuelled by the rise in people experiencing street homelessness, which was itself driven by affordability issues, as well as pandemic-related disruptions and fires in the SRO hotels in the DTES. In the lead-up to the vote, the Vancouver Police Union took the unprecedented step of endorsing Ken Sim and ABC for election. While police have never been apolitical, this was widely reported as the first time the union had openly endorsed a specific political party and was publicly condemned by numerous observers (including this chapter's authors) (Bushfield and Naylor 2022; Kozelj and St. Denis 2022).

Voters gave ABC an overwhelming mandate, with all its candidates winning by a wide margin in each race. On council, Green incumbents Carr and Fry were re-elected, along with OneCity incumbent Christine Boyle. Whether driven by concerns about public safety, frustration over the seeming inability of the previous divided council's ability to address the city's challenges effectively, or a more fundamental reassertion of the pendular logic – the NDP was by now well entrenched provincially, setting the stage for a right-of-centre counterweight at city hall – voters proved responsive to the ABC's straightforward campaign messaging.

Once again, the 2022 election included evidence of parties positioning themselves not only on a left–right economic spectrum but also on a distinct urbanist–preservationist development axis. At one end of the development spectrum was a resurrected TEAM party, now led by former NPA Councillor Colleen Hardwick and renamed TEAM for a Livable Vancouver. The party brought together ideologically diverse disaffected politicians from the NDP and the NPA and campaigned around a more cautious approach to development focused on maintaining existing neighbourhoods and communities.

At the other end of the spectrum, Mark Marissen's Progress Vancouver was built from the ashes of Hector Bremner's Yes Vancouver. As with Hardwick's TEAM, Marissen's Yes tried to show support from both sides of the traditional political divide. Progress council candidates included former BC NDP vice president and candidate Morgane Oger, as well as Mauro Francis, who had crossed over from the NPA midway through the campaign. As with Yes, Progress favoured an aggressive approach to development, using a mix of zoning reform and measures to increase both market and non-market housing supply.

Voters strongly rejected both ventures, however. Hardwick received a mere 16,769 votes and Marissen 5,830. Their council candidates performed even worse. While concerns about housing have dominated the political discourse in Vancouver for at least a decade, it is clearly not the sole driving motivation for voters at the ballot box.

Having won a clear majority not only on city council but also the park and school boards, Sim's ABC party seemed set to dominate and reshape the city's politics. The reality has been different, however, with the first two years in office proving divisive and turbulent. There have been some clear achievements, as ABC was able to quickly pass successive increases to the police budget, enabling the hiring of 100 new officers promised during the election campaign. Similarly, the plan to speed up permitting to ease construction of new housing is underway and has seen limited progress in the form of reduced permitting wait times. In other ways, though, ABC's time in power has thus far been marked by a series of unpopular policy lurches, often followed by embarrassing climbdowns. These include an abrupt (and, at the time of writing, still inconclusive) attempt to do away with the Park Board altogether, an abortive move to suspend the work of the city's ethics commissioner, and a failed attempt to reintroduce LNG heating as an option for new residential buildings in the city. Along the way, Sim's approach to the role has turned increasingly adversarial against the opposition councillors and even elements of his caucus that did not fall in line. Following Sim's surprise resurrection of his abandoned promise to abolish the elected Park Board, three ABC Park Board commissioners left the party to sit as independents, costing ABC its majority on the board. Defections on the School Board in the wake of the attempt to halt the ethics commissioner's work cost ABC its majority there as well. Even on council, a growing number of votes are seeing splits among ABC councillors, with ABCers Rebecca Bligh, Lisa Dominato, and Peter Meiszner routinely siding with the Green opposition (Bushfield 2024). Indeed, ABC expelled Bligh from the party's caucus in the run-up to a civic byelection in early 2025 (Larsen 2025). The byelection was called to replace OneCity councillor Christine Boyle following her election as MLA, and longtime Green councillor Adriane Carr following her retirement. The seats were won by OneCity candidate Lucy Maloney, and COPE candidate Sean Orr (Kshatry 2025), with ABC candidates finishing a distant sixth and seventh.

Amid all of this, council continues to advance density along the Broadway Corridor and more widely through the city under the "Vancouver Plan." Even so, housing remains deeply unaffordable in the city, and challenges related to homelessness and the poison drugs crisis remain as intractable and tragic as ever. Debates over the land – who owns it, who has a right to it, and who can develop it – remain as central and seemingly as intractable as ever.

CONCLUSION AND REGIONAL IMPLICATIONS

The last forty years have been transformative for the city economically, socially, and politically, as it has been for the broader region of Metro Vancouver. The city's physical and political landscape have both changed dramatically, though in important respects still reflect elements of the older era. Though the names have changed, power continues to alternate in the city between a right-of-centre pro-business party and a fractious assortment of left-of-centre parties that, at times, coalesce into a coherent and formidable governing alternative.

Now a globally renowned metropolis of 2.6 million people and counting, the region has multiple population centres and new high-rise developments in almost every city. Richmond, Burnaby, Surrey, Coquitlam, and other centres all face some version of the same questions of land use planning, density, and development that have confronted Vancouver city hall for decades. Some of these communities have openly embraced elements of an urbanist agenda, while others remain more deeply sceptical. As the demand for housing continues to increase, the need for greater supply will force each municipality to consider how to accommodate population growth.

Regional development and transportation plans, endorsed by mayors through regional governance structures, all call for increasing density, particularly along major transit corridors. Perhaps the strength of Metro Vancouver's unique governance model is that it provides an opportunity for a comparative approach where different jurisdictions can attempt to solve the same problems with different tools. The City of Vancouver itself, with its own charter, often serves as a closely watched laboratory for other municipalities, as changes to its governance are both highly visible and yet can be made without affecting every municipality in the province.

While housing and related urban issues remain central, the two most recent elections suggest extreme positions on questions of urban development – whether in favour of or against densification – have little traction with the electorate. Successful parties in Vancouver have tended to come from some version of the moderate middle, whether grounded in the politics of the left or right. At the same time, Vancouver's urban challenges often expose divisions within voting blocs, ultimately fracturing governing coalitions and opening space for new political actors.

So long as housing remains unaffordable for many, the city will continue to face a range of profound political, economic, and social challenges. From bike infrastructure to the urban megaproject developments and Vancouverism of the 1990s, to innovative drug policy, to Sam Sullivan's EcoDensity and Vision Vancouver's housing densification of the early twenty-first century, successive councils and mayors have lived and died by their ability to respond effectively to the complex, ideology-defying issues of land use in the city. The ability of Ken Sim's ABC to navigate such issues will define the city's politics in the coming years, with lasting implications for the future.

ACKNOWLEDGMENTS

The authors would like to acknowledge and express thanks for the support and feedback they have received in writing this chapter. Thanks to all the editors, but in particular to Jack Lucas for providing extensive and thoughtful comments on a previous version. Thanks, too, to the participants in the writer's workshop held online on 6 March 2023. Finally, many thanks to Frances Bula, who took time out from her reporting on the city to provide insightful feedback on an earlier version. Naturally all errors and omissions remain our responsibility.

NOTES

1 The term, originally from evolutionary biology (Eldredge and Gould 1972), was subsequently
   applied to politics by Spruyt (1994) and others.
2 One recent count found 2,000 homeless in the municipality and another 7,000 living precariously in
   Single-Room Occupancy hotels (City of Vancouver 2022).

3   That number may not reflect the reality of the situation for several reasons related to the census'
    limited ability to capture lived Indigenous realities.

4   When complete, the towers will house more than 6,000 rental units (Laboucan 2022).

5   For a sample of such criticism, see Sterritt (2022).

6   As with many other North American cities, the emergence and growth of Vancouver took place
    against a backdrop of racially exclusionary policies. These included everything from the often-abrupt
    annexation of land previously held by Indigenous peoples to the redevelopment of racialized
    neighbourhoods such as Hogan's Alley to recurring patterns of anti-Asian discrimination, both
    formal and informal. For an overview of such issues, see Allen (2019), Roy (1976), Roy (2003),
    Anderson (1988), Roy and Sahoo (2016), Noh et al. (1999), and Kenny (2016), *inter alia*.

7   Indeed, the history of Vancouver's mayoralty has been dominated by residents of Scottish descent.
    Kennedy Stewart was reportedly the seventh consecutive mayor with Scottish ancestry (Todd 2020).

8   That said, as of 2023, the next woman elected mayor of Vancouver will be the city's first.

9   Caught in between is a group of "house-poor" new homeowners, carrying significant mortgages and
    sensitive to every tremor and change in the housing market, as well as employment and inflation. This
    group tends to identify with the interests of other homeowners; indeed, if anything, they are even
    more sensitive to the fear of a decline in housing prices, given the implications for their net worth.
    Also in between is a final group of "comfortable renters," who, by dint of long-term tenancy and
    the compounding effects of rent control, pay rents below market rates. This group also may oppose
    densification in their neighbourhoods, given the real threat that development may pose to their
    housing situation.

10 According to one study, from 1984 to 1999, the net wealth of Vancouver homeowners increased by
    about 20 per cent to about $250,000, even as the net wealth of renters decreased by 10 per cent to
    $5,000 over the same period (Hulchanski 2001).

11 As of 2019, the median renter's income in Vancouver was just half the median income of
    homeowners (Statistics Canada 2022b). The rate of home ownership, moreover, is lower for
    younger residents across the country (Statistics Canada 2022c). That said, it is also noteworthy that
    an increasing share of Metro Vancouver's renters are becoming wealthier, with a significantly larger
    proportion of renters earning more than $100,000 in 2021 than in 2016 and a smaller share earning
    less than $30,000 – suggesting even the relatively wealthy are having trouble buying (Singh and
    O'Donnell 2022). Those unable to increase incomes may be moving away entirely.

12 It is also noteworthy that longer-term tenants in Canada are comparatively more likely to vote than
    new arrivals to a city, per McGregor and Spicer (2016), further complicating efforts to build a voting
    coalition out of the uncomfortably housed.

13 On this last point, the experience of Mark Marissen may serve as a cautionary tale. In 2018 and
    2022, the former federal and provincial Liberal strategist attempted to create a "pro-housing"
    party to tap into the frustrations of uncomfortably housed Vancouverites. In 2018, his effort was
    widely perceived as a "right-of-centre" party, and failed to gain traction among left-leaning renters
    sceptical of a market-focused approach. In 2022, Marissen's political vehicle made efforts to build
    appeal with progressive left-of-centre voters through progressive policies and the recruitment
    of NDP-affiliated politicians such as Morgane Oger to run for council, but without additional
    success.

14 Indeed, one interpretation of a recent regional referendum in which voters rejected a 0.5 per cent
    regional sales tax increase to pay for region-wide transit investments was widespread objections to
    the lack of democratic accountability for TransLink decisions (Johnson and Baluja 2015).

15 All but one municipality in BC presently uses an at-large system. The District of Lake Country
is the exception, which uses a mixed ward and at-large system (District of Lake Country 2022).
Historically, other centres have varied between at-large and wards; Surrey, for instance, only switched
to an at-large system in the 1950s. Indeed, it was that change that prompted White Rock's secession
from the larger municipality for fear of loss of influence at council.

16 The confusion can reach the conspiratorial level as in the case of the 2005 election when
independent candidate James Green received 4,273 votes for mayor. Given that 3,747 votes separated
the victorious Sam Sullivan of the NPA and Vision Vancouver's Jim Green, a few observers speculated
whether there was any coordination between Sullivan and James Green's campaign. Sullivan, for his
part, denied the allegation (CBC News 2005).

17 One partial exception may be the Vancouver Green Party, which, in its commitment to "grassroots"
democracy, has sometimes operated as a loose coalition of one-person caucuses.

18 In some cases, the relationship is formal. The Burnaby Citizens' Association, for instance, requires all
its members to be members in good standing of the provincial NDP (Burnaby Citizens Association
n.d.).

19 Geoff Meggs's career exemplifies this dynamic. Meggs worked for Premier Glen Clark's provincial
NDP government. Following the NDP's 2001 loss, Meggs moved on to work with then-mayor Larry
Campbell. Meggs was later elected as a city councillor with Vision Vancouver before resigning his seat
in 2017 to become newly elected NDP Premier John Horgan's first Chief of Staff.

20 The Georgia Viaduct was one of the few pieces of freeway infrastructure built, constructed in the
place of the expropriated Hogan's Alley, a predominantly Black neighbourhood in the east part of
downtown Vancouver (Tolfo and Doucet 2022).

21 Once again, the links between municipal and provincial parties, while informal, were profound.
Indeed, Sullivan would go on to serve as a BC Liberal MLA from 2013 to 2020.

22 Columnist turned city councillor Mike Klassen (2019) provides an example of the genre.

23 In addition to these significant financial changes, the provincial government also changed the length
of municipal government terms twice in our period of study. First, from two years to three in 1990
and then from three years to four starting in 2014. The latter change brought British Columbia
in line with the practice in the rest of Canada. (Ministry of Community, Sport and Cultural
Development 2014).

24 Calculations by Prest, based on data provided in Elections BC (n.d.).

25 The largest number of candidates for mayor in a recent election was the 58 who ran for mayor in
1996, allegedly partly due to a local freelance journalist offering to buy a pitcher of beer for anyone
who would run for mayor.

26 Vision likewise attempted to put forward its own candidate, Ian Campbell, for mayor. However,
Campbell withdrew from the race in anticipation of public disclosure of a previous domestic assault
charge (one that was stayed) (Crawford 2018).

REFERENCES

Allen, Stephanie. 2019. *Fight the Power: Redressing Displacement and Building a Just City for Black Lives in Vancouver*. Master's thesis, Simon Fraser University.

Anderson, Kay J. 1988. "Cultural Hegemony and the Race-Definition Process in Chinatown, Vancouver: 1880–1980." *Environment and Planning D: Society and Space* 6 (2): 127–49. https://doi.org/10.1068/d060127.

Anderson, Sarah. 2022. "Rent Prices in Vancouver Just Had the Most Dramatic Increase of 2022." *Daily Hive Urbanized*, August 5. https://dailyhive.com/vancouver/average-rent-vancouver-dramatic-increase.

Armstrong, David A., and Jack Lucas. 2022. "The Structure of Municipal Voting in Vancouver." *Journal of Urban Affairs*: 1–22. https://doi.org/10.31219/osf.io/2ah6e.

Balca, Dario. 2018. "Housing the Top Election Issue for Two-Thirds of Vancouverites: Poll." *CTV News*, October 18. https://bc.ctvnews.ca/housing-the-top-election-issue-for-two-thirds-of-vancouverites-poll-1.4139622.

BC Coroners Service. 2023. *Illicit Drug Toxicity Deaths in BC: January 1, 2012–December 31, 2022*. Ministry of Public Safety & Solicitor General. https://www2.gov.bc.ca/assets/gov/birth-adoption-death-marriage-and-divorce/deaths/coroners-service/statistical/illicit-drug.pdf.

Beasley, Larry. 2019. *Vancouverism*. Vancouver: On Point Press, an imprint of UBC Press. https://doi.org/10.59962/9780774890328.

Beauvais, Edana. 2018. The Grandview-Woodland Citizens' Assembly: An Experiment in Municipal Planning. *Canadian Public Administration* 61 (3): 341–60. https://doi.org/10.1111/capa.12293.

Berger, Thomas. 2004. *A City of Neighbourhoods*. City of Vancouver.

Blais, Andre. 2000. *To Vote or Not to Vote? The Merits and Limits of Rational Choice Theory*. Pittsburgh: University of Pittsburgh Press. https://doi.org/10.2307/j.ctt5hjrrf.

Braid, Kyle. 2002. *Vancouver Municipal Election 2002*. Ipsos. https://www.ipsos.com/en-ca/vancouver-municipal-election-2002.

Burnaby Citizens Association. n.d.. *Membership for 2021/2022*. Burnaby Citizens Association. Accessed June 10, 2023, from https://www.burnaby-citizens.ca/membership-for-2022.

Bushfield, Ian. 2024. "The Vibes Are Not Immaculate: Reflecting on the First Two Years Of ABC Vancouver." *Georgia Straight*, November 8. https://www.straight.com/city-culture/vibes-are-not-immaculate-reflecting-on-first-two-years-of-abc-vancouver.

Bushfield, Ian, and Mathew Naylor, hosts. 2022. "E-8 The Endorsement Show." *Cambie Report*, October 7. Audio podcast. Leg-in-Boot Media. https://www.cambiereport.ca/e-8-the-endorsement-show/.

Canadian Real Estate Association. n.d.. *National Price Map*. CREA. Accessed June 10, 2025. https://www.crea.ca/housing-market-stats/canadian-housing-market-stats/national-price-map/.

Canseco, Mario. 2022a. "*Almost Three-in-Five Vancouver Voters Want a Ward System.*" *Research Co*, June 21. https://researchco.ca/2022/06/21/vanpoli-02/.

Canseco, Mario. 2022b. "Deadlock As Voters in Vancouver Prepare for Municipal Election." *Research Co*, October 15. https://researchco.ca/2022/10/14/final-vancouver-2022/.

CBC News. 2002. "Owen Won't Seek Re-Election." *CBC News*, April 9. https://www.cbc.ca/news/canada/british-columbia/owen-won-t-seek-re-election-1.334093.

CBC News. 2005. "James Green-Jim Green Controversy Heats Up." *CBC News*, November 23. https://www.cbc.ca/news/canada/british-columbia/james-green-jim-green-controversy-heats-up-1.535462.

CBC News. 2008. "Controversial Ecodensity Charter Passes In Vancouver." *CBC News*, June 11. https://www.cbc.ca/news/canada/british-columbia/controversial-ecodensity-charter-passes-in-vancouver-1.745555.

CBC News. 2010. "Vancouver Cuts Olympic Village Social Housing." *CBC News*, April 23. https://www.cbc.ca/news/canada/british-columbia/vancouver-cuts-olympic-village-social-housing-1.882070.

City of Vancouver. n.d.-a. *2004 At-large or Wards vote*. Accessed June 10, 2023. https://vancouver.ca/your-government/2004-at-large-or-wards-vote.aspx.

City of Vancouver. n.d.-b. *City of Reconciliation*. Accessed June 10, 2023. https://vancouver.ca/people
-programs/city-of-reconciliation.aspx.

City of Vancouver. n.d.-c. *Olympic Village*. City of Vancouver. Accessed June 10, 2023. https://vancouver
.ca/home-property-development/olympic-village.aspx.

City of Vancouver. 2020. *Vancouver City Social Indicators Profile 2020*. https://vancouver.ca/files/cov/social
-indicators-profile-city-of-vancouver.pdf.

City of Vancouver. 2022. *City of Vancouver – Housing Needs Report – April 2022*. https://vancouver.ca/files
/cov/pds-housing-policy-housing-needs-report.pdf.

CityNews Vancouver. 2017. "NPA's Hector Bremner wins Vancouver Council Seat, Vows to Fix Housing
Crisis." *CityNews Vancouver*, October 14. https://vancouver.citynews.ca/2017/10/14/bremner
-wins-vancouver-council-seat-greens-vision-split-school-board/.

Condon, Patrick. 2014. Vancouver's "Spot Zoning" Is Corrupting Its Soul. *The Tyee*, July 14. https:
//thetyee.ca/Opinion/2014/07/14/Vancouver-Stop-Zoning/.

Crawford, Robyn. 2019. "Vancouver City Council Votes to Declare 'Climate Emergency'." *Global News*,
January 17. https://globalnews.ca/news/4856517/vancouver-city-council-votes-to-declare
-climate-emergency/.

Crawford, Tiffany. 2018. "Vision Vancouver Confirms Mayoral Candidate Ian Campbell had an
Assault Charge." *Vancouver Sun*, September 18. https://vancouversun.com/news/local-news/
vision-vancouver-confirms-mayoral-candidate-ian-campbell-had-an-assault-charge.

Cultus Lake Park Governance. n.d. *Cultus Lake Park*. https://www.cultuslake.bc.ca/cultus-lake-park
-governance/.

Cutler, Fred, and J. Scott Matthews. 2005. The Challenge of Municipal Voting: Vancouver 2002. *Canadian
Journal of Political Science/Revue canadienne de science politique* 38 (2): 359–82. https://doi.org/10.1017
/S0008423905040151.

De Rooij, Eline, J. Scott Matthews, and Mark Pickup. 2021. "Vancouver." *Big City Elections in Canada*.
Toronto: University of Toronto Press. https://doi.org/10.3138/9781487528577-008.

District of Lake Country. 2022. *Voter Information*. https://www.lakecountry.bc.ca/en/local-government
/voter-information.aspx.

Elections B.C. n.d. *Financial Reporting and Political Contributions Database*. Accessed June 10, 2023. https:
//elections.bc.ca/resources/financial-reporting-and-political-contributions/.

Eldredge, Niles, and Steven Jay Gould. 1972. "Punctuated Equilibria: An Alternative to Phyletic
Gradualism." In *Models in Paleobiology*, edited by Thomas J. M. Schopf, 82–115. San Francisco, CA:
Freeman Cooper. https://doi.org/10.5531/sd.paleo.7.

Fumano, Dan. 2018. "Collegiality Reigns as Vancouver's New Council Starts Work." *Vancouver Sun*,
November 6. https://vancouversun.com/news/local-news/vancouvers-new-mayor-and-council-to
-be-sworn-in-monday-afternoon.

Fumano, Dan, and Culbert, Lori. 2018. "Vancouver Mayor's Legacy on Homelessness: A Good Try, But A
Promise Not Fulfilled." *Vancouver Sun*, January 19. https://vancouversun.com/news/local-news
/robertsons-legacy-on-housing-and-homeless.

Government of British Columbia. 2023. *Decriminalizing People Who Use Drugs in B.C.* Government of
British Columbia; Province of British Columbia. https://www2.gov.bc.ca/gov/content/overdose
/decriminalization.

Hagiwara, Rosemary. 2018. "Randomized Ballot Name Order – Proposed Amendments to the
Election By-law No. 9070." *City of Vancouver*. https://council.vancouver.ca/20180606/documents
/cfsc4.pdf.

Howell, Mike. 2023. "ABC Vancouver Spent $2M to Elect Ken Sim and Running Mates." *Vancouver Is Awesome.* https://www.vancouverisawesome.com/local-news/abc-vancouver-spent-2m-to-elect-ken-sim-and-running-mates-6478587.

Hulchanski, J. David. 2001. "A Tale of Two Canadas: Homeowners Getting Richer, Renters Getting Poorer." *Centre for Urban and Community Studies Research Bulletin* 2. https://tspace.library.utoronto.ca/bitstream/1807/95888/1/Research%20Bulletins-02-A%20Tale%20of%20Two%20Canadas.pdf.

Hume, Mark. 2007. "B.C. To Buy Run-Down Hotels to House Vancouver's Homeless." *The Globe and Mail,* April 4. https://www.theglobeandmail.com/news/national/bc-to-buy-run-down-hotels-to-house-vancouvers-homeless/article4094281/.

International Olympic Committee. 2019. First Nations Stand Tall as Shining Example of Vancouver 2010 Legacy. *International Olympic Committee.* https://olympics.com/ioc/news/first-nations-stand-tall-as-shining-example-of-vancouver-2010-legacy.

Johnson, Lisa, and Tamara Baluja. 2015. "Transit Tax Shot Down in Metro Vancouver Vote." *CBC News,* July 2. https://www.cbc.ca/news/canada/british-columbia/transit-referendum-voters-say-no-to-new-metro-vancouver-tax-transit-improvements-1.3134857.

Keller, James. 2014. "Footprint of Vancouver Games Still Felt, But Impact Difficult to Measure." *The Globe and Mail,* January 17. https://www.theglobeandmail.com/sports/olympics/footprint-of-vancouver-olympics-still-felt-but-impact-difficult-to-measure/article16386868/.

Kenny, Nicolas. 2016. "Forgotten Pasts and Contested Futures in Vancouver (Passés oubliés et futurs contestés à Vancouver)." *British Journal of Canadian Studies* 29 (2): 175–97. https://doi.org/10.3828/bjcs.2016.9.

Kerr, Thomas, Will Small, Wallace Peeace, David Douglas, Adam Pierre, and Evan Wood. 2006. "Harm Reduction by A 'User-Run' Organization: A Case Study of the Vancouver Area Network of Drug Users (VANDU)." *International Journal of Drug Policy* 17 (2): 61–9. https://doi.org/10.1016/j.drugpo.2006.01.003.

Klassen, Mike. 2019. "Vancouver's New City Council Is Boring, and That's OK." *Vancouver Is Awesome,* February 26. https://www.vancouverisawesome.com/courier-archive/opinion/vancouvers-new-city-council-is-boring-and-thats-ok-3094486.

Koop, Royce, and John Kraemer. 2016. "Wards, At-Large Systems and the Focus of Representation in Canadian Cities." *Canadian Journal of Political Science/Revue canadienne de science politique* 49 (3): 433–48. https://doi.org/10.1017/S0008423916000512.

Kozelj, J., and J. St. Denis. 2022. "Police and Politics? It's Not a Good Mix, Say Critics." *The Tyee.* https://thetyee.ca/News/2022/10/10/Police-Politics-Not-Good-Mix/.

Kshatry, Shaurya. 2025. "Progressives Win Both Vancouver Council Seats in Byelection, Ruling ABC Party Loses Out." *CBC News,* April 5. https://www.cbc.ca/news/canada/british-columbia/vancouver-byelection-1.7503150 .

Kushner, Joseph, and David Siegel. 2006. "Why Do Municipal Electors Not Vote?" *Canadian Journal of Urban Research* 15 (2): 264–77.

Laboucan, Amei-Lee. 2022. "A Century After Their Village Was Burned, Sḵwx̱wú7mesh Is Rebuilding Senáḵw." *IndigiNews,* September 14. https://indiginews.com/news/a-century-after-their-village-was-destroyed-squamish-is-rebuilding-senakw.

Lamontagne, Neal. 2019. "Negotiated Value: Community Amenity Contributions and Value Capture in the City of Vancouver." *Lincoln Institute of Land Policy.* https://www.lincolninst.edu/sites/default/files/pubfiles/2019_descriptive_negotiated_value_lvc_vancouver_lamontagne.pdf.

Larsen, Karin. 2025. "Vancouver Coun. Rebecca Bligh Kicked Out of ABC Caucus." *CBC News*, February 14. https://www.cbc.ca/news/canada/british-columbia/bligh-kicked-out-of-abc -1.7460001.

Lee, Jeff. 2010. "Vancouver Mayor Gregor Robertson Caught Swearing on Mic." *Vancouver Sun*, July 12. https://vancouversun.com/news/staff-blogs/vancouver-mayor-gregor-robertson-caught-swearing -on-mic.

Lee, Jeff. 2014. "Vancouver's Fractured Left Cracks Again." *Vancouver Sun*, May 10. https://vancouversun .com/news/metro/vancouvers-fractured-left-cracks-again.

Lekovic, Katrina. 2022. *Report Back on a Citizens Assembly and Independent Election Task Force for 2022*. https: //vancouver.ca/files/cov/2022-03-08-council%20memo-report-back-on-a-citizens-assembly.pdf.

Levin, Dan. 2017. "British Columbia: The 'Wild West' of Canadian Political Cash." *The New York Times*, January 14. https://www.nytimes.com/2017/01/13/world/canada/british-columbia-christy-clark. html.

Li, Wanyee, and Perrin Grauer. 2018. "Vancouver Voters Blame Lengthy Ballot for Long Line Ups at Polling Stations for Municipal Election, Patience Tested." *StarMetro Vancouver*, October 20. https: //www.thestar.com/vancouver/vancouver-voters-blame-lengthy-ballot-for-long-line-ups-at-polling -stations-for-municipal-election/article_73d3b096-91f8-536d-97da-8053c65f082e.html.

Little, Simon. 2018. "NPA Faces Defections Over Decision to Drop Hector Bremner from Mayoral Race." *Global News*, May 10. https://globalnews.ca/news/4201523/npa-defections-hector-bremner/.

Little, Simon, and Jordan Armstrong. 2020. "Homeless Camp at Vancouver's Strathcona Park Prompts New Call for Sanctioned Tent City." *Global News*, June 26. https://globalnews.ca/news/7115080 /strathcona-tent-city/.

Lucas, Jack. 2021. "The Size and Sources of Municipal Incumbency Advantage in Canada." *Urban Affairs Review* 57 (2): 373–401. https://doi.org/10.1177/1078087419879234.

Mason, Gary. 2008. "Despite his Personal Story, Vancouver Mayor does not Inspire." *Globe and Mail*, June 10. https://www.theglobeandmail.com/opinion/despite-his-personal-story-vancouvers-mayor -does-not-inspire/article720178/.

McElroy, Justin. 2022. "Profiling Vancouver's Political Parties: ABC Vancouver, Led By Ken Sim." *CBC News*, October 5. https://www.cbc.ca/news/canada/british-columbia/abc-vancouver-ken-sim -2022-profile-1.6607033.

McGregor, Michael, and Zachary Spicer. 2016. "The Canadian Homevoter: Property Values and Municipal Politics in Canada. *Journal of Urban Affairs* 38 (1): 123–39. https://doi.org/10.1111 /juaf.12178.

Mickleburgh, Rod. 2004. "Vancouver Mayor Splits Council, Forms Own Caucus." *The Globe and Mail*, December 15. https://www.theglobeandmail.com/news/national/vancouver-mayor-splits-council -forms-own-caucus/article1008700/.

Ministry of Community, Sport and Cultural Development. 2014. "Local Elections Reform to Include Four-Year Terms." *BC Gov News*, February 25. https://news.gov.bc.ca/releases /2014CSCD0008-000215.

Nakhaie, M. Reza. 2006. "Electoral Participation in Municipal, Provincial and Federal Elections in Canada. *Canadian Journal of Political Science/Revue canadienne de science politique* 39 (2): 363–90. https: //doi.org/10.1017/S000842390606015X.

Noh, Samuel, Morton Beiser, Violet Kaspar, Feng Hou, and Joanna Rummens. 1999. "Perceived Racial Discrimination, Depression, and Coping: A Study of Southeast Asian Refugees in Canada. *Journal of health and social behavior*: 193–207. https://doi.org/10.2307/2676348.

O'Donnell, Theresa. 2022. *2021 Annual Reporting on Community Amenity Contributions and Density Bonus Zoning Contributions and Associated Report Backs*. City of Vancouver. https://vancouver.ca/files/cov/annual-report-community-amenity-contributions-and-density-bonusing-2021.pdf.

Olds, Kristopher N. 1988. Planning For the Housing Impacts of a Hallmark Event: A Case Study of EXPO 86. University of British Columbia. https://doi.org/10.14288/1.0302352.

Oppal, Wally T. 2012. *Forsaken: The Report of the Missing Women Commission of Inquiry*. https://www2.gov.bc.ca/assets/gov/law-crime-and-justice/about-bc-justice-system/inquiries/forsaken-es.pdf.

Porter, Catherine., Vjosa Isai, and Tracy Sherlock. 2022. "Lavish Money Laundering Schemes Exposed in Canada." *The New York Times*, June 15. https://www.nytimes.com/2022/06/15/world/canada/canada-money-laundering.html.

Prest, Stewart, and Ian Bushfield. 2019. "The New Urbanism: Transformation of Vancouver Municipal Politics in the 2018 Election." Paper presented at the Canadian Political Science Association Annual Conference, Vancouver, June 4–6, 2019.

Real Estate Board of Greater Vancouver. 2018. "Steady Sales and Diminished Listings Characterize 2017 for the Metro Vancouver Housing Market." *Real Estate Board of Greater Vancouver*. https://www.rebgv.org/content/dam/rebgv_org_content/pdfs/monthly-stats-packages/REBGV-Stats-Pkg-Dec-2017.pdf.

Real Estate Board of Greater Vancouver. 2020. "Home Sales Decline Below Long-Term Averages in 2019 Despite Increased Demand to End the Year." *Real Estate Board of Greater Vancouver*, January 3. https://members.rebgv.org/news/REBGV-Stats-Pkg-December-2019.pdf.

Real Estate Board of Greater Vancouver. 2022. "Metro Vancouver Home Sales Set a Record In 2021." *Real Estate Board of Greater Vancouver*, January 5. https://www.rebgv.org/content/dam/rebgv_org_content/pdfs/monthly-stats-packages/2021-December-REBGV-Stats-Package.pdf.

Reimer, Andrea. 2014. *Motion on Notice: Protocol to Acknowledge First Nations Unceded Traditional Territory*. https://council.vancouver.ca/20140624/documents/motionb3.pdf.

Roy, Patricia. E. 1976. "The preservation of peace in Vancouver: The aftermath of the anti-Chinese riot of 1887." *BC Studies: The British Columbian Quarterly* 31: 44–59.

Roy, Anjali Gera, and Sahoo, Ajaya K. 2016. "The journey of the Komagata Maru: national, transnational, diasporic." *South Asian Diaspora* 8 (2): 85–97. https://doi.org/10.1080/19438192.2016.1221201.

Singh, Kiran. 2022. "Vancouver and Surrey Might Soon Have a Person of Colour as Mayor for the First Time." *CBC News*, September 29. https://www.cbc.ca/news/canada/british-columbia/metro-vancouver-first-non-white-mayor-1.6598969.

Singh, Sandra, and Theresa O'Donnell. 2022. *City of Vancouver 2021 Census – Housing*. https://ftp.vancouver.ca/files/cov/2022-12-19-city-of-vancouver-2021-census-housing-data.pdf.

Sḵwx̱wú7mesh Úxwumixw (Squamish Nation), and City of Vancouver. 2022. *Sen̓áḵw Services Agreement*. https://vancouver.ca/files/cov/senakw-services-agreement.pdf.

Smith, Andrea B. 1981. The Origins of the NPA: A Study in Vancouver Politics 1930–1940. *University of British Columbia*. https://doi.org/10.14288/1.0095485.

Social Planning and Research Council of BC, Eberle Planning and Research, Jim Woodward & Associates Inc., Judy Graves, Kari Huhtala, Kevin Campbell, In Focus Consulting, and Michael Goldberg. 2009. *Still on our streets… Results of the 2008 Metro Vancouver Homeless Count*. Greater Vancouver Regional Steering Committee on Homelessness. https://stophomelessness.ca/wp-content/uploads/2013/09/HomelessCountReport2008Feb12.pdf.

Statistics Canada. 1981. *1981 Census of Population*. https://www12.statcan.gc.ca/

Statistics Canada. 2022a. *National Housing Day: A look at homeowners and renters*. https://www.statcan.gc.ca/o1/en/plus/2357-national-housing-day-look-homeowners-and-renters.

Statistics Canada. 2022b. *The Daily – Canadian Housing Statistics Program, 2019 and 2020*. https://www150.statcan.gc.ca/n1/daily-quotidien/220412/dq220412a-eng.htm.

Statistics Canada. 2022c. *The Daily – To buy or to rent: The housing market continues to be reshaped by several factors as Canadians search for an affordable place to call home*. https://www150.statcan.gc.ca/n1/daily-quotidien/220921/dq220921b-eng.htm.

Statistics Canada. 2023. *Census Profile*. https://www12.statcan.gc.ca/census-recensement/2021/dp-pd/prof/index.cfm?Lang=E.

Sterritt, Angela, host. 2022. "A Village Burned." *Land Back*, no. 5, December 5. Audio podcast. *CBC*. https://www.cbc.ca/news/canada/british-columbia/community/land-back-1.6561378.

Sullivan, Paul. 2002. "At last, a real fight in Vancouver." *The Globe and Mail*, October 9. https://www.theglobeandmail.com/opinion/at-last-a-real-fight-in-vancouver/article757035/.

Swift, Nick. 2004. Larry Campbell – Mayor of Vancouver. *World Mayor*. http://www.worldmayor.com/finalists2004/vancouver.html.

Taylor, Zack. 2019. *Shaping the Metropolis: Institutions and Urbanization in the United States and Canada*. Montreal: McGill-Queen's University Press. https://doi.org/10.1515/9780773558427.

Tennant, Paul. 1980. "Vancouver Civic Politics, 1929–1980." *BC Studies: The British Columbian Quarterly* 46: 3–27.

The Board of Parks and Public Recreation. 1972. *The Parks of Vancouver*. https://vancouver.ca/files/cov/parks-of-vancouver-1972.pdf.

Thomson, Matt. 2015. *Vancouver Homeless Count 2015*. City of Vancouver. https://vancouver.ca/files/cov/vancouver-homeless-count-2015.pdf.

Todd, Douglas. 2020. "Why Does Metro Vancouver Have So Many Scottish Mayors?" *Vancouver Sun*, October 30. https://vancouversun.com/opinion/columnists/douglas-todd-undeniable-scottishness-among-metro-vancouver-mayors.

Tolfo, Guiseppe, and Brian Doucet. 2022. "Livability for Whom? Planning for Livability and the Gentrification of Memory in Vancouver." *Cities* 123. https://doi.org/10.1016/j.cities.2022.103564.

Treff, Karin, & Perry, David B. 2005. "Finances of the Nation." Canadian Tax Foundation.

Tremblay, Jean-Francois. 2007. "Fiscal Federalism and Public Service Provision in Canada." *Public Policy Review* 3 (1): 51–90.

Trounstine, Jessica, and Valdini, Melody E. 2008. "The Context Matters: The Effects of Single-Member versus At-Large Districts on City Council Diversity." *American Journal of Political Science* 52 (3): 554–69. https://doi.org/10.1111/j.1540-5907.2008.00329.x.

Truth and Reconciliation Commission of Canada. 2015. *Honouring the Truth, Reconciling for the Future: Summary of the Final Report of the Truth and Reconciliation Commission of Canada*. https://nctr.ca/records/reports/.

Woodward, Jonathan. 2005. "New COPE Offshoot Enters Fall City Election." *The Globe and Mail*, August 15. https://www.theglobeandmail.com/news/national/new-cope-offshoot-enters-fall-city-election/article4119969/.

8

Calgary

Jack Lucas

Calgary is Canada's Rashomon city. For some, it is the pure embodiment of boosterism, a city that suppresses dissent beneath a frantic ethos of growth ("It's a wonder," wrote one observer, "that Calgary city council's opening prayer doesn't beg the blessings of Growth instead of God").[1] Others see a city full of fiery disagreement. Some think of Calgary as conservative, even vaguely rural – cowtown, pickup trucks, Alberta beef. Others see Instagram feeds rich with the iconography of twenty-first-century urbanism: Snøhetta architecture and Plensa sculpture, Tibetan dumplings and Peruvian chicken, vegan ice cream and late-night street markets. There is something about Calgary, something deep in its history and politics, that invites competing – even contradictory – interpretations.

Like every big city, these diverse interpretations emerge because Calgary is, in fact, many things at once, a community big enough and diverse enough to contain a multitude of experiences and lifestyles within its boundaries. Yet there *are* discernible patterns in the city's politics – patterns that confirm, but also correct, common claims about how things work in Calgary. Much has changed, I will argue, between mayor Ralph Klein's Calgary of the 1980s and mayor Jyoti Gondek's Calgary of the 2020s. But even as new political debates have emerged, common tendencies also persist, like a bass drum in a pop song, grounding the city's politics in a familiar, recognizable rhythm.

Describing this rhythm, along with the novel themes that have been layered atop it through the years, is my purpose in this chapter. I will describe two enduring cleavages in Calgary politics – a debate between "boosters" and "cutters" that would have been recognizable to the authors of the first edition of *City Politics in Canada,* along with long-standing debates between the ideological left and right in Calgary. Atop these cleavages, two additional layers of discussion and disagreement have emerged in Calgary since the early 1980s, one focused on increasing tension between municipal and provincial governments, and a second focused on non-economic policy debates such as sustainability and racial inclusion.

THE CONTEXT

Every municipal government lives inside a complex ecosystem of geographic, economic, institutional, and cultural relationships. While social scientists often describe this ecosystem in abstract language – flows, mobilities, nodes – its local effects are often clearly visible and concrete, ranging from the hyper-individual (the local resident who visits a bike-friendly city in Europe and advocates for similar policies in Calgary) to the macro-structural (a shift in global commodity prices that changes the value of commercial properties in a city). To tell the story of Calgary's politics, we must first understand something of these diverse and overlapping relationships.

Geography

The city of Calgary is situated at the confluence of the Bow and Elbow Rivers, which, after converging near the city centre, take a sharp southward turn – a geographic feature that inspired the city's traditional Blackfoot name, Mohkínstsis (meaning "elbow"). Situated in the outer foothills of the Canadian Rockies, Calgary's geographic footprint covers two major natural regions. Much of the city is foothills fescue grassland, a zone of high-elevation grassy plains with productive soil and limited forest growth. In the northwest corner, the fescue grassland gives way to foothills parkland, a rolling landscape with more plentiful aspen woodlands (Downing and Pettapiece 2006). On a clear day, the prominent peaks of the Canadian Rockies – Mt. Rae, Banded Peak, Mt. Glasgow – dominate the western horizon.

This geographic setting next door to extraordinary montane beauty lends itself to easy boosterism – especially when combined with the oft-repeated promise that Calgary's punishing winters are in fact a tolerable "dry cold." But the geographic location also has a more immediate political effect: with sparsely forested rolling terrain to the west and vast grassland plains to the east, the land that envelops Calgary offers few natural barriers to outward expansion (Guo and Fast 2019; Taylor 2014). While oceans and mountains may not serve as truly *final* limits on urban expansion (see Incheon, Korea), it is nevertheless true that mother nature whispers her constraints more quietly in Calgary than in many other Canadian cities. Having already swallowed up more than eight hundred square kilometres of grassland, few natural obstacles stand in the way of further outward development (Masson and LeSage 1994; Taylor 2014). In Calgary, decisions to limit outward expansion and prioritize population density are dictated not by cliffs or lakeshores but by political will.

Economy

Calgary's economy began as a hub for the surrounding ranching and livestock industry. After oil was discovered in Alberta, the city became a very different sort of hub: the administrative centre of the Canadian oil and gas industry, together with the financial, scientific, and technological businesses that support it (Foran 2006). As a result, Calgary is second only to Toronto in head offices in Canada (Miller and Smart 2012; Tretter 2022), and even today, despite a diverse local economy that includes technology startups, tourism, culture and film, and an "eds and meds" sector, the fossil fuels industry and its financial and technological appendages remain the heart of the local economy (Langford et al. 2016).

These economic foundations – both the early agricultural connection and the more recent oil and gas industry – continue to shape Calgary's culture and politics. The most obvious effect is the city's "cowtown" image, embodied in the annual Calgary Stampede and the tradition, initiated by mayor Don Mackay in 1950, of presenting distinguished visitors with a white cowboy hat as a gesture of hospitality and welcome (Foran 2006; Seiler and Seiler 2001). While the cowboy imagery encountered some resistance as Calgary transitioned into the oil and gas era – one alderman complained that the white hat "undermined efforts to establish Calgary as an oil and gas centre" (Seiler and Seiler 2001, 8) – the marriage of the two traditions was eased by Calgary ranchers' role as early oil sector investors (Foran 2006). These days, as thousands of Calgarians flock to the Calgary Stampede each year for free admission on "Suncor Family Day," the cultural marriage remains strong.

Beyond the cultural legacies, Calgary's role in the oil and gas sector has also had much more immediate effects on local politics – effects that result from the boom-and-bust cycles of the global energy industry (Miller and Smart 2012). In few other cities in the world is it easier to demonstrate how global economic forces, far beyond the purview of municipal councils, shape the character and content of local policy debates: an OPEC decision creates radically lower oil prices; these lower prices lead to widespread layoffs among energy sector workers in Calgary; the widespread layoffs reduce the value of commercial property in Calgary's downtown, which leads to much lower demand; assessed values of the lower-demand property begin to plummet; commercial properties outside the downtown must bear a larger portion of the non-residential tax bill; a city's local policy agenda is derailed by a tax revolt among commercial property owners. From the perspective of Calgary's mayor and councillors, the forces that control global commodity prices would hardly be further from municipal jurisdiction if they were Greek gods, but the consequences of those forces for local politics and policy are immediate and obvious.

Figure 8.1, which summarizes the number of approved building permits in Calgary between 2000 and 2021, illustrates the local consequences of this economic boom-and-bust cycle.[2] Between 2000 and 2008, the steady increase in building permits reflects an extended boom in the oil sector, followed by the sharp crash of the global financial crisis. Calgary's local economy then began another slow trip up the roller coaster, followed by another devastating crash in 2014 and 2015. A long, slow recovery then began yet again.

Population

From the standpoint of local politics, two features of Calgary's population size and demographic composition are especially notable – both of which are intimately related to the economic trends I just described. The first is the sheer *speed* of Calgary's population growth since 1980.

Figure 8.1: Building Permits in Calgary, 2000–2021

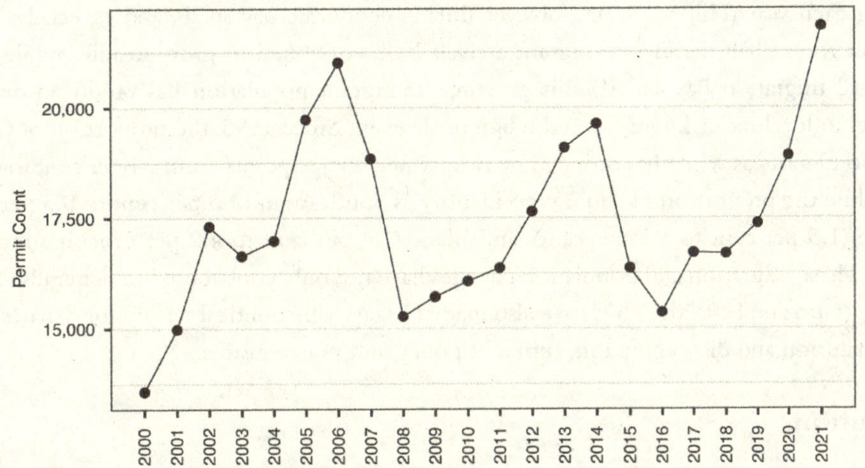

Calgary's population growth in Figure 1.1 in the introductory chapter tells a story of explosive population growth, from 592,000 in 1981 to more than 1.3 million today. No city with a comparable population in 1981 is larger than Calgary today, and just one – Mississauga – has grown at a greater rate in the same period (Feng et al. 2014). Keeping up with this explosive growth has been a major challenge, and a source of considerable political conflict, in Calgary.

The second important characteristic of Calgary's population is related to the changing *components* of population growth. Figure 1.2 (in the introduction chapter) summarizes these changes along with the other cities in this book. Each panel describes the amount of in-migration (or, if negative, out-migration) experienced each year as a result of immigration to the city from outside Canada, migration from another province, migration from elsewhere in the same province, and natural increase due to a positive births-to-deaths ratio.

While much could be said about the trends in Figure 1.2, a few patterns are especially relevant to Calgary's politics. Notice, first, that Calgary's interprovincial migration follows a roller-coaster pathway through time, one that is quite unlike most of the other cities in the figure. In many cities, such as Toronto and Montreal, interprovincial migration plays a relatively small role local population growth.[3] In Calgary, however, the upward and downward trends correspond to the city's more general economic cycles: when times are good, Canadians from other provinces flock to Calgary for new opportunities, and when times are bad, interprovincial migration dries up or even reverses direction (Hiller 2007; Langford et al. 2016).

A second important pattern in Figure 1.2 is the upward trend in Calgary's immigration panel, from about 10,000 immigrants per year in 2001 to about 15,000 per year in the immediate pre-pandemic period. While this growth is substantial, it is especially important to compare the relative size of the immigration and interprovincial migration figures. In the early 2000s, this ratio was more or less even, with similar numbers of foreign and interprovincial migrants arriving each year (Hiller 2007). However, during economic dry spells, and especially during the most recent half decade, immigrant arrivals have continued to grow steadily while interprovincial migration has stalled. This growing immigrant population has produced dramatic increases in local racial, linguistic, and religious diversity. Since 1995, the proportion of Calgarians who identify as white has dropped by twenty percentage points, from 84 per cent to 64 per cent, while the proportion of those who identify as South Asian (3.3 per cent to 9.5 per cent), Filipino (1.5 per cent to 5.5 per cent), and Black (1.4 per cent to 4.2 per cent) has increased sharply. These migration and population changes have not only contributed to demands for new housing (Guo and Fast 2019) but have also made Calgary ripe political terrain for debates about racial inclusion and discrimination, antiracist policy, and police justice.

Institutions

Calgary has a remarkable history of local institutional experimentation. In the early twentieth century, Calgary was among the first cities in Canada to adopt a near-universal franchise in municipal elections, and it was also an early adopter of at-large elections.[4] Calgary was also among the first Canadian cities to adopt partisan local elections; through most of the 1920–1960 period, city council was dominated by mayors and aldermen who had been elected as members of Labour, Social Credit, or a business-friendly "alphabet party" like the Civic Government Association (Lucas et al. 2021; Masson and LeSage 1994). Perhaps most strikingly, Calgary was just the second jurisdiction in North America to enact electoral reform, adopting a proportional representation (STV) system in 1917 and keeping the system in place until the early 1970s – until recently, the longest-lasting proportional representation system anywhere in North America (Lucas 2020).

By 1980, however, this tradition of experimentation had ended. Other cities had long since caught up to Calgary's inclusive municipal franchise (though, unlike Calgary, some retain a franchise for property-owning non-residents), and Calgary abandoned its at-large system for multi-member wards in 1961, and then single-member wards in 1977.[5] By the time of the 1980 election, when our story begins, Calgary's democratic institutions looked identical to those in most towns and cities across Canada: single-member wards, plurality voting, non-partisan elections, and a voting franchise for all local citizens.[6]

In one respect, however, Calgary retains its earlier legacy as an institutional outlier. In 1954, in response to new land development just outside Calgary's municipal boundaries, the Alberta government initiated the Royal Commission on Metropolitan Development in Calgary and Edmonton (the McNally Commission), which recommended that Calgary pursue a "unicity" development model, annexing land as needed to incorporate metropolitan growth within the boundaries of a single municipality (Ghitter and Smart 2009; Taylor 2014). This approach, the Commission argued, would enable long-term planning and ensure that Calgary would retain a healthy property tax base as the metropolitan population grew. As a result, Calgary not only became one of the earliest municipalities to adopt a unicity model of urban growth, but also came to be recognized as "perhaps the most successful example of the unicity approach in North America" (Taylor 2020, 11).

This early adoption of the unicity approach has had important effects on local politics in Calgary. The most obvious is a legacy of continued outward expansion, with especially large land annexations in 1961, 1979, and 1989 (Taylor 2014), and continued smaller annexations up to the present. The developments that populate these expanded land areas have become a source of serious political contention. Advocates point out that the new housing provides affordable residential living in communities with densities far surpassing the city's early post-war neighbourhoods, while critics point to the economic and ecological costs of continued outward development (Conger et al. 2016; Klaszus 2020; Miller 2016). In addition to these immediate policy debates, Calgary's unicity model, which incorporates a mix of inner-city housing, post-war neighbourhoods, and more recent development within a single municipal boundary, ensures that Calgary politics reflects a diverse mix of interests and ideologies in local elections and policy debates. A good deal of Calgary's conservative political reputation originates in the fact that the city has, since the Second World War, contained both its progressive-leaning downtown core and its conservative-leaning outer suburbs within a single municipal boundary.

THE PRACTICE OF POLITICS IN CALGARY

Figure 8.2 provides an overview of the careers of the eighty-five individuals who have served on Calgary city council between 1980 and 2022.[7] Each horizontal line in the figure is a separate career and begins at the time of the individual's first election; grey lines are councillors and black lines are mayors. I mark the endpoint of each career with three possible outcomes: the individual sought re-election and was defeated (a solid black vertical line); they abandoned their council position to run for mayor (a solid diamond); or they retired from council (an open circle).

For simplicity's sake, my chronological description of the practice of politics in Calgary will be organized by the city's mayors, but the careers in Figure 8.2 illustrate that long-term patterns of stability and change on Calgary city council are in fact considerably more complicated. Some elections are associated with a much larger number of career terminations than others – in the 1980s, for example, we see more turnover in 1983 and 1989 than in 1986 or 1992. Calgary's most recent election, in 2021, is especially striking, with eleven careers coming to an end in a single election. But that election is very much the exception – after most elections, the new council is a mix of old and new faces. The layered and gradual pattern of political development that I will describe may be a function, in part, of these layered political careers.

With more detailed inspection, Figure 8.2 also underscores several other important patterns. First, incumbent defeats are quite rare in Calgary – since 1980, nearly 90 per cent of incumbent

Figure 8.2: Political Careers in Calgary, 1980–2022

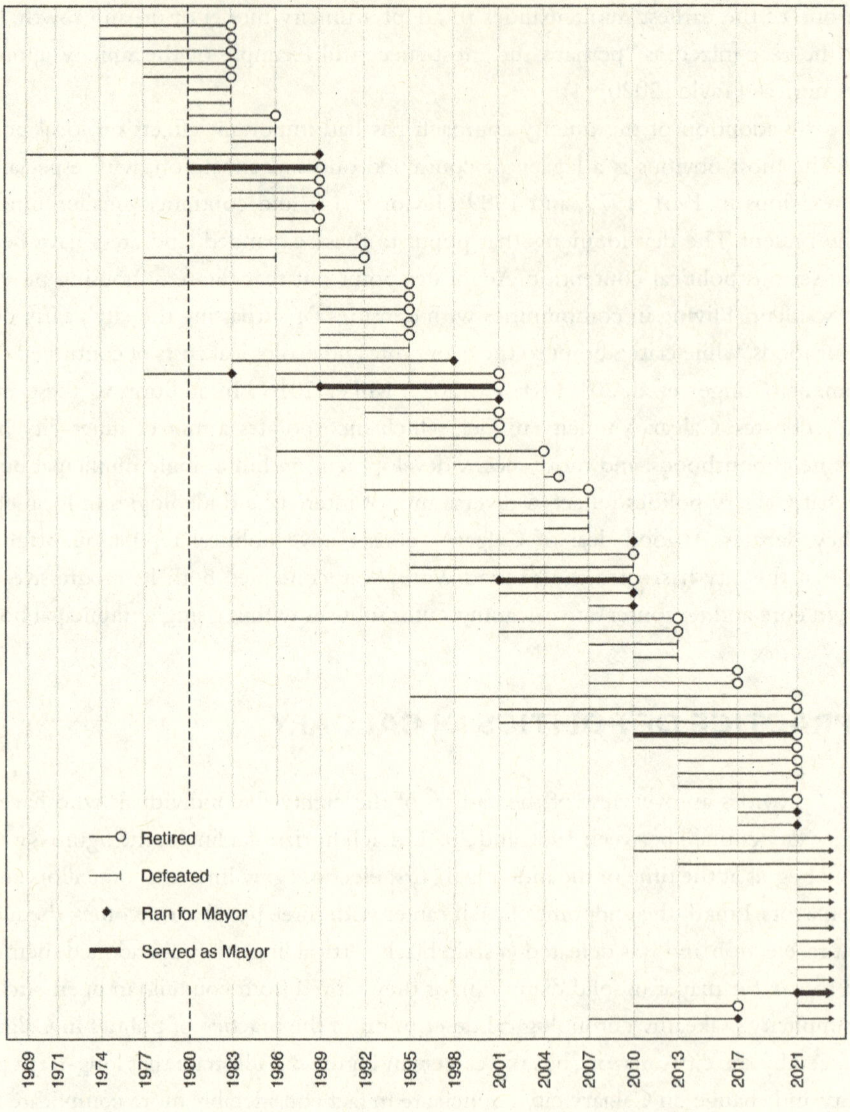

candidates have been re-elected. Incumbent candidates in Calgary's municipal elections enjoy rates of re-election far surpassing their provincial and federal counterparts – re-election rates that originate in the service work and reputation that incumbents are able to build in the community, low levels of information and coordination in Calgary's non-partisan elections, and the "self-fulfilling prophecy" that emerges when high-quality challengers, aware of the difficulty of unseating incumbents, choose to wait for open races before running (Lucas 2021a; Lucas et al. 2022).

Above all, however, the career patterns in Figure 8.2 provide little clear guidance on how we should periodize Calgary's political development since 1980. Aside from 2021, few abrupt junctures mark off obvious periods in the timeline; nor do major policy files or political debates – public transit, land-use planning and sprawl, provincial–municipal tensions, and so on – follow any obvious shared rhythm over time. Thus, in the absence of obvious disjunctures in the narrative, I have chosen to organize my presentation using a simple heuristic: Calgary's mayors. Conveniently, mayoral tenures overlap with each decade in Calgary's historical narrative: Ralph Klein (1980s), Al Duerr (1990s), Dave Bronconnier (2000s), Naheed Nenshi (2010s), and Jyoti Gondek (2020s). Despite a great deal of continuity in policy debates and council personnel, this mayoral periodization proves to be a workable guide through the past four decades of Calgary's political development.

The Klein Era

In Calgary politics, the 1980s started with a bang. The 1980 mayoral election was supposed to be an uneventful affair, featuring two well-worn candidates – Ross Alger, the incumbent, and challenger Peter Petrasuk. Don Martin, then a local columnist in Calgary, said the upcoming election was sure to be "a yawn."[8] Ralph Klein, a 37-year-old television reporter, was widely considered a second-tier candidate with little hope of success (Martin 2002).

Then Klein won – decisively – in a victory cobbled together by a ragtag band of campaign strategists, a tiny budget, and minimal policy statements beyond a promise to serve as "the people's mayor." But the populist appeal resonated powerfully, just as it would throughout Klein's political career.[9] Like other populist mayors, Klein appealed to ordinary Calgarians because, in one local reporter's words, he *was* an ordinary Calgarian: "he drinks beer, smokes other people's cigarettes, takes half-hearted stabs at exercise programs, and doesn't like getting up in the morning."[10]

Klein was very much a populist. What he was not – and this must be emphasized, given his subsequent career as a budget-slashing provincial premier – was a *conservative*. In the early 1980s, Klein was a Liberal, and he would continue to be affiliated with the Liberal Party throughout his mayoral years (he even seriously considered seeking the leadership of the Alberta Liberal Party).[11] Klein did not win in 1980 with the support of local business leaders, but instead courted support among local working people, including the labour vote.[12] "I would like to thank the people here who voted for me in the recent civic election," Klein said at a Chamber of Commerce event after the election. "I understand they're all working in the kitchen" (quoted in Martin 2002, 72).

Thrust into his new role, Klein faced several pressing challenges in the early 1980s, most of which related to the city's rapid growth. It was a period of frenzied expansion both outward, along the municipal boundaries, and upward, in the urban core – indeed, according to Don Martin, Calgary built more new office space in 1980 "than New York City and Chicago combined" (Martin 2002, 64). While progressive planning ideas had begun to arrive on the local scene by the time of Klein's election – one mayoral candidate received substantial coverage for his platform of sustainable growth and urban planning – most agreed that the city's breathless growth required immediate, extensive, developer-friendly expansion (Howard 2015; Martin 2002).

Several other policy debates also left a lasting mark. In the late 1970s, council had approved a new light rail transit (LRT) for the city, and the first line, running south from downtown, opened successfully in 1981. With a skilful combination of patient listening and decisive action, Klein oversaw the construction of two additional lines into the northeast and northwest (Hubbell 2006; Martin 2002). Further extensions to the LRT system would prove to be a major priority, and a source of disagreement, in the decades to come.

Another important event was the 1988 Winter Olympic Games. The years leading up to the Olympics had hardly been happy ones in Calgary – an economic downturn had struck the city, and the Olympic Games themselves had become embroiled in bitter controversy – but the Games proved to be a triumph, leaving the city with valuable new sporting and civic infrastructure and a cultural self-image as the city of white-hat hospitality, volunteerism, and civic pride (Hiller 1990; Martin 2002). International Olympic Committee representatives often promise that the event will leave a positive fiscal, cultural, and infrastructural legacy for its host city. Calgary ranks among the few cities for whom that promise proved true.

A populist mayor, a booming population, a joyous Olympic Games – all of this suggests an optimistic booster city untroubled by ideological disagreement. However, ideological divides were prevalent in Calgary in the 1980s. Before the 1980 election, the local newspaper reported that several political parties were expected to join the race, including a slate of candidates supported by the Calgary Labour Council and the sustainability-oriented Calgary Urban Party. Similarly, in 1983, a representative of Alberta's new Green Party announced that she was organizing a slate of municipal candidates committed to "a conserver society, nuclear disarmament, and protection of the environment."[13] In the end, none of these parties materialized – victims of internal infighting, disorganization, and a fear that voters would punish formal political parties (Masson and LeSage 1994). But the ideological disagreement that lay beneath the proposed slates persisted, along with ongoing rumours of behind-the-scenes coordination by provincial parties. During the otherwise uneventful 1983 municipal election, several conservative critics saw signs of an NDP-supported left-wing slate – an accusation that left-leaning councillors dismissed as "Red-baiting."[14] Tensions were even higher in 1986, with several ward elections described by reporters in explicitly partisan terms. Rod Sykes, Calgary's former mayor, told the local newspaper that "there has always been party involvement in city elections," explaining that the 1986 election was no different from any other.[15] Discussions of "left-wing" and "right-wing" members of council was common in local coverage in the 1980s, and most of the competitive ward races involved candidates who opposed one another from across the ideological divide.

The Duerr Era

After Ralph Klein announced that he was moving into provincial politics, local aldermen began to jostle for position in what was sure to be a competitive race for the open mayoral seat – a position whose salary, in a nod to the increasing professionalization of the city's municipal politics, had recently increased to an attractive $89,000 per year.[16] Al Duerr, a local business owner and member of council since 1983, was not considered the front runner in the race; earlier in the year, he had failed to secure the coveted quasi-incumbent status of "interim mayor" after Klein departed. Nevertheless, voters warmed to Duerr's calm demeanour and earnest, if occasionally cringeworthy, campaign – "On October 16th, let's Duerr!" exclaimed one pamphlet – and in October 1989, Duerr nosed ahead of a crowded field, winning the mayor's chair with a mere 28 per cent of the vote.[17]

While Duerr may have been pleased about the increased salary that accompanied his new role, the mayoral position had little else to recommend it through much of the 1990s. After a quiet first term, Duerr was re-elected in 1992 (with 90 per cent vote share – typical of incumbent mayors' extraordinary dominance through the 1980s, 1990s, and 2000s), and by 1993, a global economic retraction combined with continued local population growth to produce a serious local financial crisis. City council responded with property tax freezes, municipal wage freezes, and increased user fees (Howard 2015). "Life isn't easy for Mayor Al Duerr these days," wrote the local newspaper. "It's one thing to don the white Stetson and play host to Canada for Grey Cup '93. But it's quite another when he closes his office door, puts his feet up, and tries to grapple with the onslaught of job cuts, wage rollbacks, and a city facing a financial crisis."[18]

Ralph Klein, now Premier of Alberta, did little to ease Duerr's burden. Faced with his own fiscal crunch, Klein embraced the neoliberal spirit of the age, slashing provincial transfers to municipalities while simultaneously saddling them with new policy responsibilities (Miller 2007, 2016). Klein's government also amended the Municipal Government Act to abolish mandatory regional planning commissions, terminating a long-standing provincial legacy of strong regional planning (Miller 2016).

Still, despite the challenges, important new policy priorities did emerge in Calgary in the 1990s, the most important of which was a new emphasis on *sustainability*. In the 1980s, some had raised concerns about sustainable development, downtown vitality, and urban sprawl, but they had been voices in the wilderness. By the early 1990s, however, the mood had begun to shift. One early sign of change was "Calgary 2020," a "vision statement" commissioned by the city and developed by 300 local leaders and citizens (including a young Naheed Nenshi) in 1989 (Howard 2015). After stating ten "guideposts" for the city – respect for history, a culture of hard work and innovation, protection of the environment, ethnic diversity, and so on – the document outlined a vision for the city that would prioritize public and active transportation, mixed-use development, and opportunities for all. The document itself did little more than collect dust, but the vision it announced, having been developed by a diverse and prominent group of local leaders, suggested a new openness to sustainability (Howard 2015).

As the 1990s progressed, this interest continued to grow. A "Sustainable Suburbs Study," inspired by the New Urbanist thinking of the day, articulated a vision for vibrant mixed-use suburban development – a vision that took material form in Mackenzie Towne, an enormous suburban community designed by New Urbanist leader Andrés Duany (Howard 2015, 93). While the Sustainable Suburbs Study had little independent effect of its own – one planning official described the development industry's response to the study as an "unbelievable bloodbath" – the study illustrated, like Calgary 2020 before it, that sustainability had become a local priority, at least among some residents, municipal planners, and elected representatives (Miller 2016).

The Bronconnier Era

By the time Calgary's new mayor, Dave Bronconnier, was elected in 2001, Calgary had emerged from the lean years of the 1990s and was in the midst of a historic boom. Between 2002 and 2006, Calgary would be the fastest-growing municipality in the country, with similarly explosive economic expansion, including 60 per cent growth in head offices and 25 per cent growth in head office employment (Miller and Smart 2012). "If a city could be said to have a gas pedal," Bronconnier later said, "ours was to the floor."[19]

Bronconnier, a local small business owner who had been on city council since 1992, saw the traffic congestion and infrastructure demand that the boom had unleashed and pledged to get Calgary moving again. His campaign's unofficial slogan – "roads! roads! roads!" – offered little suggestion that Bronconnier would differ from the developer-friendly, expansionist, pro-growth ethos that had dominated city politics in the preceding decades (Miller and Smart 2012). On election night in 2001, few would have suspected that Bronconnier would soon become one of the city's most consequential mayors, a mayor who, together with his council colleagues, would leave behind a dramatically different political environment and policy agenda than the one they inherited.

In the first few years, Bronconnier and his colleagues focused on the original campaign promises: building new roads, widening arterial corridors, and finding new funds for upkeep and repair. This was the era of a Dave Bronconnier who "never saw a shovel handle he didn't like."[20] Over time, however, the policy agenda expanded. The most important initiative was "imagineCalgary," a public engagement project intended to develop a new vision for the city's future. The project, which engaged some 18,000 local citizens and incorporated active working groups on governance, the built environment, the natural environment, social conditions, and the economy, culminated in a report that laid out a striking new vision for the city's future, a city that "would be far more equitable, socially and economically secure, environmentally sound, and participatory" than what had come before (Feng et al. 2014; Langford et al. 2016; Miller 2016, 223). The imagineCalgary engagement process was completed in 2006, distilled into eleven core principles, and accepted by city council in 2007 as a document to be used by city staff as a guide to the city's new transportation and land use frameworks (Howard 2015).

In developing these new frameworks, which came to be called "Plan-It Calgary," the city initiated yet another intensive round of public engagement, with countless open houses and "planning summits" across the city. As it became clear that the frameworks envisioned substantial increases in density for greenfield development, along with a serious commitment to intensification in existing communities, some developers began to organize against the plan, deriding it as "social engineering" that ignored the preferences of home buyers. A group of local citizens, calling themselves "CivicCamp," were warned by supportive councillors and city officials that the entire Plan-It exercise was at risk without a strong demonstration of local support, and they mobilized in support of the new vision (Howard 2015; Miller 2016). Fearing defeat, Bronconnier met with developers and negotiated an eleventh-hour compromise, reducing the original density targets. The compromise met bitter criticism from some CivicCamp activists, including Naheed Nenshi, but many agreed that "a compromise on the plan was better than no plan at all," and the document was approved unanimously by council (Howard 2015: 156). For the first time, the City of Calgary had approved a planning document that incorporated meaningful density targets and sustainability principles and had serious regulatory teeth (Miller and Mössner 2020; Taylor 2014). The "imagineCalgary" and "Plan-It" exercises had important effects on the practice of politics in Calgary. In the immediate term, Mayor Bronconnier had ascended from road builder to city builder, a mayor who would now speak at World Urban Forum events and who would win the inaugural Canada Green Building Council Award for Government Leadership.[21] In the longer term, Plan-It would serve as the point of origin for a new and often contentious policy debate about the speed and scale of intensification in existing neighbourhoods, the costs of new development on the urban fringe, and about whose voices should count in decisions about changes to the local built environment.

Meanwhile, there was plenty more building to be done – including new LRT lines and extensions, parks and recreation facilities, police and fire stations, and even a pedestrian bridge designed by architect Santiago Calatrava, whose eye-watering price tag served for some as evidence of council's needlessly loose wallet and for others as a symbol of Calgary's emergence as a serious, attractive, world-class city.[22] The polarized response to the pedestrian bridge underscored the ongoing presence of substantial ideological disagreement in local politics. Danielle Smith, then a regular columnist at Calgary Herald, wrote before Bronconnier's election in 2001 that the city desperately needed to "elect conservative candidates" in the upcoming elections. "Forget the argument that 'there isn't a conservative or a liberal way to pave roads,'" Smith wrote. "There is – those who say otherwise are probably liberals."[23] It was hardly the first time that local leaders had called for a more conservative council – and it was hardly the last time that Calgarians would fail to heed the call.

Provincial–municipal tensions also carried on into the Bronconnier period. Whereas Al Duerr had felt powerless against Klein's austere tide, Bronconnier took a more public and confrontational approach. Tensions boiled over in the summer of 2007 when, having dismissed new premier Ed Stelmach's budget as "hocus-pocus, booga-booga economics," Bronconnier publicly demanded that Stelmach follow through on his infrastructure funding promises. Stelmach finally relented, making a $3.3 billion, ten-year commitment for infrastructure funding in Calgary (Howard 2015).

When Bronconnier announced in 2010 that he would not seek re-election, even his critics acknowledged that much had been accomplished over the mayor's three terms. One conservative activist complained about Bronconnier's record on taxes and spending but admitted his success in "improving Calgary's infrastructure and fighting crime." Ric McIver, a conservative councillor, had regularly clashed with Bronconnier but admitted that the mayor had been remarkably successful. "On any major issue that was important to him," McIver said, "he got it done during his time on city council."[24] On the eve of his retirement, the local newspaper could speak of Calgary, without apparent irony, as the "City that Dave Built."[25] It was a legacy of roads and bridges and buildings, to be sure, but it was also a legacy of sustainable planning, housing density, cultural investment, and citizen engagement. In Bronconnier's era, the potential tensions within this legacy were obscured by frantic growth and a booming economy. In the decade that followed, they would become all too clear.

The Nenshi and Gondek Era

Calgary tradition dictates that incumbent mayoral races are predictable bores, and open mayoral races are thrilling spectacles. In 2010, the city stuck to tradition. With Dave Bronconnier no longer in the mix, several high-profile candidates entered the mayoral race, and a mid-September poll, just one month before the election, showed two candidates as clear front-runners: Ric McIver, a conservative councillor, and Barb Higgins, a local broadcaster. Naheed Nenshi stood at 8.2 per cent. "Unless something fires up the electorate in the next four weeks," Leger's pollster explained, "I don't think there's going to be a big shift."[26]

Something *did* fire up the electorate – something so surprising and magical that, like Ralph Klein's 1980 victory, it still evokes a nostalgic smile among anyone who was involved in the campaign.[27] Debate continues among local politicos about just *what* it was in Nenshi's campaign that caught fire in the closing weeks. Was it the city's aversion to conservative mayors combined with Barb Higgins's disappointing performance as the obvious alternative? Was it Nenshi's whip-smart policy wonkery? His pathbreaking social media campaign? His brilliantly anti-ideological "purple army" mantra (neither red nor blue, a mix of the two)? Whatever the cause of the spectacular come-from-behind victory, it was Naheed Nenshi, on 18 October 2010, who would become the city's next mayor.[28]

Nenshi, the first Muslim ever to serve as mayor in a major North American city, was an instant celebrity. There were feature interviews on news programs across Canada and international profiles in Time Magazine and CNN. Nenshi embraced the opportunity to share his message of pragmatic city-building and inclusive growth, but he also spoke proudly of how little his background had mattered in Calgary's mayoral race. "The interesting thing for me," Nenshi said, "was that whole thing, the colour of my skin, my faith, was never part of the election. The issue of the racial background never came up; the issue of faith came up, you know, once or twice, and was quenched quite quickly."[29] According to Chima Nkemdirim, Nenshi's friend and campaign manager, the number of incidents of racist or Islamophobic language that Nenshi encountered in his 2010 campaign could be counted on one hand. By the end of Nenshi's time in office, such incidents would be occurring daily (Markusoff 2020).

In the early days, however, times were good. The 2010 council was a friendly group, generally happy to indulge the new mayor's penchant for more freewheeling council debates (Markusoff 2020). Council moved forward on projects that had begun under Bronconnier, including a major revitalization program in a long-troubled downtown neighbourhood, the East Village. After decades of feminist advocacy, council members at last agreed to change their name from "alderman" to "councillor," and Nenshi made sincere – and appreciated – gestures to reset the city's relationship with the nearby T'suu Tina First Nation (Newton 2015; Valentich 2009). Some complained that council's progress was largely symbolic, but there was little doubt that, at the very least, the tone had shifted at city hall (Markusoff 2020).

Then, in June 2013, emergency struck. In the mountains to the west, heavy rainfall and rapidly melting snow swelled Calgary's rivers, causing disastrous flooding. Mayor Nenshi quickly declared a state of emergency. What followed, amid the community evacuations, property damage, and fear, was a display of tireless leadership and community solidarity from Nenshi that few would soon forget: forty-three straight hours of social media updates, meetings with volunteers, press conferences, pep talks, and neighbourhood visits. On social media, Calgarians urged their indefatigable mayor to take a nap. Posters appeared on local buildings: "Keep Calm and Nenshi On." Newspapers across Canada filled with glowing profiles.[30] Nenshi's re-election later in 2013 might as well have been an acclamation – even his fiercest critics admitted he was unbeatable. The 2014 World Mayor Prize, which Nenshi received the following year, was icing on the cake.[31] Naheed Nenshi was on top of the world.

Quickly, however, as a result of forces far beyond Calgary city council's control, the mood began to sour. In mid-2014, global oil prices began a steep downward dive, bottoming out miles below break-even prices for Canadian oil producers. Local unemployment surged and downtown offices emptied out as the energy industry retreated into survival mode. The economic turmoil coincided with political upheaval as well: at the provincial level, the New Democratic Party ended forty-four years of Progressive Conservative rule, and at the federal level, Calgary's own Stephen Harper was defeated by a Liberal leader named Trudeau. For Calgarians, these were uncertain and unsettling times (Lucas and Santos 2021).

The local consequences were grim. A collapse in commercial property values in Calgary's downtown left a gaping hole in civic finances, and because the price collapse was geographically uneven, concentrated largely in downtown business offices, commercial property owners outside the downtown faced double digit year-over-year tax rate increases. This caused a serious fiscal challenge that would take many years, and tens of millions of dollars from municipal contingency funds, to address. As budgets tightened, tension grew. Voices of resentment and alienation grew louder, finding a receptive audience among thousands of Calgarians still looking for work. By 2017, a mayoral election that most had expected to be a standard-issue incumbent cakewalk became not only competitive but also embittered, with vile racism directed at the Nenshi campaign at a scale that no one would have imagined seven years earlier. After an uninspired campaign – Nenshi described the race as "no fun at all," and it showed – Calgary's council entered its new term in a fractious and grumpy mood (Markusoff 2020).

In the years that followed, things seemed to grow even worse. A potential bid to host the 2026 Olympic Games might have been expected to reinvigorate civic pride and enthusiasm, but intergovernmental bickering, a lack of local leadership, and a mood of rigid fiscal conservatism among the Calgary public combined to defeat a 2018 plebiscite on the potential bid (Hiller et al. 2024). In 2019, continued fiscal challenges led council to cut some $60 million from the budget, a decision that many councillors felt was necessary but profoundly demoralizing. Then came the COVID-19 pandemic, which not only sparked bitter provincial–municipal disputes but also left downtown commercial property in an even more precarious state; by the second quarter of 2022, vacancy rates in downtown commercial property had reached an astonishing 33.7 per cent. At the top of the Bow River valley, thousands of residents took advantage of the city's pandemic-era pedestrianization policy, walking and cycling on the empty street and taking sunset photographs of the city skyline and the mountains beyond. Few were aware that a third of the buildings in their photographs sat empty.

Still, despite the gloom that had settled over city council, Naheed Nenshi's final term was hardly without policy achievements. One major achievement was on secondary suites. Until the early 2000s, secondary suites (primarily basement apartments) had been a minor component of Calgary's housing stock, but the city's explosive economic and population growth created major demand for affordable rental units. In 2009, a fatal house fire in an unregulated and poorly planned secondary suite brought new attention to the issue. Nenshi and his allies on council had tried, and failed, to allow regulated citywide secondary suites, but in March 2018, he was finally able to check the box on the top campaign promise he had made in the 2010 election (Levenda et al. 2020; Markusoff 2020; van der Poorten and Miller 2017).

Other policy domains saw substantial achievements as well. The city's new central library, widely acclaimed for its aesthetic beauty and functionality, was completed on time and on budget in 2018. In public transit, the city opened new Bus Rapid Transit (BRT) lines in 2019 and, after years of debate, approved the first phase of a new north–south "Green Line" LRT in 2020. Nenshi was also able to reverse a policy decision that had occurred early in his time as mayor to remove fluoride from the city's water supply. In the closing stages of his mayoralty, Nenshi cleverly added a question on community water fluoridation to the 2021 municipal ballot, which passed with an overwhelming majority, binding the hands of the next council.

Despite the bumps and bruises, Nenshi stepped down from the mayor's chair in 2021 having not only permanently transformed the city's image beyond its borders, but having also achieved an important list of local policy goals.[32] Given the anger that had accompanied the 2017 election, as well as the debates that had arisen during the COVID-19 pandemic, it seemed that the time was finally ripe for the conservative backlash that some had been expecting in Calgary since 2013. Jeromy Farkas, a well-known conservative councillor, announced his candidacy for mayor, and a loosely organized slate of conservative candidates offered themselves in the ward races.[33] Once again, however, Calgarians chose differently, electing Jyoti Gondek, a progressive mayor, along with the most progressive council in the city's history.[34]

CALGARY'S POLITICAL CULTURE

Earlier, I suggested that Calgary's geographic position in the fescue grassland and mountain foothills has an important influence on local policy and politics. Even more important, however, is Calgary's location in *political* space as the largest city in the province of Alberta. This gives rise to several important features – as well as some important and durable misunderstandings – of Calgary's local political culture.

The most obvious consequence of Calgary's location in political space is the city's reputation as a *conservative* city. "The Calgary civic culture," writes one scholar, "is market oriented and socially and fiscally conservative" (Brunet-Jailly 2012, 296). Others describe Calgary as "a conservative city in a conservative province" (Miller and Smart 2012, 29) and as a city "renowned for its political and cultural conservatism" (Dumitrica 2014, 58). One local columnist summarized the city's reputation in more colourful terms. To outsiders, he wrote, Calgary is a "steak-eating, SUV-driving, right-wing Hicksville" (Chris Turner, quoted in Howard (2015, 7)).

Clearly, this reputation is pervasive. But the more carefully one inspects the practice of politics in Calgary, the more difficult it is to square the reputation with the reality. In every major mayoral election in Calgary since 1980, the winning mayoral candidate *defeated* at least one prominent, well-funded conservative challenger. In fact, no mayoral candidate with clear ties to provincial or federal Conservative parties has been successful in Calgary for at least four decades, and well-funded efforts to populate Calgary's city council with conservative representatives have, across several electoral cycles, largely proved unsuccessful (Lucas and Santos 2021).

Moreover, a good deal of Calgary's conservative reputation may originate in the apples-to-oranges comparison that results from the city's long-standing unicity structure. As I noted earlier, Calgary's boundaries have always included a diverse mix of inner-city and suburban residents, and to the extent that suburban dwellers tend to have more conservative preferences than those in the urban cores, this makes the average Calgarian look more conservative than the average resident in, say, Vancouver. Carve out a Vancouver-sized population from

Figure 8.3: Distribution of ideology and policy attitudes within cities

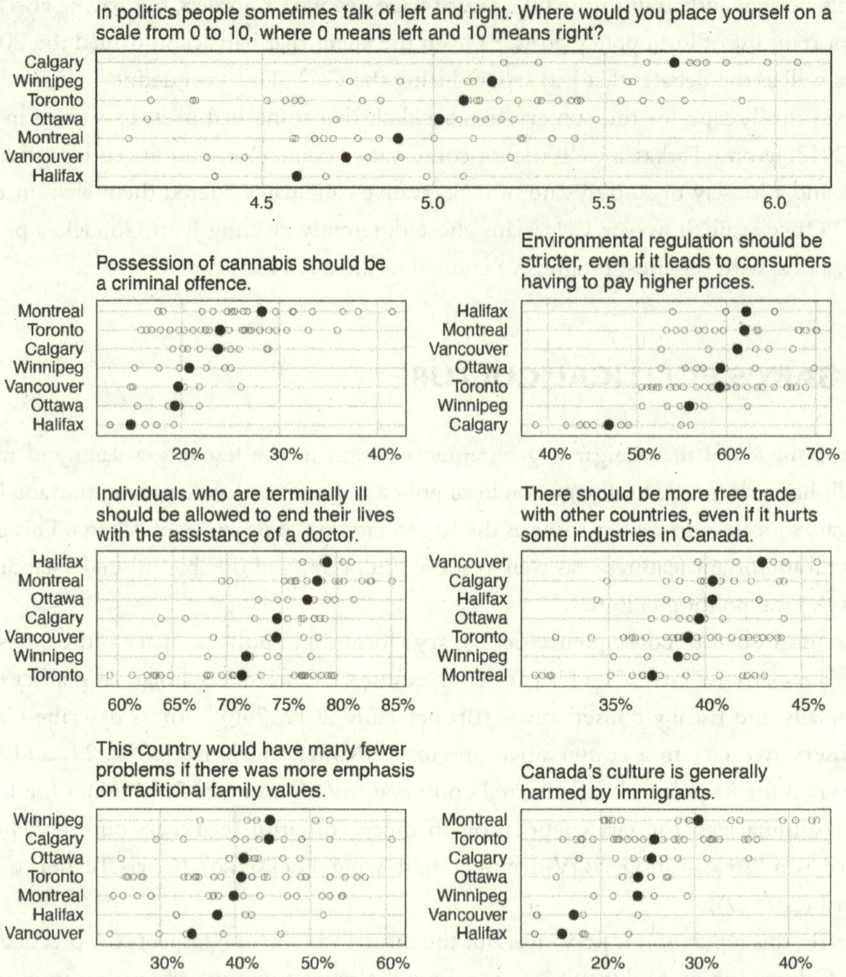
In politics people sometimes talk of left and right. Where would you place yourself on a scale from 0 to 10, where 0 means left and 10 means right?
Calgary
Winnipeg
Toronto
Ottawa
Montreal
Vancouver
Halifax
4.5
5.0
5.5
6.0
Possession of cannabis should be a criminal offence.
Montreal
Toronto
Calgary
Winnipeg
Vancouver
Ottawa
Halifax
20%
30%
40%
Environmental regulation should be stricter, even if it leads to consumers having to pay higher prices.
Halifax
Montreal
Vancouver
Ottawa
Toronto
Winnipeg
Calgary
40%
50%
60%
70%
Individuals who are terminally ill should be allowed to end their lives with the assistance of a doctor.
Halifax
Montreal
Ottawa
Calgary
Vancouver
Winnipeg
Toronto
60%
65%
70%
75%
80%
85%
There should be more free trade with other countries, even if it hurts some industries in Canada.
Vancouver
Calgary
Halifax
Ottawa
Toronto
Winnipeg
Montreal
35%
40%
45%
This country would have many fewer problems if there was more emphasis on traditional family values.
Winnipeg
Calgary
Ottawa
Toronto
Montreal
Halifax
Vancouver
30%
40%
50%
60%
Canada's culture is generally harmed by immigrants.
Montreal
Toronto
Calgary
Ottawa
Winnipeg
Vancouver
Halifax
20%
30%
40%

Calgary's urban core, and while it may not look *quite* as progressive as its west coast cousin, the resemblance might be closer than Calgary's conservative reputation would lead us to suspect.

An honest assessment of Calgary's conservatism therefore requires that we are attentive to variation not only across but also within Canadian cities. Figure 8.3 seeks to provide this sort of assessment, summarizing ideological self-placement scores and policy attitudes in Calgary and seven other cities. To construct the plot, I took advantage of the fact that Canada's federal electoral districts divide cities into smaller geographic pieces of roughly comparable population size, allowing us to explore variation in ideology and policy attitudes within each city without having to rely on municipal wards that vary dramatically in size (or, in Vancouver's case, do not exist at all).[35] Using data from a large public opinion survey, combined with a statistical technique to estimate public attitudes in specific federal electoral districts, I created district-level estimates of local residents' average self-placement on a left–right ideological scale and their positions on a number of policy issues.[36] In Figure 8.3, each panel captures the city's median score (the large black circle) along with the score for each individual district within the city (the hollow circles).

The results in Figure 8.3 help to clarify – and correct – some long-standing claims about Calgary's conservative character. Notice, in the first panel, that Calgary's median ideological self-placement score, as well as the self-placement scores for each individual district, are to the right of the other city medians. When asked to place themselves on a left–right spectrum, Calgarians are, on average, distinctly more right-leaning than their counterparts in other big Canadian cities. In other words, like other Albertans, Calgarians *are* especially likely to identify as conservative.

When we turn to issue attitudes, however, the story becomes more complicated. On a broad range of issues – including cannabis policy, medically assisted dying, international trade, traditional values, and anti-immigrant sentiment – Calgary looks much like the other cities, especially cities like Toronto or Winnipeg whose boundaries also include a mix of urban and suburban areas. On environmental regulation, however, Calgary *does* stand out from the others, with noticeably lower levels of support. Even in Calgary's downtown areas, support for environmental regulation is lower than in the average district in Toronto, Ottawa, and other cities.

Overall, the results in Figure 8.3 help to clarify Calgary's ostensible conservatism. Calgarians *are* more likely than residents in other Canadian cities to think of themselves as conservative, and they do hold distinctly conservative attitudes on some issues, particularly related to environmental policy, carbon taxes, and the oil and gas sector. This is yet another way that the city's economic environment influences its politics. On many other issues, however, Calgarians look much like their counterparts in other cities. This is even truer when we turn to specifically municipal issues. In 2021, for example, the Canadian Municipal Election Study found that Calgary residents overwhelmingly supported community water fluoridation, the Green Line LRT project, a new pro-density planning guide, reduced residential speed limits, and reallocation of $20 million from local policing to mental health and harm reduction programs (McGregor et al. 2021). On these and many other municipal issues, most Calgarians embrace a broadly progressive vision for their city's future. Thus, while Calgarians do have a distinct tendency to think of themselves as conservative, any claims about overwhelmingly conservative policy attitudes among Calgarians must be treated with scepticism – especially when those claims have to do with municipal policy.

A second important component of Alberta's political culture has figured less prominently in discussions of Calgary politics: populism. Appeals to the wisdom and good sense of the common man against the interests of a distant elite – be it Ottawa politicians, disdainful Laurentians, or global environmentalist intellectuals – have animated much of Alberta's provincial and federal politics through the years (Flanagan 2009; Sayers and Stewart 2019). There

Figure 8.4: Populism scores by district, Calgary and comparator cities

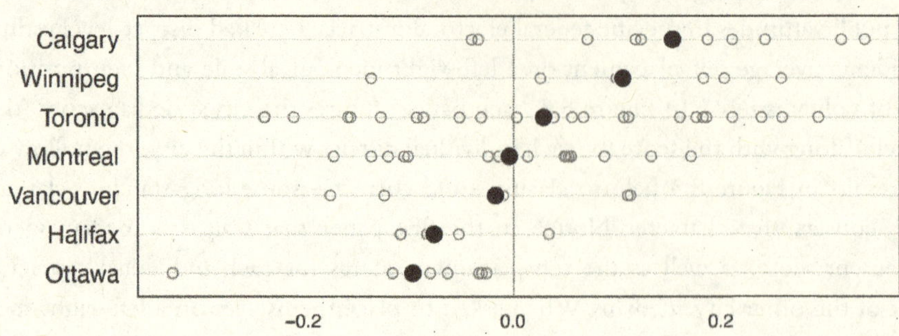

is good reason to think that Calgary – the birthplace of several populist movements, including the Reform Party of Canada – might have absorbed this populist strain into its municipal political DNA. While some have claimed that Calgary's "mayor and city councillors emerge from a business elite" (Brunet-Jailly 2012, 305), a mere glance at the city's political leaders suggests that this is transparently false: Calgary's political leadership hails far more from the petite bourgeoisie and lower-middle-class professions than from the ivory towers or downtown c-suites (Masson and LeSage 1994). Moreover, as I will discuss below, populist appeals have been central to the political leadership and electoral success of several local politicians in Calgary since 1980.

Figure 8.4, which plots populism scores for districts in Calgary and comparator cities using the same approach and layout as the previous figure, lends some support to this contention. To calculate populism scores, I relied on a set of survey questions called the "populism index," all of which measure citizens' feelings about the extent to which they feel politicians truly represent the people, politicians are trustworthy, strong leaders should be supported, and the importance of popular opinion for policy making.[37] While we must approach this measure with caution – urban political scientists have convincingly argued that local populism is as much a feature of political leadership and discourse as it is a matter of individual-level behaviour (Silver et al. 2020) – the analysis nevertheless provides a helpful comparative perspective on each city's receptivity to populist appeals.

The results in Figure 8.4 suggest that Calgarians are indeed especially likely to embrace populist attitudes. Calgary's overall average, as well as the populism scores for most local districts, are well above most other cities. Interestingly, the distribution in Calgary is very wide, with two districts at much lower populism values – the two parts of Calgary that, perhaps not coincidentally, were most strongly supportive of Naheed Nenshi (Harvard educated) and Jyoti Gondek (PhD in urban sociology). While Calgary's scores are clearly at the top of the chart, the difference is one of degree rather than kind. Unsurprisingly, given these results, recent history has shown that Calgary is hardly alone in its tradition of populist mayoral and council candidates (Erl 2021; Kiss et al. 2020; Masson and LeSage 1994).

CONCLUSION

Unlike many of the other cities in this book, Calgary's municipal government has been free of major institutional reform in the past forty years. Beneath the institutional stability, however, much has changed. The process has been one of sedimentary layering, rather than volcanic upheaval, as new practices and policy debates have been layered atop older ones without fully replacing them.

One layer that runs through the full sweep of Calgary's politics since 1980 is the ongoing challenge of economic boom and bust. Both phases of the roller coaster have their distinctive challenges: when times are bad, city council must face down the three-headed monster of fiscal restraint, social need, and a grumpy, fiscally conservative electorate. But even when times are good, there are real challenges – explosive growth means strained infrastructure, an expensive housing market, and demands for new investment from all quarters. Forty years ago, Warren Magnusson described Toronto's central ideological divide as "boosters" versus "cutters" – those who wanted to spend boldly on infrastructure, economic development, and expansion versus those who wanted the city to spend carefully and restrict itself to a modest policy agenda. "When the city had over-extended itself," Magnusson explained, "the cutters attracted a wider following," but "in times of prosperity the boosters had greater influence" (Magnusson 1983, 98). The same pattern is recognizable in Calgary: while each council has had its "cutters," the boosters tend to dominate except when the local economic situation is dire. Many important policy outcomes in Calgary – from the bitter debates about Calatrava's "Peace Bridge" to the failure of the Olympic Bid Plebiscite in 2018 – can be explained, at least in part, by pointing to the city's current position on the boom-and-bust roller coaster and the relative strength of "booster" and "cutter" voices on city council as a result of that position.

Some urban political scientists interpret the "booster versus cutter" divide as evidence of the *absence* of left–right disagreement at the municipal level.[38] In Calgary, however, ideological disagreement between recognizably left-leaning and right-leaning councillors has been an equally persistent layer in the city's politics. Every competitive mayoral election (along with most of the uncompetitive elections) has featured clear ideological competition among major candidates, and these elections are regularly described in ideological terms in local media commentary. The same is true of many council races. Candidates like Naheed Nenshi, who explicitly disavow left–right ideological disagreement, are the exception that proves the rule: Nenshi's "team purple" ethos was part of what made his campaign exciting and distinctive. Calgary's long-standing commitment to the unicity structure, its large municipal wards, and its local political culture – which, as a result of provincial and federal politics, contains more residents who consider themselves conservative than we would ordinarily expect in a Canadian city of Calgary's size – combine to generate consistent ideological disagreement in local politics (Lucas and McGregor 2020). The substance of this disagreement has varied as the local policy agenda has evolved, but the presence of a left–right cleavage has been a consistent feature of the practice of politics in Calgary through the past four decades.

Atop these foundational layers, two additional layers have been added to Calgary politics since the early 1980s. The first is municipal–provincial disagreement. Since its incorporation, Calgary's municipal government has occasionally disagreed with provincial decisions, and there have certainly been periods of serious municipal–provincial dispute prior to the 1980s (Masson and LeSage 1994). Beginning in the 1990s, however, provincial–municipal disagreement has become a more consistent and important feature of Calgary's politics: budget cuts and downloading in the 1990s; infrastructure funding in the 2000s; on-again, off-again provincial legislation to provide Edmonton and Calgary with "city charter" status in the 2010s; and bitter disagreements about pandemic policy in the 2020s. At the time of writing of this chapter, Calgary and Edmonton are facing a major impending change to local political dynamics, courtesy of a provincial government bill that will allow political parties to run in local elections and give the provincial cabinet the authority to dismiss elected councillors and order municipalities to repeal adopted bylaws (Strasser 2024). Provincial–municipal disagreement is primarily about policy autonomy and provincial funding, but it also powerfully engages deep interurban rivalries – no Calgary politician wants to be seen to be doing "Edmonton's work" on city council. This pattern reached its logical limit in 2021, when mayoral candidate Jyoti Gondek ran her campaign almost entirely against Jason Kenney, the Alberta premier at the time, rather than Jeromy Farkas, her leading opponent.[39] Much more than in the past, an important source of disagreement on municipal council in Calgary today has to do with the extent to which the city should position itself in a "resistance" role in relation to the provincial government.

A second new layer that has been added to Calgary's politics has been the increasing presence of *non-economic* dimensions of policy disagreement. The most important of these has been an emphasis on *sustainability* in local policy decisions, particularly in debates on land use planning and growth. As I noted earlier, sustainability became an important part of the local conversation in the 1990s and was then integrated into actual local policy in the first decade of the twenty-first century with the plans that emerged from imagineCalgary and Plan-It. Since then, questions of sustainability have become crucial to the practice of politics in the city. Today, policy issues are regularly debated not only in terms of their economic costs and benefits but also their implications for long-term sustainability. City council's 2021 decision to declare a "climate emergency" – Mayor Jyoti Gondek's first act in office – is just the latest example of an established tradition in which sustainability has become a crucial component of many local policy debates.

More recently, non-economic debates about *social justice* have also become increasingly prominent.[40] Discussions of Indigenous reconciliation, racism in local policing, and policy initiatives to support historically marginalized and equity-deserving groups are now frequent on Calgary city council. Moreover, policy discussions that might once have been structured entirely by "booster versus cutter" arguments – the size of the police budget, the names of new bridges and parks, budget support for cultural events and local organizations – are now debated in terms of social justice, antiracism, and the so-called woke versus anti-woke divide. This new politics – what some call "identity politics" – goes well beyond symbolic representation and powerfully shapes council's policy agenda and budget priorities.

This multi-layered political environment – boosters and cutters, left and right, economic and non-economic – speaks to the complexity of local politics across Canada today. Against the "city limits" theorists, who insist that municipal policy making is invariably focused on "developmental" issues (Peterson 1981), Calgary's political development reveals the consistent presence of deeply ideological policy disagreement, and demonstrates how quickly a seemingly "developmental" issue, like economic development or tax rates, can be framed in non-developmental, ideological terms. Against "nationalization" theorists, on the other hand, who argue that municipal politics is becoming little more than a proxy war for larger national cleavages (Hopkins 2018), Calgary also demonstrates that local politics is about much more than national debates and is clearly shaped by its local economic, institutional, and cultural context. Urban political scientists must navigate between these extremes, acknowledging that Calgary's politics – like other cities – is firmly enmeshed within a larger political ecosystem, while also accepting that creative local political leaders like Ralph Klein or Naheed Nenshi can build local coalitions, at least for a time, that would be difficult to hold together in other cities or at other levels of government.

Above all, however, Calgary's political development offers a strong caution against excessively "localist" theories of urban politics – theories for which the practice of politics that is visible in a city must be traced to *causes* in that same city (Brenner 2009). There is no doubt that much about Calgary's politics must be explained in terms of the personalities and passions of the people who have governed the city, the local activists who work tirelessly to raise the profile of policy issues, and the relationships of trust (or distrust) among local policymakers at a particular point in time (Feng et al. 2014). There is plenty of local drama and individual agency in Calgary's story. Even so, our mental model of politics in Calgary must not stop at the city's boundaries. It must necessarily be a *multilevel* model, one in which broader forces continually shift the balance of probabilities in the local political arena: the probability that a candidate will win in a local election, a policy issue will rise or fall on the local agenda, or a new housing project will be approved at the city's outer fringe. Some of these forces are slow-moving and, in the absence of a comparative perspective, barely perceptible, such as the effect of Calgary's natural geography on its spatial development. Others, like the price of oil, are dramatic and rapid. Either way, they are only visible – and Calgary's politics only intelligible – from a multilevel point of view.

NOTES

1 Ron Wood, quoted in Howard 2015: 64.
2 "Building Permits," Calgary Open Data. https://data.calgary.ca/Business-and-Economic-Activity/Building-Permits/c2es-76ed/data.
3 Halifax's trajectory, a random walk about zero followed by a sharp upward trend, probably captures pandemic-era migration to Nova Scotia.
4 The city moved to a dual franchise, with voting rights for property owners *and* local residents, in 1915; see Statutes of Alberta 1915, c.26. At-large elections were adopted in Calgary after a plebiscite

to abolish wards passed narrowly in 1913. See City of Calgary Council Minutes May 26, 1913, October 27, 1913.

5 Calgary's very eventful process of adopting wards in 1961 is described in Masson and LeSage (1994); additional detail is available in Calgary Public Library Clippings Files ("Calgary Election Ballots").

6 In 2025, as a result of legislative changes by the Government of Alberta, political parties will run once again in Calgary's municipal elections, and for the first time, party and slate labels will be listed on the municipal ballot. This change, if it persists, is likely to be consequential for the practice of politics in Calgary. Three parties have already been announced in anticipation of the 2025 election.

7 My source for this plot is the Canadian Municipal Elections Database; see Lucas et al. (2021).

8 Calgary Public Library Clippings Files, 1980 Election; *Calgary Herald*, September 20, 1980.

9 Calgary Public Library (hereinafter CPL) Local History Vertical Files: 1980 Election; *Calgary Herald*, October 16, 1980.

10 CPL Vertical Files: 1983 Election; *Calgary Herald*, October 16, 1983.

11 References to Klein's Liberal Party affiliation are easy to find in contemporary media coverage. See CPL Vertical Files: 1980 Election; *Calgary Herald*, September 25, 1980, and 1986 Election; *Calgary Herald*, October 14, 1986, as well as Martin (2002).

12 CPL Vertical Files: 1980 Election; *Calgary Herald*, October 14, 1980.

13 For 1980, see CPL Vertical Files: 1980 Election; *Calgary Herald*, October 20, 1979, and March 4, 1980; for 1980, see 1983 Election; *Calgary Herald*, July 6, 1983.

14 CPL Vertical Files: 1980 Election; *Calgary Herald*, September 17, 1983; September 20, 1983.

15 CPL Vertical Files: 1986 Election; *Calgary Herald*, October 14, 1986.

16 CPL Vertical Files: 1989 Election; *Calgary Herald*, January 11, 1989, September 10, 1989, October 7, 1989. For the more general professionalization of municipal councils in Alberta in the 1980s, see *Calgary Herald*, September 10, 1989; and Masson and LeSage (1994).

17 The pamphlet is available in CPL Pamphlets PAM 352.23216. For the other details about Duerr and the interim mayor position, see CPL Vertical Files: 1989 Election; *Calgary Herald*, January 5, 1989; January 11, 1989; October 7, 1989; October 17, 1989.

18 CPL Vertical Files: Duerr, Al; *Calgary Herald*, November 27, 1993.

19 Quoted in Howard (2015, 109). Bronconnier was referring to the height of the boom, in 2006, but as Howard rightly suggests, his comment captured the more general sentiment.

20 *Calgary Herald*, February 14, 2010, A14.

21 *Calgary Herald*, February 14, 2010, A4.

22 *Calgary Herald*, February 24, 2010, A14.

23 *Calgary Herald*, June 4, 2001, A11. In fact, those who say otherwise are probably conservatives: see Lucas (2021b).

24 *Calgary Herald*, February 24, 2010, A1.

25 *Calgary Herald*, February 24, 2010, A14.

26 *Calgary Herald*, September 19, 2010, A6.

27 During the 2017 Calgary election, in an interview I conducted with a long-time strategist who had been involved in Klein's 1980 victory, a brief mention of the 1980 mayoral election literally brought nostalgic tears to the eyes of the grizzled campaign veteran.

28 *Calgary Herald*, February 24, 2010, A4; April 7, 2021, A4.

29 TVO The Agenda, May 28, 2021.

30 See, for example, *Calgary Herald*, June 13, 2013; *Canadian Press*, June 28, 2013; *Montreal Gazette*, July 8, 2013.

31 CBC News Calgary, February 2, 2015.

32 *Calgary Herald*, April 7, 2021, A4.

33 CBC News Calgary, November 19, 2019, October 9, 2021.

34 Like Nenshi, Gondek denied that ideology was especially relevant in municipal politics, calling herself a "radical centrist," but her council votes, criticisms of the United Conservative Party (UCP) government, and consistent advocacy for racial justice, feminism, and Indigenous reconciliation have clearly positioned her as a progressive. See Lucas et al. 2023.

35 The population size of Canada's federal electoral districts varies a great deal across Canada, but in the eight cities in my comparison, the population variation is less stark, ranging from a low of about 75,000 (Central Nova) to a high of 121,000 (Ottawa South). 65 per cent of these districts have populations between 90,000 and 110,000, and 98 per cent have populations between 80,000 and 120,000.

36 The survey dataset is the 2019 Canadian Election Study (Stephenson et al. 2021), and the statistical technique is multilevel regression and poststratification (MRP). For more information on MRP, see Lucas and Armstrong (2021). To identify districts within each city, I used district and municipality boundary files to identify all instances in which at least half of a federal district's land mass was contained within the city's boundaries. More restrictive criteria yield similar results.

37 α =0.73. These questions are drawn once again from the Canadian Election Study, 2019. I created a latent populism measure from these items using a Bayesian factor analysis model and then measured district-level populism scores using MRP.

38 To be clear, this is not Magnusson's argument; his account of city politics in Toronto attends very carefully to left–right divides in Toronto politics.

39 Jyoti Gondek's campaign manager described this strategy in detail in a podcast recorded after the election: https://www.decidecampaigns.com/news/the-gondek-campaign-the-podcast.

40 Where social justice is understood, following Joseph Heath, to capture "what individuals actually wind up receiving, in the form of income, status, health, leisure, and other 'good things in life'."

REFERENCES

Brenner, Neil. 2009. "Is There a Politics of 'Urban' Development? Reflections on the US Case." In *The City in American Political Development*, edited by Richardson Dilworth, 121–40. New York: Routledge.

Brunet-Jailly, Emmanuel. 2012. "Civic Culture in Calgary: The Oil and Developers' Land." In *Comparative Civic Culture: The Role of Local Culture in Urban Policy-Making*, edited by Laura A. Reese and Raymond A. Rosenfeld, 295–323. Farnham: Ashgate.

Conger, Brian W, Bev G Dahlby, and Melville L McMillan. 2016. "Municipal Revenue Generation and Development in the Calgary and Edmonton Metropolitan Regions." *School of Public Policy Research Papers* 9 (39): 1–35. https://doi.org/10.55016/ojs/sppp.v9i1.42612.

Downing, D.J., and W.W. Pettapiece. 2006. *Natural Regions & Subregions of Alberta*. Edmonton: Government of Alberta.

Dumitrica, Delia. 2014. "Politics as 'Customer Relations': Social Media and Political Authenticity in the 2010 Municipal Elections in Calgary, Canada." *Javnost - The Public* 21 (1): 53–69. https://doi.org/10.1080/13183222.2014.11009139.

Erl, Chris. 2021. "The People and The Nation: The 'Thick' and the 'Thin' of Right-Wing Populism in Canada." *Social Science Quarterly* 102 (1): 107–24. https://doi.org/10.1111/ssqu.12889.

Feng, Patrick, Ben Li, and Cooper H. Langford. 2014. "300 People Who Make a Difference: Associative Governance in Calgary." In *Governing Urban Economies*, edited by Neil Bradford and Allison Bramwell, 184–202. University of Toronto Press. https://doi.org/10.3138/9781442617223-010.

Flanagan, Thomas. 2009. *Waiting for the Wave: The Reform Party and the Conservative Movement*. Montreal: McGill-Queen's University Press. https://doi.org/10.1515/9780773575271.

Foran, Max. 2006. "Lassoed and Branded: The Calgary Exhibition and Stampede and the City of Calgary, 1889–1976." *Urban History Review* 34 (2): 30–42. https://doi.org/10.7202/1016011ar.

Ghitter, Geoff, and Alan Smart. 2009. "Mad Cows, Regional Governance, and Urban Sprawl: Path Dependence and Unintended Consequences in the Calgary Region." *Urban Affairs Review* 44 (5): 617–44. https://doi.org/10.1177/1078087408325257.

Guo, Jiaao, and Victoria Fast. 2019. "Insights on Urban Density." *Spatial Knowledge and Information Canada* 7 (3): 1–7.

Hiller, Harry H. 2007. "Gateway Cities and Arriviste Cities: Alberta's Recent Urban Growth in Canadian Context." *Prairie Forum* 32 (1): 47–66.

Hiller, Harry H. 1990. "The Urban Transformation of a Landmark Event: The 1988 Calgary Winter Olympics." *Urban Affairs Quarterly* 26 (1): 118–37. https://doi.org/10.1177/004208169002600106.

Hiller, Harry H., Jack Lucas, and R. Michael McGregor. 2024. "Why Do Local Residents Oppose Olympic Bids? An Electoral Perspective from Calgary 2026." *Event Management* 28(5): 749–61. https://doi.org/10.3727/152599524X17067412396200.

Hopkins, Daniel J. 2018. *The Increasingly United States*. Chicago: University of Chicago Press.

Howard, Tom. 2015. "From Risky Business to Common Sense: Sustainability, Hegemony, and Urban Policy in Calgary." PhD thesis, Vancouver: University of British Columbia.

Hubbell, John. 2006. "Light Rail Transit in Calgary The First 25 Years." In *Joint International Light Rail Conference*, 19. St. Louis, Missouri.

Kiss, Simon J., Andrea M. L. Perrella, and Zachary Spicer. 2020. "Right-Wing Populism in a Metropolis: Personal Financial Stress, Conservative Attitudes, and Rob Ford's Toronto." *Journal of Urban Affairs* 42 (7): 1028–46. https://doi.org/10.1080/07352166.2019.1657021.

Klaszus, Jeremy. 2020. "Paying for It: Calgary's 14 New Communities." *The Sprawl*, February 10. https://www.sprawlcalgary.com/sprawlcast-calgarys-14-new-communities.

Langford, Cooper H., Ben Li, and Camille D. Ryan. 2016. "Innovation from an Oil and Gas Platform: Calgary." In *Growing Urban Economies*, edited by David A. Wolfe and Meric S. Gertler, 157–78. University of Toronto Press. https://doi.org/10.3138/9781442629455-010.

Levenda, Anthony M., Noel Keough, Melanie Rock, and Byron Miller. 2020. "Rethinking Public Participation in the Smart City." *The Canadian Geographer /Le Géographe Canadien* 64 (3): 344–58. https://doi.org/10.1111/cag.12601.

Lucas, Jack. 2020. "Reaction or Reform? Subnational Evidence on PR Adoption from Canadian Cities." *Representation* 56 (1): 89–109. https://doi.org/10.1080/00344893.2019.1700154.

Lucas, Jack. 2021a. "The Size and Sources of Municipal Incumbency Advantage in Canada." *Urban Affairs Review* 57 (2): 373–401. https://doi.org/10.1177/1078087419879234.

Lucas, Jack. 2021b. "The Ideological Structure of Municipal Non-Ideology." *Urban Affairs Review*, August, 107808742110383. https://doi.org/10.1177/10780874211038321.

Lucas, Jack, R. Michael McGregor, and Aengus Bridgman. 2023. "Spatial Voting in Non-Partisan Cities: A Case Study." *Electoral Studies* 82 (April): 102599. https://doi.org/10.1016/j.electstud.2023.102599.

Lucas, Jack, and David A. Armstrong. 2021. "Policy Ideology and Local Ideological Representation in Canada." *Canadian Journal of Political Science* 54 (4): 959–76. https://doi.org/10.1017/S0008423921000652.

Lucas, Jack, and R. Michael McGregor. 2020. "Are City Elections Unique? Perceptions of Electoral Cleavages and Social Sorting Across Levels of Government." *Electoral Studies* 66 (May): 1–8. https://doi.org/10.1016/j.electstud.2020.102165.

Lucas, Jack, R. Michael McGregor, and Kim Lee Tuxhorn. 2022. "Closest to the People? Incumbency Advantage and the Personal Vote in Non-Partisan Elections." *Political Research Quarterly* 75 (1): 188–202. https://doi.org/10.1177/1065912921990751.

Lucas, Jack, Reed Merrill, Kelly Blidook, Sandra Breux, Laura Conrad, Gabriel Eidelman, Royce Koop, Daniella Marciano, Zack Taylor, and Salomé Vallette. 2021. "Women's Municipal Electoral Performance: An Introduction to the Canadian Municipal Elections Database." *Canadian Journal of Political Science* 54 (1): 125–33. https://doi.org/10.1017/S000842392000102X.

Lucas, Jack, and John Santos. 2021. "Calgary." In *Big City Elections in Canada*, edited by Jack Lucas and R. Michael McGregor, 33–52. Toronto: University of Toronto Press. https://doi.org/10.3138/9781487528577-005.

Magnusson, Warren. 1983. "Toronto." In *City Politics in Canada*, 94–139. Toronto: University of Toronto Press. https://doi.org/10.3138/9781487575908-004.

Markusoff, Jason. 2020. "Naheed Nenshi Falls Back to Earth." *Maclean's*.

Martin, Don. 2002. *King Ralph: The Political Life and Success of Ralph Klein*. Key Porter Books.

Masson, Jack K., and Edward C. LeSage. 1994. *Alberta's Local Governments: Politics and Democracy*. Edmonton: University of Alberta Press.

McGregor, R. Michael, Cameron D. Anderson, Éric Bélanger, Sandra Breux, Jack Lucas, J. Scott Matthews, Anne Mévellec, Aaron A. Moore, Scott Pruysers, Laura B. Stephenson, and Erin Tolley. 2021. "The Canadian Municipal Election Study." *Frontiers in Political Science* 3 (September): 1–6. https://doi.org/10.3389/fpos.2021.745331.

Miller, Byron. 2007. "Modes of Governance, Modes of Resistance: Contesting Neoliberalism in Calgary." In *Contesting Neoliberalism: Urban Frontiers*, edited by Helga Leitner, Jamie Peck, and Eric S. Sheppard, 223–49. New York: Guilford Press.

Miller, Byron. 2016. "Sustainability Fix Meets Growth Machine." In *Governing Cities Through Regions: Canadian and European Perspectives*, edited by Roger Keil, Pierre Hamel, Julie Anne Boudreau, and Stefan Kipfer, 213–38. Waterloo: Wilfrid-Laurier University Press.

Miller, Byron, and Samuel Mössner. 2020. "Urban Sustainability and Counter-Sustainability: Spatial Contradictions and Conflicts in Policy and Governance in the Freiburg and Calgary Metropolitan Regions." *Urban Studies* 57 (11): 2241–62. https://doi.org/10.1177/0042098020919280.

Miller, Byron, and Alan Smart. 2012. "Ascending the Main Stage? Calgary in the Multilevel Governance Drama." In *Sites of Governance*, edited by Martin Horak and Robert Young, 26–52. Montreal: McGill-Queen's University Press. https://doi.org/10.1515/9780773586918-003.

Newton, Randi. 2015. "Exploring the Factors That Support Cooperative and Equitable Municipal-First Nation Relationships: A Case Study of the City of Calgary and the Tsuut'ina Nation." Master's thesis, Queen's University, Kingston.

Peterson, Paul. 1981. *City Limits*. Chicago: University of Chicago Press.

Sayers, Anthony M., and David K. Stewart. 2019. "Out of the Blue: Goodbye Tories, Hello Jason Kenney." In *Orange Chinook: Politics in the New Alberta*, edited by Duane Bratt, 399–423. Calgary: University of Alberta Press. https://doi.org/10.1515/9781773850276-019.

Seiler, Robert M., and Tamara P. Seiler. 2001. "Ceremonial Rhetoric and Civic Identity: The Case of the White Hat." *Journal of Canadian Studies* 36 (1): 29–49. https://doi.org/10.3138/jcs.36.1.29.

Silver, Daniel, Zack Taylor, and Fernando Calderón-Figueroa. 2020. "Populism in the City: The Case of Ford Nation." *International Journal of Politics, Culture, and Society* 33 (1): 1–21. https://doi.org/10.1007/s10767-018-9310-1.

Stephenson, Laura B., Allison Harell, Daniel Rubenson, and Peter John Loewen. 2021. "Measuring Preferences and Behaviours in the 2019 Canadian Election Study." *Canadian Journal of Political Science* 54 (1): 118–24. https://doi.org/10.1017/S0008423920001006.

Strasser, Scott. 2024. "Province Tables Bill Allowing Political Parties in Calgary, Edmonton Elections." *Calgary Herald*, April 25. https://calgaryherald.com/news/politics/ric-mciver-outlines-changes-to-laea-embargoed.

Taylor, Zack. 2014. "Alberta Cities at the Crossroads: Urban Development Challenges and Opportunities in Historical and Comparative Perspective." *School of Public Policy Research Papers* 7 (12): 1–43. https://doi.org/10.2139/ssrn.2474377.

Taylor, Zack. 2020. *Theme and Variations: Metropolitan Governance in Canada.* Toronto: Institute on Municipal Finance and Governance, Munk School of Global Affairs and Public Policy.

Tretter, Eliot. 2022. "Petrocity." *The International Journal of Architectonic, Spatial, and Environmental Design* 16 (1): 55–70. https://doi.org/10.18848/2325-1662/CGP/v16i01/55-70.

Valentich, Mary. 2009. "Calgary's Resistance to Changing from 'Alderman' to 'Councillor'." *Theory in Action* 2 (1): 66–85. https://doi.org/10.3798/tia.1937-0237.08029.

van der Poorten, Kylee, and Byron Miller. 2017. "Secondary Suites, Second-Class Citizens: The History and Geography of Calgary's Most Controversial Housing Policy; Secondary Suites, Second-Class Citizens." *The Canadian Geographer/Le Géographe Canadien* 61 (4): 564–78. https://doi.org/10.1111/cag.12425.

Conclusion: Change and Continuity in Canadian City Politics

Martin Horak, Jack Lucas, and Zack Taylor

Canada's big cities are very different places today than they were forty years ago, when Sancton and Magnusson's *City Politics in Canada* was published. They have experienced major population growth, driven primarily by immigration, and they are far more ethnoculturally diverse than they were in the late twentieth century. The economic importance of manufacturing has declined, and big cities have instead become hubs of the post-industrial "knowledge economy." Income inequality has increased, and socio-spatial polarization between wealthy and poor neighbourhoods has grown. As the data in our opening chapter show, the speed and intensity of these transformations differ from one city to another. In some cities, such as Calgary and Toronto, the changes have been rapid and dramatic; in others, such as Halifax and Winnipeg, they have been more incremental. But these are differences of degree, not of kind. Like their counterparts in other wealthy democracies, big Canadian cities in the twenty-first century are dynamic, growing, globally connected post-industrial metropoles.

In contrast to these sweeping social and economic transformations, the institutional foundations of local government in Canada's big cities have changed little. To be sure, the boundaries of some urban municipalities have expanded due to provincially imposed amalgamations, with profound implications for local politics that we will discuss later in this chapter. In legal and fiscal terms, however, the main story is one of continuity. Big-city local governments remain constitutionally subordinate to provincial governments. Their core responsibilities still involve managing property development, providing services to property, and supplying collective infrastructure such as parks, transit, and libraries. The fiscal health of urban local governments has fluctuated over the last forty years, and many experienced a run of lean years around the turn of the millennium. But their sources of revenue in the 2020s are basically the same as in the 1980s: property taxes and other property-based revenues dominate, supplemented by user fees and intergovernmental transfers, mainly from the provinces.

The evolution of city politics in Canada, vividly captured in our seven city chapters, mirrors this juxtaposition of institutional continuity and social transformation. The key policy issues that animate politics in our cities – transportation, planning and development, housing, local public services – have changed very little. In this sense, Sancton's claim that municipal politics in Canada is primarily about property remains as true today as it was then. However, the terms of debate and conflict about local political issues *have* changed – sometimes dramatically so. Middle-class homeowners, celebrated as grassroots heroes in the 1970s for stopping development, are now often vilified as conservative NIMBY actors who stand in the way of sustainable and socially equitable development. Debates about density, transit, and active transportation are now often animated by new arguments about environmental sustainability. As cities have become more diverse and unequal, questions of cultural and social inclusion have become increasingly important in debates about local service provision. Affordability concerns have become central to local debates about housing, especially in the last few years. The issues may be the same, but the policy debates about those issues look very different.

In this final chapter, we seek to make sense of this mix of continuity and change, bringing together insights from our chapters on Montreal, Toronto, Halifax, Ottawa, Winnipeg, Vancouver, and Calgary to address some of the big questions about city politics in Canada, which we introduced at the beginning of this volume. We begin with the question of cleavages: What are the main axes of division or conflict that characterize city politics across our cases, and how have these changed since the late twentieth century? We argue that there have been important shifts in this regard. The mid-century development cleavage is gone, and a new dominant development cleavage pits older middle-class homeowners against a loose coalition of proponents of what we call "twenty-first century urbanism." Notwithstanding common claims that Canadian city politics is "non-ideological," we have also seen the emergence of a durable ideological cleavage between an urban right that espouses low taxes and business principles in local government, and a left that embraces a variety of environmental and social goals. Finally, long-standing ethnic, linguistic, and religious group identity cleavages in city politics have faded, and despite (or perhaps because of) the sociocultural hyper-diversity of many large Canadian cities, ethno-racial cleavages are generally not dominant in our city politics.

Despite these common trajectories, the practice of politics differs markedly across Canadian cities. For instance, since the 1997 amalgamation, decision making in Toronto has become increasingly centralized; by contrast, our authors characterize Montreal as "a textbook example of decentralization and the distribution of political spaces for representation and participation" (56). In some cities, such as Ottawa and Toronto, the ideological cleavage coincides with a spatial divide between the city core and its outlying areas; in others, such as Vancouver, the spatial articulation of this divide is much less evident. In the next section of this chapter, we turn our attention to making sense of this local variation. We highlight the role of three important factors: (1) the amalgamations of Halifax, Toronto, Ottawa, and Montreal, which transformed local political dynamics in those cities by internalizing previously external political divides; (2) variation in the structure of local representative and participatory institutions, which shape how residents can access local government; and (3) the different growth trajectories of our cities, which shape the resources of local governments and influence which political issues are most salient.

In our introductory chapter, we suggested that a local, comparative, and developmental perspective on city politics in Canada, which has been largely absent from recent research, can illuminate phenomena that might otherwise remain obscured. In the final section of this chapter, we engage with this claim, exploring not just how such work can contribute to the development of urban political theory, but also the questions that our work in this book raises for future research. We argue that students of Canadian city politics are particularly well placed to contribute to our understanding of the ongoing development of Canada's multilevel democracy, political power in the urban political economy, the dynamics and political consequences of spatial cleavages, and the promise and limits of participatory mechanisms as tools for enhancing the performance and legitimacy of democratic governments. As this book is being completed, the dominant issue on the national public agenda is the unaffordability of housing. We end this chapter, and the book, by proposing that the politics of housing is fruitful empirical terrain for theoretical development.

EVOLUTION OF POLITICAL CLEAVAGES IN CANADIAN CITIES

Transformation of the Development Cleavage

Over forty years ago, Andrew Sancton wrote that "virtually all conflict in Canadian urban politics can be located on a pro-development-anti-development spectrum" (Sancton 1983, 295). The fiercest political battles in Canadian cities of the 1960s and 1970s pitted the development establishment – large local developers, boosterist politicians, and modernist urban planners – against residents of inner-city neighbourhoods, who rallied under the banner of "reform" to oppose the dominant development agenda. In hindsight, it has become clear that this cleavage in Canadian cities was a product of a particular historical period. Mid-century urban redevelopment practices often displaced residents and even demolished large parts of established neighbourhoods to make way for urban expressways, high-rise housing, commercial spaces, and office towers. The reformers were a loose alliance of middle-class and working-class residents whose agenda was focused on stopping this redevelopment "machine."

As Magnusson's incisive 1983 analysis notes, the reform movement was from the outset driven by internal tensions between middle-class homeowners who valued neighbourhood preservation and community amenities, and working-class advocates for redistribution (Magnusson 1983a, 31–7). These tensions soon splintered reform activism, and only a few Canadian cities ever experienced reform-led councils. Magnusson ultimately concludes that the sound and fury of the reform movement did not amount to much: "[I]t would be difficult to say that municipal government was dramatically different in 1980 from what it had been in 1960. Most of the conditions that had led municipal councils to be supportive of urban growth and to aid the private developer were still extant" (Magnusson 1983a, 35).

It is true that growth and development have remained central goals in local politics, as our city chapters clearly show. However, the reform movement did accomplish something different and consequential: it shifted decision-making power away from planners and developers and towards residents, particularly homeowners. Largely as a result of reform activism, resident participation in planning and development approvals was institutionalized in provincial and local regulations across much of Canada in the 1970s and 1980s (Hodge and Gordon 2014, 132). This in turn pushed urban development practices to align with what American scholars Altshuler and Luberoff (2004) call the "do no harm" principle. Major new development now avoided established residential neighbourhoods and was channelled either *upward* into compact, transit-oriented, high-density development nodes, or *outward* into suburban sprawl.

The economic transformation of Canadian urban cores helped to stabilize this new development compromise even as city populations continued to grow. Deindustrialization in the 1980s and 1990s left behind significant expanses of derelict land in core urban areas, giving local politicians the space to channel development pressures away from existing neighbourhoods. Indeed, over the past forty years, many of the landscapes of the old industrial economy – Vancouver's False Creek, Toronto's waterfront, Montreal's Old Port, and old warehouse districts in numerous cities – have become focal points for new residential and commercial development. At the same time, in municipalities that encompass suburban and rural areas within their boundaries – such as Calgary, Winnipeg, and post-amalgamation Halifax and Ottawa – large-scale new development has occurred mainly on the suburban and exurban fringes.

The mid-century pro- versus anti-development cleavage is no longer visible in Canadian city politics. Yet conflict over development remains common – whether it be struggles between neighbouring homeowners and condo developers over the height and traffic impacts of intense residential infill projects, battles over the routing of rapid transit lines, or debates about separated bike lanes. In many cities, development conflicts have increased in intensity since the turn of the millennium as populations have continued to increase, and governments have moved to address urban infrastructure needs that had been neglected during the fiscally lean 1990s. In this context, a new development cleavage has emerged over the last twenty years, one that involves a striking inversion of players in relation to the earlier historical period.

On one side of the new cleavage are middle-class homeowners – the same group that spearheaded the 1970s reform movement. Now, as then, their focus is on opposing new development that threatens to disrupt established neighbourhoods. Yet while in the 1960s and 1970s they were the challengers in city politics, they are now very much part of the establishment. On the other side is a diverse collection of both grassroots and elite actors – environmental and transit activists, housing advocates, (some) developers, planners, and others – who support some version of what we could call "twenty-first century urbanism": the promotion of higher-density, mixed-use neighbourhoods, transit and active transportation, and strong public amenities.

This new development cleavage is a significant force, but it is not as clearly defined as its counterpart in the 1960s and 1970s, in part because it does not straightforwardly pit vulnerable neighbourhood-level Davids against powerful private-sector Goliaths. Indeed, the contemporary development cleavage often pits middle-class children against their middle-class home-owning parents. What is more, while in the 1960s and 1970s city planners were villains that the neighbourhood movement sought to dethrone, they are today seen as heroes, at least by those who embrace twenty-first century urbanism's discourses of density and sustainability. Like mid-century reformism before it, twenty-first century urbanism has a complex relationship to class-based interests. As we will see in the next section, many of its advocates are self-consciously progressive, yet a growing body of research shows that twenty-first century urbanism has – intentionally or unintentionally – supported a new wave of gentrification that has displaced working-class and marginal residents in core urban neighbourhoods (Catungal et al. 2009; Lehrer et al. 2010; Quastel et al. 2012). In addition, the cleavage between NIMBY homeowners and twenty-first century urbanists operates in a transformed development environment, one in which international property investors compete with local actors in many segments of the real estate market (August and Walks 2018). This is especially true in Toronto and Vancouver, where urban residential real estate has become a safe harbour for global capital (Ley 2020; Gordon 2022).

For all its complexity – perhaps in part because of it – the contemporary development cleavage is not as dominant as its earlier counterpart, and many important local political issues are not clearly structured by it. In fact, in many ways, the contemporary cleavage over the development of the built environment can be seen as one dimension of a broader underlying division that has grown in visibility and salience in Canadian city politics since the 1990s: an ideological divide between urban conservatives and urban progressives.

The Ideological Cleavage

There is very little talk of "left" and "right" in the first edition of *City Politics in Canada*. In the book's concluding chapter, Sancton described urban politics not in terms of ideological, geographic, or class conflict, but rather in terms of competing versions of a single dominant phenomenon: boosterism. "The reality," wrote Sancton, "is that Canadian municipal politicians have little choice but to build their political careers on policies designed to make their city more prosperous, more appealing, and more pleasant than other competing cities" (1983, 293). Sancton was not alone in emphasizing the absence of ideology from urban politics. In *City Limits*, published just two years earlier, American urban politics scholar Paul Peterson (1981) provided a theoretical foundation for much the same argument, suggesting that local politics is necessarily devoted to "developmental" policies that aim to improve the city's economic position and, therefore, in his view, the entire community, rather than policies that might divide local residents along ideological lines.

Yet the rich developmental narratives in the first edition of *City Politics in Canada* also make it clear that ideological divides *were* present in many cities, at least at an earlier stage in their history. Many historical political conflicts described in the book have an obvious ideological flavour: the Vancouver and District Labour Council's role in the development of the Vancouver party system in the 1960s (Gutstein 1983, 204–5); active Cooperative Commonwealth Federation (CCF) candidates in Edmonton in the 1940s (Lightbody 1983, 261); ideologically charged provincial–municipal disagreement in Ottawa in the 1970s (Andrew 1983, 156–7); and, perhaps most strikingly, Winnipeg's 1920 municipal election, where former strike leaders faced off against business interests in the most contentious and closely watched municipal election in Canada's history.

In the early 1980s, left–right conflict in Canadian cities may well have reached a low ebb, yet such conflict was historically the norm rather than the exception. This reality is obscured in part by the long-time insistence of many local politicians and pundits – especially those on the ideological right – that local politics is "non-ideological" (Lucas 2023). While the fierce class divides of the interwar years have not returned to Canadian city politics, the chapters in *this* edition of *City Politics in Canada* suggest that left–right conflict has returned, albeit in a reconfigured form that is no longer as clearly grounded in class-based interests.

This evidence for a left–right divide in contemporary city politics is clearly visible in our authors' descriptions of municipal elections: Vancouver's "right-wing" parties and candidates facing off against an assortment of "left-wing" alternatives; Calgary's "conservative" mayoral candidate, Jeromy Farkas, challenging the centre-left progressive, Jyoti Gondek; the "plateau progressivism" of Projet Montréal versus the centre-rightism of Denis Coderre; John Tory's toryism against Olivia Chow's orange army; and so on. Even where ideological competition is less clearly labelled, it is by no means invisible. In Ottawa, for example, pundits made a great deal of the internal strife between the "strong left" federal Liberals and the "centre-left" provincial Liberals that manifested in the competition between the two leading mayoral candidates in 2022. Similarly, in Halifax in 2020, the mayor's only serious challenger, councillor Matt Whitman, built his campaign on an ideologically charged, socially conservative platform. When we look beyond elections and dig into the actual policy disagreements described in our chapters – housing affordability in Vancouver, homeless shelters in Ottawa, Cornwallis commemoration in Halifax – the divides between ideological left and right are equally clear.[1]

Over the long run, then, the relative importance of ideological disagreement in Canadian city politics has been subject to ebbs and flows. In the nineteenth and early twentieth centuries, disagreements about urban infrastructure, land speculation, and political corruption were often fierce, but they reflected little obvious ideological structure (Nelles and Armstrong 1976; Weaver 1977). In the interwar period, the increasing strength of the urban labour movement, combined with agrarian and progressive critiques of the wider Canadian status quo, made for a more class-oriented and openly ideological urban politics (Bright 1998; Epp-Koop 2015; Tennant 1980). After the Second World War, as provincial governments took over responsibility for social policy and health care, leaving municipalities to focus their attention on the challenges of explosive urbanization, the disagreement between the corporate conservatism of the development industry and the conservationist conservatism of the middle-class urban reformers became the more prominent cleavage. Since then, however, the ideological tide has rolled in yet again.

While this nutshell narrative of nearly two centuries of urban political history is necessarily stylized, it nevertheless illuminates some important aspects of left–right politics in our cities, which can help us to understand its re-emergence in the past four decades. First, arguments about the "ideological" or "non-ideological" character of Canadian urban politics are likely to be unhelpful when divorced from their developmental context; the persuasiveness of each "side" of this argument will depend on the period under consideration. Second, and even more importantly, the long-term historical narrative also suggests important clues about *why* left–right conflict may ebb and flow in Canadian urban politics.

One critical ingredient is the changing position of municipalities in the intergovernmental division of labour. It is no coincidence that left–right disagreement became more salient in Canadian cities in the early twentieth century as cities accumulated responsibilities in health care and social welfare (Banting 1987; Guest 1997; Lucas 2018). Nor is it a coincidence that the tenor of local disagreement shifted in the post-war years as provincial welfare states emerged. As we have seen in previous chapters, in the 1990s the intergovernmental division of labour changed again with federal and provincial retrenchment and downloading. In many large cities, these changes produced significant fiscal challenges. At a time when neo-liberal ideas were on the rise in Canadian politics more generally, these conditions produced a new urban conservatism focused on fiscal restraint, prioritization of services to property over new amenities, and a preference for market-based approaches to development and service delivery. Public–private partnerships, tax restraint (recall, for instance, Winnipeg's *thirteen straight years* of tax freezes from the late 1990s into the 2010s), and other fiscally conservative policies in turn contributed to the rise of new problems, including underinvestment in urban infrastructure and cutbacks to local supports for poor and vulnerable residents. This in turn generated resistance and response from voices on the economic left in these cities, thus contributing to a reinvigoration of ideological conflict.

A second important contributor to the re-emergence of left–right politics is the evolving character of Canadian ideological debates more generally. City politics plays out within a multilevel policy context, but it also exists within a broader ecosystem of *political* debate. Canada's interwar period involved contentious ideological and class divides in city politics, in part because it was a time of strong ideological contention in general, as new parties and social movements – labour activists, agrarian radicals, progressive reformers, suffrage advocates, social credit visionaries – roiled the Canadian political system. When these contentious years gave way to a less openly conflictual post-war politics – the "Tweedledum and Tweedledee" years variously celebrated and bemoaned by the intellectual class – ideological conflict was dampened at the municipal level as well. Since the early 1980s, clear ideological divides between Canada's major political parties have again become a durable feature of federal party competition (Cochrane 2015), and when voters and political candidates begin to "think ideologically" with respect to federal politics, they are more likely to do so with respect to municipal politics as well.

As we have seen, the changing ideological context of Canadian politics contributed to the emergence of a new urban conservatism in the 1990s. It also contributed in important ways to the transformation of the urban left. The rise of "post-materialist" values in advanced democracies in recent decades has led to the increased salience of issues such as the environment and climate change, accommodation of cultural diversity, and racial justice (Inglehart 1990). These issues are closely identified with "new left" politics at the federal and provincial scales, as well as with contemporary "progressive" urban politics. This has given municipal policy issues that might once have been considered developmental or even "technical" in character a more obvious ideological flavour: a policy decision about hiring practices on the police force, now animated by how well the officers reflect the diversity of the wider community; procurement of a new transit bus, now focused on the relative carbon footprint of electric, hybrid, and diesel vehicles; a bridge construction project, now focused on opportunities to use low-carbon concrete. The shifting dimensions of ideological debate in advanced democracies across the globe have thus meant that once-quiet arenas of municipal policy are now frequently subject to much more animated ideological disagreement.

In contrast to the urban left of the early twentieth century, the new urban left in Canada's cities is not as clearly grounded in class interests. Its social foundations more closely resemble those of the 1970s reform movement, in that it involves an alliance of middle-class urbanites and advocates for historically marginalized groups. However, leftward shifts among highly educated urban knowledge workers, combined with the increased importance of social and cultural issues in contemporary ideological disagreement, make this alliance somewhat less tense than it was in the 1970s. The contemporary urban left in Canada prioritizes a variety of goals. These include environmental sustainability and quality of life, embodied in the twenty-first century urbanist principles of density, strong public amenities, and accessible urban design; but also equity and redistribution as they connect to diversity and racial justice, with a particular focus on discrimination and equity in regulation and policing of the urban population. Across our cities, activists and political representatives on the left advocate for culturally sensitive and responsive land development, multilingual and culturally sensitive service delivery, racial reconciliation and equity in policing, and increased spending on housing and other supports for "equity-deserving" groups, such as new Canadians, racialized minorities, and urban Indigenous populations. Given the limited functional responsibilities and fiscal powers of local governments, however, the policy outcomes have tended to involve various regulatory or symbolic equity and reconciliation initiatives, rather than significant new programs that redistribute public resources towards marginalized populations.

Interestingly, the contrasting sources of the new urban right and left – the urban right rooted in the fiscal conservatism and market liberalism of the 1990s, and the urban left grounded in post-materialist concerns with environmental sustainability and social justice – have given ideological disagreement a somewhat asymmetric quality in many cities. On the right, borrowing from long-standing municipal tradition and more recent neo-liberal ideas, many insist that municipal politics is *not* ideological, but rather a "business enterprise" that ought to be managed on sound business principles. On the left, activists and elected representatives respond by insisting that policy decisions that *seem* neutral or technical are in fact deeply connected to equity, equality, and justice. The result is a peculiarly municipal "meta-debate" about the character of politics, itself divided along left–right ideological lines (Lucas 2023).

A final ingredient that shapes municipal ideological disagreement is one that Magnusson and Sancton's rich historical narratives in *City Politics in Canada* made abundantly clear: the role of local political elites in *constructing* the cleavages that animate local politics. In the interwar period, for example, a great deal of fierce ideological and class debate focused on local democratic institutional reform (Bright 1998; Lucas 2020), as local labour activists reframed long-standing policy debates in class-based, ideological terms, and local business elites responded in kind. In more recent decades, such creative political action continues to shape urban politics. At times, it can serve to dampen ideological disagreement. In Calgary, for example, the success of Naheed Nenshi's "purple coalition" in 2010 meant that the city council's 2010–2014 term had less ideological structure than may have been the case under a different mayor. John Tory's ideologically cross-cutting coalition in Toronto in 2014, positioned between Ford on the right and Chow on the left, had a similar effect (McGregor et al. 2021). In other cases, creative local action can heighten ideological disagreement, as previously low-conflict domains become infused with new arguments – as in the case of the heightened ideological salience of subway construction and garbage pickup under Toronto's Rob Ford. These acts of political entrepreneurship, scattered throughout our case chapters in this book, also help to shape the ebb and flow of ideological conflict in Canadian city politics.

Group Identity Cleavages in Canadian City Politics

Numerous Canadian cities were divided by language or religion in earlier times. For example, in their 1983 chapter on Halifax, Cameron and Aucoin identify informal power-sharing among Catholics and Protestants as a key feature of local politics into the 1960s (170–171); French and Irish Catholics also cooperated in nineteenth-century Quebec City (Grace 2003). In the nineteenth and twentieth centuries, religious organizations such as the Orange Order actively worked to preserve Protestant control over civic administration to the detriment of Catholic and Orthodox minorities in Ontario and prairie cities (Palmer 1974; Smyth 2015).

Montreal, Moncton, and Quebec City have long featured some degree of conflict and informal power-sharing among English- and French-speaking residents, sometimes cross-cutting and reinforcing the Protestant–Catholic divide. Linguistic conflict in Montreal was managed by the formation of separate English-majority suburban municipalities (Sancton 1985). As salient as these cleavages have been, they have all but disappeared as important features of Canadian urban politics.

Since the 1980s, immigration from all corners of the world has made the largest Canadian cities hyper-diverse. The populations of Toronto and Vancouver are now roughly half foreign-born and half "visible minority," and Calgary, Montreal, and Ottawa are not far behind (see Figure 1.3 in the introductory chapter). Increased ethnocultural diversity might be expected to heighten the salience of group-identity politics in local electoral politics, but in fact, the opposite has occurred. With few exceptions, our chapters identify little evidence of ethnic, linguistic, or religious group-identity conflict in local elections in recent decades. Why is this the case?

The first reason is jurisdictional. A key focus of linguistic and religious conflict – primary and secondary education – largely plays out in other venues: locally within separately elected school boards, and provincially as provincial governments exert the greatest influence over curriculum and resource allocation. A second reason may have to do with the time lag in the political incorporation and mobilization of recent immigrants. The local religious and linguistic conflicts of earlier eras typically involved two or more long-standing resident groups. By contrast, the new immigrants who have transformed the face of Canadian cities over the past generation typically take a number of years to become politically active at similar levels as long-standing residents (Wallace et al. 2024).

A further reason lies in the way that increasing ethnocultural diversity has interacted with existing electoral systems and patterns of political mobilization. Historical patterns of settlement in early twentieth century established substantial and enduring concentrations of various ethnic groups – Italians, Greeks, Portuguese, and so on – in most of our cities. This process has accelerated as the overall level of immigration to Canada and immigrants' tendency to concentrate in the country's largest cities have increased (Teixeira et al. 2012). In this context, ward systems have occasionally facilitated ethnic bloc voting and the election of ethnic group members to councils (Siemiatycki and Saloojee 2002).

However, the hyper-diversity of immigration to Canadian cities has tended to dampen intergroup divides. A contrast with the American experience can clarify the logic of this apparent paradox. Many large American central cities have also experienced increased ethnocultural diversification through immigration in recent decades, but long-entrenched patterns of racial division in local politics have tended to subsume the diverse demands of different ethnocultural groups into political conflict among a limited number of ostensibly monolithic groups – whites, Blacks, and Latinos (Hajnal and Trounstine 2014). In Canadian cities, by contrast, racial division has historically been less politicized, so as ethnocultural groups become politically mobilized,

Table 9.1: Descriptive representation on Canadian city councils, November 2023

	Visible minority council members	% Visible minority council members	% Visible minority population (2021)	Women on council	% Women on council	Woman mayor?	Visible minority mayor?
Toronto	9/26	34.6%	55.7%	11/26	42.3%	Y	Y
Halifax	2/17	11.8%	16.8%	9/17	52.9%	N	N
Ottawa	3/25	12.0%	32.5%	9/25	36.0%	N	N
Winnipeg	3/15	20.0%	34.4%	5/15	33.3%	N	N
Vancouver	3/11	27.2%	54.5%	5/11	45.5%	N	Y
Calgary	5/15	33.3%	41.4%	5/15	33.3%	Y	Y
Montreal	10/65	15.4%	38.8%	34/65	52.3%	Y	N

Note and sources: Percentage of visible minority residents in the municipal population are taken from Statistics Canada data table 98-10-0599-01. Categorization of visible minority and woman representatives was based on a review of photos and other information (e.g., profile texts) on municipal websites. Categorization was independently performed by three people and median values were taken in cases where the counts did not match.

they are less likely to coalesce into larger "racial" blocs. In this context, increasing diversity has decreased the political salience of ethnic group identity, because in most cities (and in many cases, even in individual wards), there are simply too many groups for any one or two to be electorally decisive. In addition, research suggests that other, cross-cutting cleavages – such as ideology – are more salient than ethnic, linguistic, racial, and religious cleavages in contemporary Canadian urban politics (Doering et al. 2021; Lucas and McGregor 2021).

The weakness of ethnic group-based voting in Canadian urban politics contributes to a notable deficit in descriptive representation (see Table 9.1). In at least three of these cities – Toronto, Montreal, and Calgary – the most recent round of general elections was widely cited in the local media as a watershed year for electing a greater proportion of racialized councillors, but there remains a deficit in descriptive representation in all cases.

While electoral conflict and political representation are thus not organized along group lines, we do see considerable evidence of issue-based activism by ethnic and other identity groups in our cities. Black and Indigenous residents have mobilized against discriminatory policing in many Canadian cities (Maynard 2017), including those that we have profiled in this book, and in Halifax, African Nova Scotians have demanded recognition and compensation for city's post-war demolition of their historic neighbourhoods (Nelson 2000). We also see activism by and on behalf of undocumented migrants, in the form of "sanctuary cities" policies (Jeffries and Ridgley 2020). While sometimes criticized for being tokenistic or stigmatizing (Calderón-Figueroa et al. 2022), local administrations in many cities have also been responsive to the interests of specific minority communities, for example through multicultural programming, which has lessened the need for ethnicity-based organizing around policy concerns (Siemiatycki 2011).

THE PRACTICE OF CITY POLITICS: SOURCES OF LOCAL VARIATION

Common Trajectories, Local Differences

So far, we have identified some shared trends across our case cities. Over the last forty years, Canada's biggest urban areas have shifted decisively towards post-industrial economies. They have all continued to grow, albeit at different rates, and all of them have become more ethno-culturally diverse as well as more economically unequal. The core institutional foundations of local government have changed little since the 1980s, but patterns of political cleavage have evolved significantly: the dominant development cleavage has shifted, and now primarily pits middle-class homeowners against advocates of twenty-first century urbanism; the historical left–right ideological divide has become increasingly salient; and, conversely, group-identity politics has become submerged and less pronounced.

Despite these common trends, the character of city politics and local political conflict varies greatly across our seven cities. In some cities, such as Winnipeg, Halifax, and Toronto, our authors describe local politics as elite-centred and inaccessible to ordinary citizens and grassroots groups; in others, most notably Montreal, but also to some extent Calgary, local politics appears to have become more open to citizen engagement in recent years. Equally striking, in some of our case cities – most notably Toronto, Ottawa, and Halifax – political cleavages are geographically articulated, such that political debates often pit central city dwellers against suburbanites, but in other cities, this dynamic is weaker or absent.

In addition, there is no straightforward relationship between the ideological leanings of local populations and the policy agendas and priorities of local political leaders. For instance, according to the survey results presented in Lucas's chapter on Calgary (see Figure 8.3), the average Halifax resident is more left-leaning than the average Montrealer; yet key concerns of the "new urban left," such as sustainable transportation and affordable housing, clearly have a more prominent place on the political agenda in Montreal than in Halifax. Likewise, the same survey suggests that Calgary residents lean more to the right than Winnipeg residents, yet Lucas shows that in Calgary concerns such as environmental sustainability and reconciliation with Indigenous peoples have gained a significant place on the policy agenda, whereas Stelman and Moore show that in Winnipeg, successive councils have been resolutely focused on keeping taxes low and facilitating private development.

What, then, explains such local variation in the practice of city politics across our cases, despite their broadly similar developmental trajectories? It is neither possible nor advisable for us to propose a comprehensive theory, especially based on only seven cases. However, if we take our seven cities together, some important sources of local variation become apparent. We will focus on three: municipal boundaries, local institutions, and city–regional economic trajectories. These sources of variation interact with each other and with additional, idiosyncratic elements, thus creating unique local contexts that shape the practice of politics in distinct ways in each city.

Municipal Boundaries and Spatial Conflict

A wave of amalgamations across several provinces around the turn of the millennium dramatically expanded the boundaries of Halifax (1996), Toronto (1997), Ottawa (2000), and Montreal (2002). These changes are discussed in detail in the introduction (13–16), as well as in the individual city chapters, so we need not re-tell the story here. For our present purposes, it is important to note the impact of these amalgamations on city politics. What comes across most clearly from our case chapters is that amalgamations brought together pre-existing political communities with different political cultures and preferences, thereby internalizing new sources of political conflict. Intermunicipal conflicts became intramunicipal conflicts. In his conclusion to the 1983 *City Politics in Canada* volume, Sancton wrote that "there is more competition among Canadian cities than within them" (294). Today, this claim is much harder to sustain, at least in amalgamated cities.

The amalgamations of the late 1990s and early 2000s varied in scope. Some effectively erased the difference between the local and the metropolitan scales of government. Post-amalgamation Ottawa and Halifax are metropolitan-scale municipalities that include core neighbourhoods, new suburbs and exurbs, and large swaths of rural land. Post-amalgamation Toronto is not nearly so geographically extensive, and the new city remains one of 25 municipalities in the Greater Toronto Area. Nonetheless, it is by far Canada's largest single municipality, and it includes both the old urban core and large areas of post-war suburbs. Montreal's amalgamation – followed in 2006 by partial de-amalgamation – was relatively more modest, and the city remains only one entity in a complex and multitiered web of local governments in the metropolitan region; however, it still expanded the central city's population by about 50 per cent, incorporating new constituencies with geographically distinct preferences into the politics of the central city.

In Toronto, Ottawa, and Halifax, amalgamations brought neighbourhoods in the urban core, which were accustomed to receiving a wide range of services and collective amenities, together with post-war suburban neighbourhoods, where local government was seen as a limited entity focused on providing property-related services. In both Ottawa and Halifax, rural areas accustomed to receiving limited services at a minimal rate of taxation were also brought into the fold. As Horak and Taylor note, the first decade after Toronto's amalgamation was consumed with conflict over the harmonization of services and tax rates, generally up to the levels of the most generous former jurisdiction. In Ottawa and Halifax, conflict is partially mitigated by applying differential property tax rates to zones defined by the level of services provided.

Insofar as they brought together areas of the city whose residents have different material interests, relationships with local government, and policy preferences, amalgamations intensified allocational politics – that is, location-based conflict over the geographic distribution of public resources. In their work on spatial cleavages in urban politics, Doering et al. (2021) identify a dominant "lifestyle"-based cleavage in Toronto, which pits residents of the old city core against those in the post-war suburbs. The "war on the car" and the "15-minute city" have become political rallying points in Toronto, Montreal, and other large Canadian cities, which highlight the clash between the divergent interests of suburban car commuters and residents of dense, mixed-use neighbourhoods in urban cores. Cycling, rapid transit, and traffic calming infrastructure are pitted against downtown highway access and parking availability.

While Doering, Silver, and Taylor do not describe this geographical cleavage in ideological terms, it is clear from the chapters in this book that there is often an ideological dimension, because proponents of the new urban left and what we have called "twenty-first century urbanism" tend to live in urban cores, whereas suburban and rural populations lean more to the right. In addition, in city politics, ideological conflict over taxes and services is very much tied to the physical characteristics of neighbourhoods. Given these interconnections, place labels such as "downtown" and "suburb" often become heuristics for left–right conflict, and vice versa.

In some cities, most notably Toronto and Ottawa, local politicians – especially mayoral candidates – have actively mobilized these coinciding spatial and ideological cleavages to build and maintain winning coalitions. The result has been a volatile, conflictual, and at times dysfunctional local politics. The fierce debates over the representation of core, suburban, and rural areas on Ottawa city council, as well as the recurring and often raucous geopoliticization of transit policy in Toronto, are among the clearest manifestations of this strategic mobilization of the territorial cleavage in local electoral politics.

Representative Institutions and Public Engagement

Beyond elections, the extent to which politics and decision making are accessible to and inclusive of residents varies widely across our cities. In part, this variation is a function of the structure of the local institutions of representative democracy.[2] In some cities, political representation is highly concentrated, while in others, it is more dispersed, making it easier for residents to access and influence their local elected officials (Taylor forthcoming). Toronto's 3 million citizens are represented by 25 councillors and a mayor; by contrast, Montreal's 1.8 million residents are represented by a total of 103 elected officials – 65 city council members, some of whom also sit on one of the 18 borough councils, and an additional 38 representatives who only serve at the borough level. In addition, some cities have concentrated executive structures that distance decision making from public influence. The mayors of Montreal, Vancouver, and – most recently – Toronto and Ottawa have significant executive powers. But even in cities that adhere to the classic Canadian model of the "weak mayor," a concentration of executive authority can emerge. For instance, while Winnipeg's mayor has few executive powers, they name the members of the powerful executive policy committee (EPC), which, as Stelman and Moore's chapter makes clear, sets the policy agenda largely independent of the council as a whole, resulting in a "depoliticized" system of government that is highly insulated from public input between elections.

As we noted earlier, the reform movement of the 1960s and 1970s institutionalized public participation in development approvals in urban Canada, making homeowners a major force in Canadian city politics. Yet beyond the realm of property development, the availability of institutionalized public engagement mechanisms – participatory budgeting processes, consultative bodies, citizen initiative powers, and so on – varies widely across our cities. Montreal again appears to be at one end of the spectrum in this regard. As Bherer, Breux, and Van Neste show, in recent years Montreal has developed a wide variety of mechanisms for resident engagement, ranging from the ability of residents to initiate public consultation on a policy issue through petition to "neighbourhood tables" for discussion of hyper-local issues, as well as a variety of participatory budgeting and citizen-led policy initiatives at both the city and the borough levels. This remarkable proliferation of participatory opportunities has been driven by Projet Montréal's, the governing party since 2017, which is rooted in the city's tradition of activist local politics.

No other city among our seven has anywhere close to Montreal's range of institutionalized participatory opportunities. Indeed, the authors of our chapters on Winnipeg, Toronto, and Halifax all comment on the scarcity of such opportunities, with Finbow noting that the lack of engagement mechanisms contributes to "a general sense of disconnect between decision makers and social constituencies" in Halifax (119). However, even in cities where participatory opportunities are weakly institutionalized, it is not uncommon for administrators in particular policy fields to engage and connect with the populations that they serve. Such sectoral engagement can be an important input for policy, and it has contributed to policy innovation in a variety of settings, ranging from sustainability planning in Calgary and multicultural services in Toronto, to race relations initiatives in Halifax and urban design policies in Vancouver.

From a developmental perspective, the contrasting approaches to government–citizen relations we have discussed – centralization versus decentralization of representative institutions, insulation from versus. engagement with residents – represent different responses to the same basic problem: the emergence of an increasingly complex local political environment, which leads to the proliferation of new policy demands. In post-amalgamation Toronto, the provincial government and local politicians have repeatedly centralized representative decision making to make the governance of this huge and diverse city more manageable; in Winnipeg, the consolidation of power in the EPC from the 1970s to the early 2000s was part of an effort by successive suburban-dominated councils to restrict the political agenda to items that were consistent with their interests. By contrast, as Calgary has become more socially diverse, it has gradually institutionalized more participatory policymaking mechanisms, eroding the previously dominant role of corporate notables in city politics. And in Montreal, which has a long tradition of activist local politics, the creation of a multilayered local authority through the restructuring of the early 2000s provided opportunities for a new progressive political party – Projet Montréal – to rapidly grow and deepen participatory mechanisms.

Economic Trajectories and Policy Choices

Even though our seven cities have shared a basic trajectory from industrial to post-industrial economy since the 1980s, specific patterns of population growth and economic development have differed. For instance, Vancouver and Toronto are both central cities in metropolitan areas that have experienced sustained, rapid growth, and they have been the leading sites for foreign investment in Canadian property markets (Ley 2020). Calgary has experienced the most rapid population growth of all of our case cities, but this has happened in the context of a cyclical boom-and-bust local economy that is tied to the oil industry. Winnipeg has, overall, been the slowest growing of our seven cities, but the pace of its population growth has accelerated since the early 2000s as it has become a significant immigrant destination.

Urban political economy research suggests that politicians in cities with a strong "market position" – that is, cities that are attractive to investors and new residents – have a broader menu of policy choices than politicians in slow-growing or declining cities (Savitch and Kantor 2002). In the Canadian context, where local governments rely heavily on property-based taxation, fast growth allows local governments to increase their resource base – and therefore their policy choice – by increasing the property tax base; but it also brings characteristic local policy problems, such as the need to build new infrastructure and the problem of rising housing costs. By contrast, slow growth or economic stagnation limits local government resources and focuses political attention on efforts to attract new investment and residents (Leo and Anderson 2006).

Our case chapters document the connections between economic trajectories and policy choices. For instance, a healthy local tax base has allowed the left-wing mayors and councils that periodically govern Vancouver to spend generously on public amenities and services, even as housing affordability remains a serious and apparently intractable problem. In his chapter on Calgary, Lucas emphasizes the effect of the boom-bust cycle on city politics and finances, noting that periods of boom are accompanied by public spending on new infrastructure and services, whereas periods of bust bring fiscal conservatism and belt-tightening to the fore. Meanwhile, the slower-growth city of Winnipeg has long practiced a highly constrained local politics, characterized by low taxes and a focus on basic services and accompanied by persistent boosterist efforts to attract new investment.

Yet not all cities neatly fit the expected patterns. Slow-growth Montreal, for instance, has recently been a hotbed of policy innovation and local spending on "progressive" policy initiatives. In Toronto, there is an enduring sense of fiscal crisis despite decades of steady – and often rapid – population growth. Winnipeg has retained the single-minded focus on a minimalist, low-tax local government even as growth has accelerated over the past decade and social needs in the urban core have deepened. Clearly, local institutions in all their variety mediate the relationship between economic trajectories and policy choices. In Winnipeg, the institutional entrenchment of a political executive that is insulated from the rest of council has helped to "lock in" a focus on low taxes and basic services. Conversely, Montreal's complex and multilayered institutional structure of governance has provided political opportunities for political activists to put progressive issues on the local policy agenda despite economic constraints. And in all Canadian cities, a revenue system that is ill-suited to the needs of a major metropolis makes it difficult for the city to transform economic growth and dynamism into increased local government capacity.

Local Variation: The Limits of Comparative Explanation

Beyond the three elements that we have discussed above, other factors also shape the dynamics of city politics in particular places, but these factors defy easy comparison and categorization. Many of our authors emphasize the importance of mayoral leadership. Some mayors, such as Toronto's Rob Ford, mobilize division and discontent, while others, such as Calgary's Naheed Nenshi, attempt to build bridges between diverse constituencies and perspectives. Some mayors, such as Halifax's Mike Savage or Toronto's John Tory, are brokers who negotiate among ideas and priorities that are not necessarily their own, while others, such as Winnipeg's Glen Murray or Toronto's David Miller, advance big new policy ideas with varying degrees of success. Whether local political parties exist, and if so, how they are organized, also impacts the practice of local politics. However, among our cities, political parties only exist in Montreal and Vancouver, and they operate very differently in the two cities, so it is difficult to generalize meaningfully about their impacts. While we can learn much from comparing the ways in which municipal boundaries, the structure of local institutions, and other factors push city politics in different directions, it is also true that, viewed holistically, city politics in any one place is sui generis, since it is rooted in the unique, historically evolved interaction of structural conditions, interests, institutions, and political culture (Dilworth 2020).

CANADIAN CITY POLITICS AND URBAN POLITICAL THEORY

At the outset of this volume, we argued that we have much to learn from a local-comparative-developmental perspective on Canadian city politics, which has been largely neglected in recent scholarship. We hope that the reader will agree that our city studies shed new light on developmental trends and diverse trajectories in Canadian city politics over the last forty years. But what can our approach contribute to the development of *theory* in the study of urban politics and, for that matter, in political science more generally? As Taylor and Eidelman (2010) have argued, there has been little in the way of endogenous theory development in the study of Canadian city politics. Canadian scholars have usually drawn either on frameworks and approaches developed by American urban politics scholars – community power, urban regime analysis, public choice, the growth machine approach, and so on – or on frameworks that emerge from the study of national politics – such as theories of policy process or political behaviour.

Such theoretical borrowing has often advanced our understanding of Canadian city politics. For instance, recent research that borrows concepts and methods from national-level electoral and legislative studies has shown – as the studies in this volume also suggest – that politics in Canadian cities is often ideologically structured, despite frequent claims to the contrary. However, the studies in our volume also show that Canadian city politics is distinct. It is neither national politics writ small, nor is it American urban politics with different place names, although it has elements in common with both. The studies presented in this volume suggest several avenues of inquiry that can take advantage of this combination of commonality and distinctiveness in ways that not only allow us to better understand the dynamics of Canadian city politics, but also to contribute to theory development in political science more generally.

The City in Multilevel Democracy

Our perspective on city politics in this book has been intentionally and self-consciously local. Yet, as Andrew Sancton reminds us, local governments "will always only be *part* of a multi-level system of government for cities" (2008, 32). As Sellers et al. (2020) put it in their *Multilevel Democracy*, "local institutions have persisted as inextricable components of national democratic political systems. The constraints as well as possibilities of democratic rule are impossible to understand without attention to how these institutions operate" (4). The locally focused accounts in this book demonstrate the extent to which the last forty years of city politics in Canada have been shaped by the actions of other levels of government. The budget cutbacks and welfare state retrenchment of the 1990s created a new operating environment for local politics, one that encouraged the development of contemporary urban conservatism, but that also saw the emergence of problems and challenges that animate much contemporary urban progressivism. As the chapter authors have described, provincially imposed restructuring of local government boundaries, systems of local representation, and fiscal and legal arrangements has fundamentally transformed politics in several of our cities. And the capacity and willingness of local governments to respond to local policy demands, especially any redistributive demands, have depended on the availability of support from other levels of government.

To understand the politics of Canadian cities therefore requires recognition of the city's embeddedness in a multilevel system of governance and politics (Taylor 2019, 2024). These systems have changed considerably over the time period considered in this book and will continue to do so. Provincial interventions have provoked considerable political reaction from local politicians and publics, and these changes have altered the venues and incentive structures of local decision making. Yet our case chapters have also shown that local political institutions often resist easy change, and patterns of political conflict develop in locally path-dependent ways, despite the occasional punctuation of top-down provincial intervention. Canadian urban politics scholars are in a strong position to exploit cross-national, cross-provincial, and cross-city variation to theorize the causes and effects of our cities' entanglements with politics at multiple scales.

Urban Political Economy

The concept of power is both central to political science and marginal to currently dominant approaches in the discipline. The political economy tradition of urban political research was long animated by questions about power – Who governs? How is power exercised in urban politics? But even in urban politics, the concept has recently receded from view. However, comparative–developmental work on city politics can bring it back to centre stage. For instance, the evolutionary overview of city politics in this book presents a mystery with respect to urban political power in Canada: the near-disappearance of developers and local business elites as visible wielders of policy power. While these actors do occasionally feature in our city narratives, they rarely appear as agents who shape city policy agendas. This is a remarkable contrast to the accounts in the 1983 edition of *City Politics in Canada*, in which property-related elites are portrayed as a dominant force – in many cases, *the* dominant force – in city politics.

What accounts for this disappearing act? Various possibilities are worthy of exploration. Perhaps the institutionalization of citizen participation in the 1970s, along with the rise of new urban interests and demands in the decades since, have served as effective counterweights. Conversely, perhaps, as critical political economists might suggest, the legitimacy of private business power has become naturalized in a neo-liberal era, such that it is no longer politically controversial. It may be that the delocalization of capital through processes of corporate consolidation and transnationalization have downgraded the importance of the local business actors that were so central to urban regime and growth machine analyses. Or perhaps the development compromise that emerged in the 1980s – in which homeowners ruled established neighbourhoods, while developers got their way on brownfield and greenfield sites – has contained the influence of landed elites within politically acceptable bounds. Research that systematically tackles this puzzle promises to shed light on the evolving dynamics of city politics in Canada and advance our theoretical understanding of the essential, but currently marginalized, concept of power in urban political studies.

Space, Place, and Cleavage

Students of Canadian city politics are also well positioned to theorize and develop empirical understandings of the spatial cleavages that divide cities. As we saw, the provincial imposition of municipal mergers in several Canadian cities at the turn of the millennium has created large and diverse urban polities in which some cleavages are spatially articulated. While research on the content and dynamics of these spatial cleavages has now begun to emerge (Doering et al. 2021; Moore 2021; Taylor and Armstrong 2024), much more can be learned about the sources and political effects of such spatial political cleavages. For instance, some local political leaders in Canadian cities have actively exploited spatial divisions for electoral gain, while others have sought to bridge them. At a time when spatial cleavages are increasingly salient in national politics in Canada and beyond (Rodríguez-Pose 2018; Ford and Jennings 2020; Taylor et al. 2023), investigating how contrasting elite appeals to place-based cleavages affect mass political attitudes, group identities, and policy processes may allow students of Canadian urban politics to make important contributions to a broader emerging area of inquiry.

Institutions and Democratic Engagement

The urban political arena is also fertile territory for studying the relationships between institutions and democratic engagement. Local governments in Canada's big cities are at the front lines of responding to a demographically changing society, one that is constantly generating new policy demands and concerns. While a vigorous literature has already emerged on local elections, resident engagement between elections in Canadian cities has not received as much scholarly attention. Localists regularly argue that the smaller scale of local governments makes them more accessible to citizens, but there are good reasons to be suspicious of the general validity of this argument (Latimer 2023). Most of the municipalities profiled in this book are hardly "small" in scale, and there is strong evidence that the participatory planning and development mechanisms institutionalized in the 1970s systematically privilege the voices of older, white, middle-class homeowners (Einstein et al. 2019; Einstein et al. 2024).

Yet, as we saw earlier, Canadian cities structure participation opportunities in very different ways. Elected officials are more accessible to residents in some cities than others, and some municipalities have focused significant energy on expanding and diversifying participatory opportunities, whereas other have not. This variation gives scholars an opportunity to investigate how institutions matter for democratic engagement. Reviewing our empirical chapters, it seems that, at least from the perspective of our authors, cities that have more numerous and diverse participatory mechanisms are ultimately better governed. In those places where local government is remote and inaccessible, our authors tend to describe city politics as being in "crisis" or unable to respond to important citizen demands; conversely, in places where governing processes are more decentralized and engagement opportunities have proliferated, our authors' assessments of the quality of local democratic outputs are more positive.

Our findings, then, suggest that municipalities that have complex and diverse channels of democratic engagement may better placed to respond to socially diverse and divided urban populations. This idea is intuitively appealing. Yet our case chapters were not designed to investigate it systematically, and it is possible that it says more about the normative biases of urban political researchers than the realities of city politics. After all, the powers and resources available to local governments are constrained. Even in places that are more permeable to citizen demands, the scope for redistributive policies is limited, and the demands and interests of marginalized populations are often met with responses that are more symbolic than substantive in character. Whether or not participatory mechanisms such as those in Montreal can precipitate a deeper shift in the dynamics of urban policymaking is a key question for future research, one that can illuminate the potential and the limits of democratic process innovations in the face of broader structural constraints.

Housing Politics as an Empirical Terrain for Theory Development

We have highlighted the embeddedness of Canadian cities in broader systems of governance and the market economy, and have pointed to place-based cleavages and democratic institutions as important objects in need of theorization. To provide a more concrete illustration of these themes, we conclude with a brief vignette on housing policy – perhaps Canada's most salient local political issue at present – as an example of how the theoretical agendas we have described intersect with empirical realities.

Canada's housing crisis has – for the first time in a long time – strengthened the policy voice of property developers in city politics. The homebuilding industry has framed the crisis as one that is driven by insufficient supply (Moffatt 2021), and it has advocated for a range of policy changes at both the local and provincial levels. At the same time, the housing crisis has amplified the voices of affordable housing and homeless advocates on the political left. These include numerous academic housing scholars, who have long argued that the non-market and affordable rental sectors are under-funded (Dantzler 2022). Affordable housing advocates frame the housing crisis as one of distorted demand, fuelled by the liberalization of Canadian housing markets and investment rules in the early 2000s (August 2020), as well as by many years of low interest rates and generous federal mortgage supports. Accordingly, they advocate for policy solutions that disincentivize the use of housing as an investment vehicle, and channel more government resources and incentives into the affordable rental and non-market sectors.

In response, local and provincial governments have switched development paradigms. After several decades of increasing regulation to promote residential intensification, transit-oriented development, and environmental sustainability goals, provincial and local governments have moved towards facilitating the rapid construction of new homes, to a degree not seen since the post-war suburban boom. This new paradigm threatens to destabilize the development compromise forged in the 1970s, which has long directed new construction away from established residential neighbourhoods. Given the centrality of property issues to local politics and the political mobilization of homeowners, it is quite likely that these changes will generate intense local conflicts at multiple scales – block, neighbourhood, and city – conflicts that may reshape local politics in Canadian cities for years to come. The ideologically polarized debate about the housing crisis, which pits supply-side advocates on the right against demand-management advocates on the left, may very well sharpen and clarify the ideological stakes in Canadian city politics, reinforcing the political salience of the left–right ideological cleavage.

The housing crisis and responses to it also highlight the multilevel dimension of urban politics and governance in Canada. In order to stimulate housing development, some provincial governments have unilaterally intervened in local affairs in ways that would have been unthinkable a decade ago, while the federal government has leapfrogged over provincial governments to establish new direct relationships with municipalities. The challenges of coordinating policy responses across government levels in a system where political and administrative links across these levels are weak and unstable are manifest. Governments pull in different policy directions, even as actors "venue shop" for advantage. All of these dynamics demand theorization and analysis.

Clearly, much remains to be done. Our goal in this book has been to present a holistic and developmental study of the practice of local politics in Canada's big cities. Building on Magnusson and Sancton's pioneering work, we have situated city politics within its changing social, economic, and institutional contexts – a perspective that takes seriously both structural forces and the agency of local political actors. In doing so, we hope to reinvigorate political scientists' interest in the causes and consequences of local democracy and local political contention, both present-day and historical, in Canadian cities.

NOTES

1   Recent research in Canada has used surveys of municipal politicians and the general public to show that policy attitudes on these and other issues have obvious ideological "structure" – meaning that a politician or municipal resident's attitude on one policy issue (such as active transportation) tends to be related to their attitude on other policy issues (such as tax policy or policing) in ways that suggest a coherent "left" and "right" in municipal policy disagreement.
2   For more information, see the Canadian Municipal Attributes Portal (C-MAP), a comprehensive database of institutional arrangements in all Canadian municipalities with over 25,000 residents (Horak and Taylor 2023).

REFERENCES

Altshuler, Alan A., and David E. Luberoff. 2004. *Mega-Projects: The Changing Politics of Urban Public Investment*. Lanham, MD: Rowman & Littlefield.

Andrew, Caroline. 1983. "Ottawa-Hull." In *City Politics in Canada*, edited by Andrew Sancton and Warren Magnusson, 140–65. Toronto: University of Toronto Press.

August, Martine. 2020. "The Financialization of Canadian Multi-Family Rental Housing: From Trailer to Tower." *Journal of Urban Affairs* 42 (7): 975–97. https://doi.org/10.1080/07352166.2019.1705846.

August, Martine, and Alan Walks. 2018. "Gentrification, Suburban Decline, and the Financialization of Multi-Family Rental Housing: The Case of Toronto." *Geoforum* 89: 124–36. https://doi.org/10.1016/j.geoforum.2017.04.011.

Banting, Keith. 1987. *The Welfare State and Canadian Federalism*. 2nd ed. Montreal: McGill-Queen's University Press.

Bright, David. 1998. *The Limits of Labour: Class Formation and the Labour Movement in Calgary, 1883–1929*. Vancouver, BC: University of British Columbia Press.

Calderón-Figueroa, Fernando A., Daniel Silver, and Olimpia Bidian. 2022. "The Dilemmas of Spatializing Social Issues." *Socius* 8: 23780231221103059. https://doi.org/10.1177/23780231221103059.

Catungal, John Paul, Deborah Leslie, and Yvonne Hii. 2009. "Geographies of Displacement in the Creative City: The Case of Liberty Village, Toronto." *Urban Studies* 46 (5-6): 1095–114. https://doi.org/10.1177/0042098009103856.

Cochrane, Christopher. 2015. *Left and Right: The Small World of Political Ideas*. Montreal: McGill-Queen's University Press.

Dantzler, Prentiss A. 2022. "Housing Affordability, Market Interventions, and Policy Platforms in the 2022 Ontario Provincial Election." *Sociology Compass* 16 (11): 1–14. https://doi.org/10.1111/soc4.13046.

Dilworth, Richardson. 2020. "Comparative Case Study Methods in Urban Political Development." *Social Sciences* 9 (10). https://doi.org/10.3390/socsci9100183.

Doering, Jan, Daniel Silver, and Zack Taylor. 2021. "The Spatial Articulation of Urban Political Cleavages." *Urban Affairs Review* 57 (4): 911–51. https://doi.org/10.1177/1078087420940789.

Einstein, Katherine Levine, David M. Glick, and Maxwell Palmer. 2019. *Neighborhood Defenders: Participatory Politics and America's Housing Crisis*. New York, NY: Cambridge University Press.

Einstein, Katherine Levine, Maxwell Palmer, Ellis Hamilton, and Ethan Singer. 2024. "The Gray Vote: How Older Home-Owning Voters Dominate Local Elections." https://sites.bu.edu/kleinstein/files/2024/03/Gray_Vote.pdf.

Epp-Koop, Stefan. 2015. *We're Going to Run This City: Winnipeg's Political Left after the General Strike*. Winnipeg: University of Manitoba Press.

Ford, Robert, and Will Jennings. 2020. "The Changing Cleavage Politics of Western Europe." *Annual Review of Political Science* 23 (2020): 295–314. https://doi.org/10.1146/annurev-polisci-052217-104957.

Gordon, Joshua C. 2022. "Solving Puzzles in the Canadian Housing Market: Foreign Ownership and De-Coupling in Toronto and Vancouver." *Housing Studies* 37 (7): 1250–73. https://doi.org/10.1080/02673037.2020.1842340.

Grace, Robert J. 2003. "A Demographic and Social Profile of Quebec City's Irish Populations, 1842–1861." *Journal of American Ethnic History* 23 (1): 55–84. https://doi.org/10.2307/27501378.

Guest, Dennis. 1997. *The Emergence of Social Security in Canada*. 3rd ed. Vancouver, BC: UBC Press.

Gutstein, Donald. 1983. "Vancouver." In *City Politics in Canada*, edited by Andrew Sancton and Warren Magnusson, 189–220. Toronto: University of Toronto Press.

Hajnal, Zoltan, and Jessica Trounstine. 2014. "What Underlies Urban Politics? Race, Class, Ideology, Partisanship, and the Urban Vote." *Urban Affairs Review* 50 (1): 63–99. https://doi.org/10.1177/1078087413485216.

Hodge, Gerald, and David L. A. Gordon. 2014. *Planning Canadian Communities: An Introduction to the Principles, Practice, and Participants*. 6th ed. Toronto: Nelson Education.

Inglehart, Ronald. 1990. *Culture Shift in Advanced Industrial Society*. Princeton, NJ: Princeton University Press.

Jeffries, Fiona, and Jennifer Ridgley. 2020. "Building the Sanctuary City from the Ground Up: Abolitionist Solidarity and Transformative Reform." *Citizenship Studies* 24 (4): 548–67. https://doi.org/10.1080/13621025.2020.1755177.

Latimer, Trevor. 2023. *Small Isn't Beautiful: The Case Against Localism*. Washington, DC: Brookings Institution Press.

Lehrer, Ute, Roger Keil, and Stefan Kipfer. 2010. "Reurbanization in Toronto: Condominium boom and social housing revitalization." *disP - The Planning Review* 46 (180): 81–90. https://doi.org/10.1080/02513625.2010.10557065.

Leo, Christopher, and Kathryn Anderson. 2006. "Being Realistic about Urban Growth." *Journal of Urban Affairs* 28 (2): 169–89. https://doi.org/10.1111/j.0735-2166.2006.00266.x.

Ley, David. 2020. "A Regional Growth Ecology, a Great Wall of Capital and a Metropolitan Housing Market." *Urban Studies* 58 (2): 297–315. https://doi.org/10.1177/0042098019895226.

Lightbody, James. 1983. "Edmonton." In *City Politics in Canada*, edited by Andrew Sancton and Warren Magnusson, 255–90. Toronto: University of Toronto Press.

Lucas, Jack. 2018. "Toward Delegation: Social Policy Centralization in Toronto, 1870–1929." *Journal of Policy History* 30 (2): 272–300. https://doi.org/10.1017/S0898030618000076.

Lucas, Jack. 2020. "Reaction or Reform? Subnational Evidence on P.R. Adoption from Canadian Cities." *Representation* 56 (1): 89–109. https://doi.org/10.1080/00344893.2019.1700154.

Lucas, Jack. 2023. "The Ideological Structure of Municipal Non-Ideology." *Urban Affairs Review* 59 (1): 275–93. https://doi.org/10.1177/10780874211038321.

Lucas, Jack, and R. Michael McGregor. 2021. "Conclusion." In *Big City Elections in Canada*, edited by Jack Lucas and R. Michael McGregor, 213–29. Toronto: University of Toronto Press.

Maynard, Robyn. 2017. *Policing Black Lives: State Violence in Canada from Slavery to the Present*. Fernwood Publishing.

McGregor, R. Michael, Aaron A. Moore, and Laura B. Stephenson. 2021. *Electing a Mega-Mayor: Toronto 2014*. Toronto: University of Toronto Press.

Moffatt, Mike. 2021. "Baby Needs a New Home: Projecting Ontario's Growing Number of Families and their Housing Needs." *Smart Prosperity Institute*. https://institute.smartprosperity.ca/sites/default/files/Baby-Needs-a-New-Home-Oct-1.pdf.

Moore, Aaron A. 2021. "Motivations for Mobilization: Comparing Urban and Suburban Residents' Participation in the Politics of Planning and Development." *Urban Affairs Review* 58 (4): 1124–51. https://doi.org/10.1177/10780874211016939.

Nelles, H.V., and Christopher Armstrong. 1976. "The Great Fight for Clean Government." *Urban History Review* (2): 50–66. https://doi.org/10.7202/1019530ar.

Nelson, Jennifer J. 2000. "The Space of Africville: Creating, Regulating and Remembering the Urban 'Slum'." *Canadian Journal of Law and Society* 15 (2): 163–85. https://doi.org/10.1017/S0829320100006402.

Palmer, Howard. 1974. "Nativism in Alberta, 1925–1930." *Historical Papers/Communications Historiques* 9 (1): 183–212. https://doi.org/10.7202/030782ar.

Peterson, Paul. 1981. *City Limits*. Chicago: University of Chicago Press.

Quastel, Noah, Markus Moos, and Nicholas Lynch. 2012. "Sustainability-As-Density and the Return of the Social: The Case of Vancouver, British Columbia." *Urban Geography* 33 (7): 1055–84. https://doi.org/10.2747/0272-3638.33.7.1055.

Rodríguez-Pose, Andrés. 2018. "The Revenge of the Places that Don't Matter (and What to Do About it)." *Cambridge Journal of Regions, Economy and Society* 11 (1): 189–209. https://doi.org/10.1093/cjres/rsx024.

Sancton, Andrew. 1983. "Conclusion: Canadian City Politics in Comparative Perspective." In *City Politics in Canada*, edited by Warren Magnusson and Andrew Sancton, 291–317. Toronto: University of Toronto Press.

Sancton, Andrew. 1985. *Governing the Island of Montreal: Language Differences and Metropolitan Politics, Lane Studies in Regional Government*. Berkeley: University of California Press.

Sancton, Andrew. 2008. *The Limits of Boundaries: Why City-Regions Cannot Be Self-Governing*. Montreal: McGill-Queen's University Press.

Savitch, Hank V., and Paul Kantor. 2002. *Cities in the International Marketplace: The Political Economy of Urban Development in North America and Western Europe*. Princeton University Press.

Sellers, Jefferey M., Anders Lidström, and Yooil Bae. 2020. *Multilevel Democracy: How Local Institutions and Civil Society Shape the Modern State*. Cambridge, UK: Cambridge University Press.

Siemiatycki, Myer. 2011. "Governing Immigrant City: Immigrant Political Representation in Toronto." *American Behavioral Scientist* 55 (9): 1214–34. https://doi.org/10.1177/0002764211407840.

Siemiatycki, Myer, and Anver Saloojee. 2002. "Ethnoracial Political Representation in Toronto: Patterns and Problems." *Journal of International Migration and Integration* 3 (2): 241–73. https://doi.org/10.1007/s12134-002-1013-8.

Smyth, William J. 2015. *Toronto, the Belfast of Canada: The Orange Order and the Shaping of Municipal Culture*. Toronto: University of Toronto Press.

Taylor, Zack. forthcoming. "Strange Multiplicity: Diverse Patterns of Governance for Canadian Metropolitan Areas." In *Local Government Structure and Intergovernmental Relations: An Urban-Rural Perspective*, edited by Karl Kössler. London, UK: Palgrave Macmillan.

Taylor, Zack. 2019. *Shaping the Metropolis: Institutions and Urbanization in the United States and Canada*. Montreal: McGill-Queen's University Press.

Taylor, Zack, and David A. Armstrong. 2024. "Political Divisions in Large Cities: The Socio-Spatial Basis of Legislative Behavior in Chicago and Toronto." *Journal of Politics* 87 (2): 496–509. https://doi.org/10.1086/732976.

Taylor, Zack, and Gabriel Eidelman. 2010. "Canadian Political Science and the City: A Limited Engagement." *Canadian Journal of Political Science* 43 (4): 961–81. https://doi.org/10.10170S0008423910000715.

Taylor, Zack, Jack Lucas, David A. Armstrong, and Ryan Bakker. 2023. "The Development of the Urban-Rural Cleavage in Anglo-American Democracies." *Comparative Political Studies*. https://doi.org/10.1177/00104140231194060.

Teixeira, Carlos, Wei Li, and Audrey Kobayashi, eds. 2012. *Immigrant Geographies of North American Cities*. Toronto: Oxford University Press.

Tennant, Paul. 1980. "Vancouver Civic Politics, 1929-1980." *BC Studies* 46: 3–27. https://doi.org/10.14288/bcs.v0i46.1054.

Wallace, Rebecca, R. Michael McGregor, and Erin Tolley. 2024. "Why Don't More Immigrants Vote in Local Elections?" In *Political Engagement in Canadian City Elections*, edited by R. Michael McGregor and B. Stephenson Laura, 27–57. Montreal: McGill-Queen's University Press.

Weaver, John C. 1977. "Tomorrow's Metropolis" Revisited: A Critical Assessment of Urban Reform in Canada, 1890–1920." In *The Canadian City: Essays in Urban History*, edited by Gilbert A. Stelter and Alan F.J. Artibise, 393–418. Toronto: McClelland and Stewart.

Afterword

Warren Magnusson

One of the most disturbing myths about politics is that there must be some place where all the big decisions are made: in the United States, maybe the White House or Congress; in Canada, maybe the prime minister's office (PMO) or Cabinet. We all sense that the real centres of power might be elsewhere: in the backrooms, corporate headquarters, or wherever. This can lead to strange conspiracy theories.

Part of the point of *City Politics in Canada*, in both its editions, is to suggest that the major municipalities, and not just the federal government and the provinces, are centres of power, whose politics must be understood if we are to make sense of things. Of course, a focus on the big cities does not tell us everything we need to know; the world we live in is much more complex than that. Neither power nor politics is centred in just one place in a city, province, or country. As I have argued elsewhere (Magnusson 2011), the dispersal of power and politics is "like a city" in its complexity. Making sense of this complexity is a challenge, to put it mildly.

Each chapter in this volume responds to that challenge in a different way, but what they have in common is a focus on the municipal governments of the cities concerned. The same was true of the first edition (including my own contributions). In part this was, and is, a remedial response to a Canadian politics literature that largely ignores municipalities. Municipal politics is important too. Perhaps people can understand that more easily if they focus on our biggest cities, many of which have bigger populations than most of our provinces. We are immediately confronted with a problem, however. Some cities, like Halifax, Calgary, and Winnipeg, encompass all or almost all of their metropolitan areas. Others, like Vancouver and Montreal, have much more limited jurisdiction. Vancouver is hardly much bigger than Surrey in terms of population, and both are within the Metro Vancouver Regional District, which has many important powers. Ottawa is an even more complicated case because the metropolitan area spills over into Quebec, whose municipal institutions differ from Ontario's. At least in Vancouver and Montreal, there is a single provincial government that can take responsibility for metropolitan problems. The same is true of Toronto, although in that case, it is very difficult to draw a boundary around the metropolitan area because it is so huge.

The editors raise many good questions about municipal politics. I would like to add a few. How does municipal politics relate to activity in what we often call "civil society"? If there are regional institutions, how do they relate to municipal politics? What are the implications of different institutional designs? Are there systematic variations in municipal politics according to the area of the country in which the municipality is situated? Are there differences between municipal politics in big cities, small and large suburbs, small cities, towns, and rural areas?

The editors note how things have changed since the first edition of this book came out in the 1980s. One aspect of this is how neighbourhood preservationists, who could style themselves as "reformers" half a century ago, now seem like old-fashioned conservatives, out of touch with the realities of the contemporary housing crisis. (My daughter often reminds me of this.) It seems that there always has been a housing crisis: my parents had many tales of the difficulties of finding housing after the Second World War (I was an early walker, and my mother attributed this to the cold floors in the unwinterized cottage they were forced to rent when I was a baby); their parents suffered much more during the Great Depression (my paternal grandfather lost the house he had built himself). Going still further back in time, we learn of horrific slums in cities of all sorts. Isn't a crisis in housing just a standard feature of urban politics?

Another standard feature is the environmental crisis. Half a century ago, we might have thought of that mainly in terms of pollution. Now it is mainly a matter of climate change, although of course pollution is a major culprit in that regard. The difference is that so many activities we once thought were innocent – harvesting wood, raising cattle, heating our homes with natural gas, and so on – now seem highly problematic. Most of us live in cities, and those of us who do not still live urban lives. The impact of our urban way of life is everywhere, so that what is happening in the farms and forests – and also in the rivers and deserts, as well as in the great oceans around us – is as much determined by the way we manage urban life as what happens in housing markets or transportation systems. If that is so, do we still have reason to focus on cities in our efforts to understand our present lives? I think we do, because almost all of our problems are bound up with urbanism as a way of life.

Part of the rationale for local government is that it allows for a fuller and more effective democracy. Faced as we now are with a disaster that most of us who were alive forty years ago could hardly have imagined, we must know that decisive action at all levels is required of us. People must act decisively as individuals, in their local communities, and more broadly. A vibrant, highly localized democracy is the key to this. The challenge and the opportunity are especially marked in big cities, many of which have responded with various elements of what the editors call twenty-first century urbanism: higher densities, more transit, increased walkability, safer cycling, reduced waste, better recycling, water conservation, reduced dependence on fossil fuels, and so on. Bolder action is required, but that depends on deepening democracy, which has to begin where people live.

In many respects, our world is like a huge city plagued by innumerable problems, from horrific violence to desperate poverty and environmental disasters. We can begin to address these problems if we focus on our own neighbourhoods and reach out from there.

Contributors

Laurence Bherer is professor of political science at the University of Montréal. Her research focuses on political participation, local democracy, and political engagement. She is the coauthor, most recently, of *Le Québec en mouvements: Continuité et renouvellement des pratiques militantes* (Les Presses de l'Université de Montréal 2023) with Pascal Dufour and Geneviève Pagé.

Sandra Breux is professor and Canada Research Chair in municipal elections at the Institut national de la recherche scientifique, where she is cohead of the Smart Pedestrians Cities Laboratory and scientific director of the Villes Régions Monde (VRM) Network. She is also codirector of the Canadian Municipal Barometer project.

Ian Bushfield is the executive director of the BC Humanist Association and a frequent commentator on British Columbia and Vancouver politics. He cohosts the PolitiCoast and Cambie Report podcasts.

Robert G. Finbow is Eric Dennis Memorial Professor of political science and deputy director of the Jean Monnet European Union Centre of Excellence at Dalhousie University. A specialist in international political economy, he recently edited a collection titled *CETA Implementation and Implications: Unravelling the Puzzle* (McGill-Queen's University Press 2022) and is the author of *The Limits of Regionalism: NAFTA's Labour Accord* (Routledge 2017).

Martin Horak is associate professor of political science at the University of Western Ontario. He is associate director of Western's Centre for Urban Policy and Local Governance, past director of Western's Local Government Program, and a pillar lead of the Canadian Municipal Barometer project.

Jack Lucas is professor of political science at the University of Calgary. He is codirector of the Canadian Municipal Barometer project and author, most recently, of *Ideology in Canadian Municipal Politics* (University of Toronto Press 2024), which won the Seymour Martin Lipset Best Book Award from the American Political Science Association.

Warren Magnusson was professor emeritus of political science at the University of Victoria. A political theorist with a particular interest in the urban and the local as sites of politics and government, he was the author of *Politics of Urbanism: Seeing Like a City* (2011) and *The Search for Political Space* (1996). With Andrew Sancton, he coedited *City Politics in Canada* in 1983. He passed away in April, 2025.

Aaron A. Moore is professor of political science at the University of Winnipeg, where he specializes in urban politics and local governance. He is the author of *Planning Politics in Toronto: The Ontario Municipal Board and Urban Development* (University of Toronto Press 2013) and, with R. Michael McGregor and Laura B. Stephenson, *Electing a Mega-mayor: Toronto 2014* (University of Toronto Press 2021).

Stewart Prest is lecturer in political science at the University of British Columbia.

Andrew Sancton is professor emeritus of political science at the University of Western Ontario. During his long and distinguished career, he published numerous books and articles on metropolitan governance, the governance and politics of Montreal, municipal amalgamations, and electoral boundary making. With Warren Magnusson, he coedited *City Politics in Canada* in 1983. He received the Mildred A. Schwartz Lifetime Achievement Award from the American Political Science Association's Canadian Politics section in 2023.

Ursula Stelman is former director of community services at the City of Winnipeg and lecturer in the University of Western Ontario's local government program. After a long career in municipal administration, she completed her doctoral research from the University of London, on organizational performance and policy implementation in African cities in 2011.

Zack Taylor is associate professor of political science at the University of Western Ontario. He was the founding director of Western's Centre for Urban Policy and Local Governance and is a fellow at the University of Toronto's Institute on Municipal Finance and Governance. His book *Shaping the Metropolis: Institutions and Urbanization in the United States and Canada* was published by McGill-Queen's University Press in 2019.

Luc Turgeon is professor at the School of Political Studies at the University of Ottawa, where he is affiliated with the Centre de recherche interdisciplinaire sur la diversité et la démocratie, le Groupe de recherche sur les sociétés plurinationales, and the Research Centre on the Future of Cities.

Sophie L. Van Neste is associate professor and Canada Research Chair in urban climate action at the Institut national de la recherche scientifique. She is principal investigator of Labo Équité Climat and leads the Villes Climat Inégalités project. With Patrice Melé and Corinne Larrue, she is the coauthor of *Transitions socioécologiques et milieux de vie: Entre expérimentation, politisation et institutionnalisation* (Les Presses de l'Université de Montréal 2024).

Index

Political Development:
Comparative Perspectives